Biological Perspectives on
Language

MIT Press Studies in Neuropsychology and Neurolinguistics
David Caplan, series editor

Psychobiology of Language, edited by Michael Studdert-Kennedy, 1983

Biological Perspectives on Language, edited by David Caplan, André Roch Lecours, and Alan Smith, 1984

Biological Perspectives on Language

edited by
David Caplan
André Roch Lecours
Alan Smith

The MIT Press
Cambridge, Massachusetts
London, England

This book was set in Times Roman
by The MIT Press Computergraphics Department and printed and bound
by Halliday Lithograph
in the United States of America.

Library of Congress Cataloging in Publication Data

Main entry under title:

Biological perspectives on language

(MIT Press studies in neuropsychology and neurolinguistics)

Articles first presented at Third Conference of Centre de recherches en sciences neurologiques, Université de Montréal, May 1981.

Includes bibliographies and index.

1. Neurolinguistics—Congresses. 2. Language disorders—Congresses.
I. Caplan, David, 1947– . II. Lecours, André Roch, 1936– .
III. Smith, Alan. IV. Université de Montréal. Centre de recherches en sciences neurologiques. Conference (3rd : 1981 : Université de Montréal)
V. Series.
QP399.B55 1984   153.6   83–26520
ISBN 0–262–03101-9

Contents

Contents

Series Foreword

The MIT Press Studies in Neuropsychology and Neurolinguistics present theoretical and empirical research on the neural mechanisms underlying language and cognition and their pathologies. These studies focus on the application of concepts and methods of the study of normal language and cognition to the investigation of their disturbances and on the use of increasingly detailed observational techniques in the clinical and basic neurosciences to the neural bases of language and cognition (normal and disturbed). The scope is broad, including humans and nonhumans, development, and related issues. The series is intended to make available significant new work in neuropsychology and neurolinguistics, to disseminate new approaches, ideas, and results, and to catalyze new research.

The members of the advisory board for this series are Maureen Dennis of The Hospital for Sick Children in Toronto, Patricia Goldman-Rakic of Yale University, Steven Hilliard of the University of California at San Diego, Marc Jeannerod of INSERM at Lyon, Scott Kelso of Haskins Laboratories, André Roch Lecours of the University of Montreal, John C. Marshall of Oxford University, Larry Squires of the University of California at San Diego, and Edgar B. Zurif of the City University of New York and Brandeis University.

David Caplan

Introduction

This volume is an edited collection of papers presented at the Third Conference of the Centre de Recherches en Sciences Neurologiques of the Université de Montréal, at which some 20 scientists whose work deals with language and its biological basis presented general reviews of their own work and related subjects to an audience of over 300 people active in a wide variety of disciplines and at various scientific levels. The organizers of the conference, aware of the audience, requested that each invited speaker present an overview of his or her subject as well as results of special importance and the speaker's own work. This volume thus contains, for the most part, specialized and technical material set in context in a way intended to make it comprehensible to the nonspecialist.

The theme of the conference, reflected in the title of this volume, was biological perspectives on language. It is hard to think of a theme of greater generality, or one that could include a larger number of conceptualizations of problems and research areas or admit of more methodological approaches. The concepts that guided the choice of speakers and contributors were, however, reasonably restricted, if still broad. Profoundly influenced by contemporary linguistic analyses, the conference organizers and volume editors were chiefly concerned with bringing together workers whose presentations and reflections would bear on hypotheses emanating from these analyses. Among the questions considered here are the following: Is the structure of the language code formally similar to or dissimilar from that of representational systems found in other areas of human cognition? Is there any functional ability in nonhumans that involves representations of a sort similar to those constituting language and that could be considered an evolutionary precursor of language abilities? What brain regions are involved in representing language structures and in the processing of these structures in language use, and how do these regions vary with genetic and environmental factors? What is the evolutionary history of these areas? (The last question can receive an answer regardless of the role these areas play in the functional capacities of other organisms.) Are these brain regions those that are involved in the representation and processing of language in development? What do we know about the physiology of these areas? How do we conceptualize the question of how physiological events could be related to mental activities such as language processing, and how might we model such physiological events?

A strong position on several of these questions has emerged from recent linguistic work. In discussions of the implications of linguistic results it is often claimed that the structures that constitute language are formally unlike those that underlie other functional abilities. This claim has most frequently emerged from studies of syntactic structure, but it is supported by analyses of phonological systems. A further claim is that the structures of language are not similar to those we need to postulate to account for behaviors of other species. The strongest inference that can be drawn from these statements is that essential aspects of language—significant parts of syntactic and phonological systems, at the least—have no evolutionary history but are uniquely human mental properties. If this is the case, adherence to a materialist doctrine that attributes mental structures to neural structure and function requires that some neural structures and/or functions be unique to humans and devoted to language representations and/or processing, and that, whatever their nonhuman homologues, these neural structures or functions be, in some measure, new in evolutionary history. Many aspects of this formulation are controversial, from the version of materialism that might be appropriate for the relation of mental and physical predicates to the empirical justification of the linguistic and nonlinguistic analyses on which the initial premises of the argument are based. Fodor, for one, has argued against the necessity (and even the desirability) of what he terms "type reductionism" and holds out the possibility that a very weak form of materialism (his "token physicalism") is adequate for purposes of the philosophy of science. Token physicalism, however, denies any causal role to neural structures in the determination of mental structures and, if correct, would imply that no conclusion about neural structures or their evolution could be drawn from analyses of language or other mental systems. The question has been much debated in recent years. In the second domain mentioned above, some researchers have argued that some aspects of linguistic structure are reflections of more general, pragmatic factors affecting communication and have questioned the isolation of the mental structures of language from those required for other functions. We are aware of the controversy surrounding the "linguistic" perspective on the biology of language. Our convictions, despite these controversies and despite differences among us over these issues, are that this formulation can give structure to the questions enumerated above and that one "perspective on the biology of language" (within which several "biological perspectives on language" are possible) consists of clarification, debate, and investigation of these questions and the answers to them afforded by recent linguistic studies.

The chapters in this volume deal with these questions quite directly. All of them present approaches to the biological basis for language and therefore bear on the characterization of the organic structures responsible for language, their ontogenetic development, and their evolutionary history and role. We have included brief introductions to the three parts of the volume; the purposes of these are to highlight issues presented in individual chapters, to indicate connections and gaps between concepts and claims advanced in different chapters, and to relate the ensemble of chapters in each part to these basic themes.

The linguistic structures and the language-related processes dealt with in this volume can, for the most part, be termed central. That is, they are fairly abstract representations, such as syntactic structures, phonological representations, and semantic readings. There is a considerable literature that deals with more "peripheral" aspects of language: acoustic-phonetic structures and the perception and production of these structures. By and large, the answers afforded to the basic biological questions that emerge from studies of the latter are the opposites to those advanced by linguists interested in the central aspects of language. A not unreasonable hypothesis is that the central, abstract aspects of language, both formal and semantic, and their accompanying neural substrate, are evolutionarily unique structures that are superimposed on, make use of, and are in some measure constrained by their utilization of peripheral effector and receptor systems that have evolved more gradually and whose basic biological functions lay and lie in other spheres. Whatever the truth of the matter, the approaches represented in this volume are representative of one focus of interest and need to be compared and integrated with studies of other aspects of the question. We hope that the articles collected provide timely introductions to a related set of questions about the neural basis of language.

David Caplan
André Roch Lecours
Alan Smith

Biological Perspectives on
Language

Part I

Issues at the Interface of Neurological and Psychological Approaches to Language

These four chapters focus on the interaction of psychological and linguistic studies of language, on the one hand, and neurological approaches to language on the other. They present different aspects of this interaction and raise different methodological and theoretical points.

In chapter 1, Caplan develops the notion of a mental organ for language. The argument, illustrated by simple analyses of syntactic structures and their parsing, is that language consists of structures of a highly specific sort and that the psychological processes involved in recognizing (and otherwise utilizing) these structures are also highly specialized. Caplan's claim is that at least some of these structures and processes are unlike those found in the description of other mental activities—that is, that they are autonomous in the sense that they are unique to language and its use. Caplan acknowledges that the employment of these structures requires interaction with many other mental capacities. One psychological question, then, is how this interaction is structured: How do psychological capacities, such as the ability to retain verbal material in a short-term "working" memory or the ability to deploy attention to one or another aspect of a task, which may be conceived of as applying to functional domains outside of the recognition and production of the structures of language *per se*, interact with the psychological routines that appear to be specialized for the processing of the structures of language? There has been considerable debate about this matter in both experimental and theoretical work in recent years. The corresponding structural linguistic question can also be posed: If linguistic representations involve different notational systems and reflect different aspects of language structure, such as syntax and phonology, how do these representations interact to yield a full description of a linguistic structure? Here there is a great deal of agreement about the answer. Most theories of language structure assume and justify an architectural arrangement of the major components of a grammar, which specifies particular points at which one type of representation can be referred to by rules of another component and further specifies the internal structure of components of a grammar so that processes belonging to a given component are scheduled and constrained by the internal architecture of that component. With respect to linguistic structures, then, language appears to be quite "modular" down to a fairly detailed level of structure if some contemporary theories are correct. Its psychological processing may be modular in the same sense.

Caplan's point is that these facts about language and its processing—the uniqueness to language of certain psychological processes (autonomy) and their isolation from other cognitive structures and processes (modularity)—suggest a possible comparable structural and physiological analysis of the neural mechanisms that serve language. This need not, of course, be the case. Computers maintain separation of programs running simultaneously and of programming languages, despite the absence of hardware specialized for particular programs or languages; indeed, within very general limits, programs of various types can be run over the same physical elements. Whether the neural elements related to language and their physiology are distinct from those responsible for other psychological functions is an empirical question. The thesis that language is autonomous and modular with respect to its structure and its psychological processes makes the autonomy and the modularity of the neural mechanisms for language a consideration; if language were neither autonomous nor modular at the functional level, the question of these features of its neural basis would not arise.

Caplan stresses that the level of description of neural structures and operations to be related to language is of critical importance in evaluating these questions. Correlations of language functions with gross areas of the brain, established by the analysis of aphasic symptoms and syndromes and by more direct observational methods where possible, is likely to be misleading. Lesions in a particular area of the brain may affect a variety of psychological functions, each related to different aspects of the cellular structures and physiological operations of neural elements in that area. In such cases, particular pathological processes may yield more specific functional impairments than others, and the study of functional deficits seen after different types of injury may reveal an organic modularity that is not discernible when only one process is studied. This observation is particularly relevant to the analysis of aphasia, which has been heavily influenced by a desire to resolve a 125-year-old debate about the empirical validity of the view that specific language functions are localized in particular regions of the brain. The study of circumscribed stable lesions (essentially, strokes in a stable state) provides the best test of particular hypotheses of this nature. If the theoretical question guiding research is no longer restricted to the question of localization of a language function but extends to whether a neural structure related to a language function is autonomous and modular, the impetus to restrict study to vascular cases is removed and in its place the study of a variety of etiologies for aphasia may become necessary.

There is another aspect to the neural question emphasized by Caplan: Not only may a gross analysis of the brain be misleading with respect to the autonomy and modularity of neural mechanisms related to language; additionally, a more detailed analysis is required for more than a correlational relationship between neural structures and language functions. Such correlations as exist between gross areas of the brain and language functions must be due to the composition of these regions of the brain. This is a point with which Geschwind is in strong agreement. The principal theme of Geschwind's chapter is that much more can and must be made of neural facts, both in the formulation of neurolinguistic theories and in the development of theories in linguistics and psycholinguistics themselves. Part of Geschwind's concern is that descriptions of normal and (more recently) of pathological language have been developed without consideration of the neural organization that underlies language. Increasingly, these considerations are related to advances in our understanding of the neural structures and operations Caplan points to as relevant to language: cellular and subcellular structures and physiological events. Geschwind cites the literature on the neural basis for plasticity as one area in which progress at the cellular level has been made.

A good part of Geschwind's chapter is devoted to an argument based on his reading of the history of aphasiology and neuropsychology. He considers that advances in the understanding of language-brain relations made in the late nineteenth century were lost during a period of the present century. Elsewhere Geschwind has commented on political events and social forces that he believes contributed to the ignoring of these earlier advances. These social forces are not now in play, but there is a striking similarity, in Geschwind's view, between some present-day intellectual currents and those present in the period following the First World War. Geschwind has suggested that an approach to the aphasias heavily influenced by psychological theories of the day developed (is developing) that did (does) not make contact with more neuroanatomical work that characterized an earlier period. He cautions against this tendency on three scores. First, work totally divorced from neural concerns cannot lead to the development of organic treatments for the diseases that cause aphasia and thus fails to pursue an important avenue of potentially applied research. Second, by overlooking the work of earlier periods, researchers are in danger of "rediscovering the wheel." Geschwind cites Zurif's scholarship on the appreciation and characterization of the nature of comprehension in agrammatic aphasics as an example of an area in which previous investigators' discoveries have been ignored and forgotten. Geschwind seems to imply that it is the lack of concern for neurological aspects of the problem of aphasia that has led to this unawareness of earlier contributions in which functional analyses were often presented within a set of concerns more neurological than those that seem to Geschwind to motivate some current work. It is perhaps a related point in Geschwind's paper that Caplan's approval of Morton and Mehler's thesis that the recovery of language functions after early left-hemisphere injury is a phenomenon not seen in other aspects of neural function is misguided and that a greater appreciation of neurological issues (for instance, the facts of development or recovery of functions associated with the cerebellum in cases of early cerebellar lesion) would put this particular issue in a very different light. Third, Geschwind maintains that theories of language and psycholinguistics should be constrained by and responsive to neurological facts; he suggests that the behavioral sciences and psychology—in particular, the under-standing of the behavioral effects of disease—advance mainly during times when the organic bases of the phenomena under study are well understood.

At one level, it is impossible to disagree with Geschwind's arguments. The development of organic therapies for the aphasias will require study of their organic determinants; recognition of the origins of con-temporary ideas and familiarity with phenomena in a wide range of functional areas that might bear (in this case, do bear) on a stated hypothesis are, of course, highly desirable aspects of scholarship; there have been important connections between advances in the understand-ing of psychological phenomena and the understanding of their organic bases. However, Geschwind's views raise a number of basic questions. It is not always the case that progress in the understanding of a function attends upon advances made with respect to its organic basis. Psycho-logical studies of the performances of aphasic patients from Jackson's on have yielded insights into the factors regulating performance of patients, often without appreciation of a neural correlate of the psy-chological function. The theories of language structure and processing that have been developed in contemporary linguistics and psychological work fall into this category. However desirable it may be for neural theories and observations to speak to and suggest aspects of the or-ganization and use of language structures, neurological observations have had little impact on the development of theories of language.

In part this is probably a result of the difficulties of interdisciplinary research, which are particularly felt in a case where knowledge about the nervous system and theories of language are both in a state of rapid expansion; in part the problem is that not enough is known about the way neural mechanisms would support the sorts of representations and functions that contemporary psychology and linguistics are developing. Consider a specific example of a neural analysis, one that has influenced aphasiological thinking for two decades: Geschwind's hypothesis that the connections of the inferior parietal lobe allow for the development of cross-modal associations between nonlimbic stimuli and that this is the basis of the ability to name objects. Whatever the truth of this theory, it achieves a correlation between neurological structures and psychological functioning in which the two levels of description are operationally related. When we consider, however, other aspects of the same process of relating meanings to words, the relationship between the psychology and the neurology of the process becomes obscure. What is the neural basis—structural or physiological—for word-to-word associations, for establishing the meanings of abstract terms, or for the extraordinary ease with which children at certain ages acquire vocabulary? These, and a plethora of similar questions derived from language phenomena currently under study, have no answers at present. Indeed, the direction that research into answers to these questions can take is primarily suggested by the mechanisms postulated by psychology, such as semantic and other "nets" underlying word associations.

Morton's interesting reconstruction of the basic typologies of models of language processing that were articulated in the earliest period of research on aphasia reaches an even stronger conclusion. In his view, perfectly acceptable models of the psychology of language and of language disorders were abandoned because their neural bases were not established or were found to be other than expected. Morton argues that psychological models must be developed and evaluated independent of their neuropsychological aspects. In view of the degree to which current neurological understanding constrains psychological modeling at the level of detail found in modern psycholinguistic theories, Morton's point is pertinent to today's work. Restriction of psychological models to a level of descriptive detail that can be explicated in terms of neural theory would impose a very severe and undesirable restriction on such investigations. The functional taxonomy of the dyslexias in terms of pathologies of particular reading routes, which has grown up without the satisfying correlation that "pure alexia" and "alexia with agraphia" have to neural loci, exemplifies the development of functional analyses that cannot yet be tightly constrained by neural facts. This does not mean, however, that no constraints on psychological theories, or no guidance of these theories, can come from neurological considerations. Heterogeneity of patient performances within what is initially taken to be a single syndrome may be due to differences in lesions, for instance. Certain aspects of the performances of brain-injured patients may reflect systems that are anatomically isolated from more efficient systems normally in use. The performances of deep dyslexics, for instance, has been claimed to represent right-hemisphere reading. If this claim can be maintained, it will allow the development of a model of reading that would not have to account for the types of errors seen in this syndrome as the basis for normative, left-hemisphere-based reading.

Thus, while there seems at present to be no clear way in which present knowledge of the organization of the nervous system highly constrains the details of psychological models of language structure and processing (but see part III for work that bears on this task), there are nonetheless a number of neuropsychological and neurolinguistic considerations that can be helpful in theory construction in psycholinguistics.

Nottebohm's chapter, which concludes part I, illustrates how detailed studies of functional capacities and neural mechanisms can be related and presents many such relations. One will serve to illustrate the type of correlation that has been discovered: Female canaries do not sing, but if treated with testosterone and subjected to ovariectomy in adult life they develop song repertoires as complex as those of the average male. Anatomical studies of these testosterone-treated female birds show significant increases in size of the nuclei responsible for song in the male, HVc and RA. Microscopic studies of these nuclei indicate that dendritic trees increase in the treated females, and further studies indicate that radiolabeled thymidine is incorporated into neurons in these nuclei. The latter observation is one of the few examples for neurogenesis in adult life. Its functional significance is still uncertain, though; although there is comparable intake of ^{3}H-labeled thymidine in ovariectomized adult female birds treated with cholesterol (rather than testosterone), they do not come into song and dendritic aborization does not increase. What role neurogenesis plays in functional plasticity in adult life is a topic with great potential therapeutic significance, and it points to neural mechanisms underlying adult learning that have not been considered previously. Many aspects of the functional capacities of the songbirds have yet to be related to neural structures and mechanisms. What determines whether a bird is an imitator or an improvisor in vocal learning? What determines the critical period for song acquisition? Nottebohm's work provides answers to some questions and suggests lines of research into others.

What relevance do these studies have to language and its neural basis? Nottebohm considers similarities and dissimilarities between the bird and the human. In both, there are effector areas of the brain which are controlled by other, specialized areas. These latter areas may play important roles in both reception and production of song and language. Both systems show hemispheric dominance. Here, however, the systems differ. The human system does not seem to involve a duplication of language capacities or of potential for these capacities in the right hemisphere, as judged by recovery after aphasia and by split-brain studies, whereas the right hemisphere of the canary can master song after resection of the "song center" (the HVc) on the left. The differences in plasticity may be related to the kinds of neural events (neurogenesis and dendritic increases) Nottebohm has investigated.

We do not need to consider birdsong a language to learn something about neural mechanisms that may be relevant to human language from its study. This is true in general; we may learn something about the brain and obtain ideas about how the human brain supports language from studies of the neural basis for virtually any functional capacity. But birdsong is a functional ability that resembles language in several respects; it shows vocal learning, it has distinct communicative functions, it has an internal structure such that lesions can affect its complexity, and it is lateralized to one hemisphere. Thus, it affords the opportunity to investigate the neural mechanisms underlying many functional capacities that seem particularly relevant to language. The evolutionary connection between the two systems will be clearer as the similarities and dissimilarities in mechanisms are clarified.

Part I is intended to introduce topics, hypotheses, and examples of studies relating language and the brain. Geschwind emphasizes the need for such studies, Morton's chapter illustrates the value of studying language in brain-injured populations, Caplan presents hypotheses on the organization of language and the brain, and Nottebohm provides an example of how detailed structure-function correlations can be investigated in species for which the experimental methodology is less limited than for humans. Parts II and III explore many of these questions in more detail.

Chapter 1

The Mental Organ for Language   David Caplan

One of the most productive concepts in contemporary psychology is that cognitive function is the result of a number of autonomous processes, each operating in a particular intellectual sphere, whose individual natures and interactions determine the cognitive capacity of an organism and thereby influence its behavioral repertoire. One area of cognition in which this concept has emerged most clearly is that of human linguistic abilities. Investigators have spoken of the existence of a "language-responsible cognitive structure" (Klein 1977) and a "mental organ for language" (Chomsky 1977)—terms intended to imply that the human capacity to utilize language is one of the functional entities that make up human mental endowment. I shall consider some of the issues raised by the notion of an autonomous capacity for language and illustrate some of the properties of language structure and processing that characterize this hypothesized autonomous component of the human mind.

Among the features of language which investigators have recognized and which have led them to postulate a mental organ for language are the following three:

- Language as we conceive of it pretheoretically is unique to humans. Though many animals have systems of signal and sign display that allow for restricted types of communication, for expression of individual, social, and species identity, and for other functions served by language, our impressions are that what humans call language is not found in comparable form in other species, and that human language and other systems differ in qualitative rather than simply quantitative ways. The qualitative differences are felt to exist at the levels of formal structure, semantic power, and social and personal utility.
- Language is universally present in humans. Except in cases of neurological impairment and failure to be exposed to linguistic environments,

all humans develop language in auditory-oral modes. Humans born deaf and raised in communities and families of the deaf develop language in manual-visual modes, and the language they develop is, within limits imposed by the mode of expression, highly similar in form and probably in semantic power to the auditory-oral language that hearing humans achieve.

• Humans learn language without explicit instruction. In comparison with the highly structured input that characterizes educational efforts in other domains (consider simple arithmetic), linguistic input to children is quite unstructured and poorly reinforced. Furthermore, children appear to acquire language in regular sequences, despite diverse exposure conditions and sequences of exposure, and they acquire knowledge about language that goes well beyond mastery of the actual exemplars to which they are exposed. It would appear, therefore, that language acquisition does not utilize environmental input in the same way as is the case in other domains in which "learning" takes place.

These three observations ("claims" might be a better word) are central premises on which the concept of a mental organ for language has been based. I shall not review the data that support them; rather, I accept them as a point of departure for the characterization of a postulated mental organ for language. We can conceive of an explanation for these three observations if we postulate that humans are naturally equipped to develop and utilize what we call language, that this "being equipped" is an inherent capacity of humans that develops with the maturing nervous system, and that other species lack this "equipment." Klein (1977) presents four postulates that spell out this position more explicitly:

• What accounts for a person's capacity for language is that he possesses a language-responsible cognitive structure. Although the structure of the LRCS may vary somewhat from speaker to speaker, certain universal features are common to the LRCS of each normal individual.

• Considerable uniformity exists in the onset time and the course of language acquisition among normal individuals. These universal features of language acquisition are controlled to a significant extent by maturational changes in the LRCS.

• The physical realization of the LRCS is a brain structure. This realization is sufficiently uniform among neurologically normal individuals that an (idealized) description of the LRCS as a neural structure is possible.

• The possession of the LRCS is a species-specific trait; that is, universal features of the LRCS (both structural and developmental) are controlled by genetic factors characteristic of every normal member of the human species.

These postulates, which embody much current thinking about the biology of language, present a general characterization of the concept of a mental organ for language that links the view of the species-specific cognitive ability to a particular material basis (a neurological "realization" based on a genetic endowment) and to a strongly nativist psychology.

Accepting this characterization as a point of departure, I propose to focus on work that provides a more detailed characterization of the postulated mental organ for language. I shall emphasize the issue of the autonomy of linguistic systems from other cognitive systems. Evidence that language faculties constitute an autonomous aspect of human mental life would be, *ipso facto*, evidence that, if there is a mental organ in the sense discussed by Klein, it has particular highly restricted domain of application—namely, what we might intuitively appreciate and ultimately theoretically define to be human language.

Moreover, the notion of the domain specificity, or autonomy, of a mental organ for language invites us to phrase specific hypotheses about phylogeny and ontogeny. Phylogenetically, we might consider that what separates humans from other species is our possession of just this particular domain-specific mental capacity, at least in its entirety. We might look for separate evolutionary lines, as in the vocal learning seen in birdsong, or intelligent behavior we can discover by observation or experimentation in other primates, and compare the resulting functional abilities with those of language. It may turn out to be the case that, when characterized in detail, various aspects of the behavioral repertoires of other species are in fact highly similar to parts of a language faculty. This might provide us with animal models for the physical basis of at least parts of a language faculty. Ontogenetically, we could consider that what it is that develops under genetic control of neurological maturation, with appropriate environmental exposure, is just the restricted cognitive capacity that we associate with language. Other aspects of the ontogeny of cognition and intelligent behavior could be the result of very different psychological processes and might follow quite different principles of development.

What, then, are some of the features of a postulated mental organ for language? Current linguistic and psycholinguistic work suggests that we conceive of a mental organ for language as consisting of two separate related elements: a set of knowledge structures and a set of procedures that utilize these knowledge structures. I shall deal with each of these features in turn.

There are several features of linguistic knowledge that have been emphasized in the modern literature and that we may take as ontological claims about this aspect of cognitive psychology. Some are psychological, such as the observations that, though we are aware of certain aspects of the structure of language (such as the existence of discrete elements which we term words), much linguistic knowledge is unconscious, or the claim that important aspects of this knowledge are innate and serve as a basis of the child's ability to identify and assimilate features of the particular language to which it is exposed. Some of these features are structural; "core" linguistic knowledge is highly detailed, has structural features not found in other domains of the intellect, and is itself modular in the sense that it consists of structural components, each representationally distinct and coherent, which interact in highly constrained ways to yield the full array of structures that make up linguistic knowledge. These latter structural features—modularity, uniqueness, and representational coherence—are those that lead to the conclusion that, at the level of knowledge structures, a mental organ for language is an autonomous aspect of human intelligence.

To exemplify one analysis that presents hypotheses about the type of knowledge that might be represented in a mental organ for language, we may consider some quite simple and uncontroversial facts about English syntax, the analysis of which has provided some of the most interesting contemporary hypotheses about certain aspects of language structure. The phenomena in question are restrictions on the possible movement of constituents of sentences in English, a domain of empirical observation since the earliest days of transformational theory. The following analysis is highly simplified and omits many discrepant and technical details, but it does illustrate at least some of the basic properties of the knowledge structures in a postulated mental organ for language.

Among the data to which any theory of grammar must be responsive are facts about English such as that only the first of the three following sentences is acceptable.

1. The boat that you believe John painted is a yawl.
2. *The boat that you believe the claim John painted is a yawl.
3. *The boat that you asked who painted is a yawl.

It is clear that sentences 2 and 3 are not ruled out on semantic grounds; if they were well formed, their meanings would be perfectly clear. They must be unacceptable because of their syntactic structure.

Simplifying considerably, we may say that sentences 2 and 3 are unacceptable in English because the noun phrase *the boat*, which serves as the subject of the principal verb of the main clause, is also the direct object of the verb in the embedded clause and cannot serve both these functions simultaneously in structures such as those found in sentences 2 and 3. Given the acceptability of sentence 1, it cannot simply be the case that English does not allow one noun phrase to fulfill these two thematic functions simultaneously; rather, there must be some feature of English grammar that is relevant to the distinction between the first sentence and the other two and that is derived from the syntactic structures of sentences in which a single noun phrase fulfills both these thematic roles.

It minimally follows that in order to be able to describe and ultimately explain facts such as these we need a system for representing the relevant aspects of the structure of English sentences. A wide variety of such descriptive frameworks have been proposed; I shall present one in a very simplified way. This system postulates that English contains structures called clauses, which themselves consist of an introductory complementizer and a propositional content (Bresnan 1972). We may assign each of these elements distinctive labels, using the symbol S (which we may think of as sentence) for the propositional content of a clause, and the symbol S̄ for the entire clause, including its complementizer. In other words, we are assuming that the basic rules of English contain the following two phrase-structure rules:

4. $\bar{S} \rightarrow COMP + S$
5. $S \rightarrow NP + VP$.

Within this framework, movement rules place a subset of words (*who, which, that,* and so on) in a complementizer position.

We are now in a position to assign highly simplified syntactic structures to the relevant portions of sentences 1–3:

6. The boat $[_S[_{COMP}$ which that$][_S$ you believe $[_S[_{COMP}$ that$][_S$ John painted t$]]$ is a yawl.
7. The boat $[_S[_{COMP}$ which that$][_S$ you believe $[_{NP}$ the claim $[_S[_{COMP}$ that$]$ $[_S$ John painted t$]]]]]$ is a yawl.
8. The boat $[_S[_{COMP}$ which that$][_S$ you asked $[_S[_{COMP}$ who$][_S$ t_1 painted $t_2]]]]$ is a yawl.

The symbol t stands for trace and marks the original position in the embedded clause of the noun phrase *the boat*, which has been moved to the complementizer position and replaced with the word *which*. The resulting complex complementizer *which that* in sentences 6–8 is ultimately reduced to the single complementizer *that* in sentences 1–3.

Given syntactic structures with this level of detail, we are in a position to identify the differences between sentence 1 and sentences 2 and 3. This statement requires that we identify the symbol called a bounding node, which has the property of constraining movement rules such as the one that moves *the boat* to the complementizer position. The elements in movement rules cannot be separated by more than one bounding node (Chomsky 1977). There is evidence that the node NP is the bounding node in all languages. In addition, most languages have a second bounding node that relates either to clauses or to propositions. In English the bounding node is related to the propositional content of the clause, S. In sentence 1 (structure 6), *which* moves to each COMP in turn and is never more than one bounding node from its previous position. In sentence 3 (structure 8), the lower COMP is filled, so *which* must move directly to the higher COMP. In sentence 2 (structure 7), the lower COMP is separated from the higher COMP by both NP and S. Thus, *which* would have to move over two bounding nodes at some point in the generation of sentences 2 and 3. Thus, sentences 2 and 3 violate the conditions on movement transformations in English.

Other languages appear to have different choices of bounding nodes. Rizzi (1978) has argued that in Italian the bounding node is S̄ rather than S, a situation that leads to the acceptability of sentences 1 and 3 and the unacceptability of sentence 2. Moreover, the acceptability or unacceptability of structures such as 1–3 is critically dependent on the fact that the constituent noun phrase, *the boat*, is moved without leaving an overt marker in its original position. There are languages (such as Hebrew) in which such a marker is found in the form of a resumptive pronoun, a different process than movement; in such languages, structures such as 2 and 3, with the appropriate resumptive pronoun in the embedded clause, are acceptable.

The analysis of constraints on movement transformations, and many other analyses couched in a similar vocabulary, have suggested properties of a mental organ for language. It seems clear that the types of knowledge structures that are a prerequisite for the statement of conditions on movement transformations are unconscious aspects of our knowledge of English. It is clear that children receive no specific instruction about such structures and constraints, and it seems highly improbable that any general learning strategy would arrive at an appreciation of such structures. On the contrary, it has been argued that the process of language acquisition is understandable only if we postulate that the child comes to the identification and analysis of certain "core" features of sentences such as 1 equipped with a rich conceptual apparatus in which only a small number of parameters, such as the choice of S or S̄ as bounding node, need to be specified (Chomsky 1977, 1982).

This analysis, and others like it, may also serve to illustrate and justify the concepts of the autonomy and the modularity of linguistic structures.

The issue of autonomy of the representations that constitute knowledge structures in a postulated mental organ for language involves the uniqueness and internal coherence of the formal representations needed to express linguistic knowledge. It is quite clear that contemporary cognitive psychology does not make use of substantive symbols or of the organization of such symbols, which appear in the description and explanation of linguistic regularities and which, by hypothesis, are the knowledge structures represented in a mental organ for language. The statement of conditions on movement transformations that I have just presented, even in this schematic form, is possible only within a very highly articulated framework of syntactic structure, which does not appear in existing theories of other cognitive abilities. The mental organ for language would thus appear to be separate from other faculties with respect to its particular knowledge structures, in the sense that it uses structures not found elsewhere.

The converse of this statement is less clear. There do seem to be plausible candidates for overlap between structures found in theories of other aspects of mental life and linguistic representation. For instance, a variety of logical notations have found a role in capturing linguistic regularities as well as in describing human reasoning, prototype theory seems to apply to linguistic as well as nonlinguistic cognitive categories, and the devices of recursion and transformation utilized in many versions of syntax are borrowed from mathematical formalisms and may find a place in other theories of cognition. Though a certain degree of caution is indicated in accepting the conclusion that these substantive and formal devices are identical in the domains of linguistic and nonlinguistic knowledge, we must certainly accept the possibility that language may borrow from a variety of formal and substantive sources and achieve representational autonomy only in the sense that it adds to the formalisms utilized elsewhere and not in the sense that none of its formal and substantive devices have parallels in other systems.

The final feature of linguistic representations that I wish to emphasize is their modularity. I have noted above that the statement of conditions on movement transformations requires a certain set of syntactic representations. It is equally important in the present context that it does not require other aspects of linguistic information. Thus, for instance, although some forms of lexical information (that is, information that depends on particular words) can influence phenomena such as the conditions on movement and on other transformations, there seems to be no case in which the actual phonological sound of a word is relevant to the statement of such conditions. In fact, the sound pattern of individual words is a highly complex linguistic structure. The fact that it does not influence other complex phenomena in natural language suggests that each of these phenomena depends on a separate system, and that these systems combine to yield the totality of linguistic structures only by the input and output to and from various subsystems of linguistic representations.

It should be borne in mind that the syntactical example I have chosen is but one of a larger number of similar examples of the autonomy and the modularity of linguistic representations. Though the study of syntax is the area in which these features of linguistic representations have been most frequently emphasized with respect to their implications for a postulated mental organ for language, exactly the same points emerge from studies of thematic relations (Bresnan 1982), metrical structures in phonology (Liberman and Prince 1977), morphology (Leiber 1980), and many other areas. If there is a mental organ for language, and if it does contain knowledge structures of the sorts I have very sketchily outlined here, it is indeed a rich system.

The second part of a postulated mental organ for language consists of a set of procedures that utilize linguistic structures of the sort I have sketched. Here, as in the area of knowledge structures themselves, the question of a mental organ for language is closely related to the question of the autonomy of these procedures. In very general terms, what we would like to know is whether the procedures that utilize linguistic knowledge are specific to this purpose or whether they are particular applications of mental procedures that operate on other sorts of mental representations. It is far beyond the scope of this chapter to attempt a plausible characterization and taxonomy of psychological procedures, and I shall again present simply one example of a psycholinguistic process (sentence parsing) for which the delineation of general features of the process itself and analyses bearing on its particular nature are fairly well developed and directly relevant. Like my presentation of the analysis of syntactic structures relevant to conditions on movement transformations, the example that follows omits much conflicting evidence and technical detail but, I hope, will serve the purpose of illustration.

Sentence parsing, the assignment of a syntactic structural description to an utterance, is increasingly being understood as one of a number of so-called "on-line" tasks that involve linguistic representations. A number of general characteristics of on-line processes have been suggested (Marslen-Wilson and Tyler 1980). On-line processes are unconscious, not permeable by systems of belief (Pylyshyn 1981), rapid, obligatory, and dependent on information in the physical signal ("bottom-up" in this particular sense), and they interact with other processes in constrained and efficient ways. Sentence parsing falls into the class of on-line processes when seen in these terms. The fragment of an analysis of a sentence parser that I shall present is due to Fodor and Frazier (see Frazier and Fodor 1978 and Fodor and Frazier 1980) and illustrates some of the specific properties that a sentence parser may have.

Consider the following sentences:

9. John bought the book for Susan.
10. John bought the book that I had been trying to obtain for Susan.

One's first interpretation of sentence 9 is that Susan is who John bought the book for, and one's first interpretation of sentence 10 is that what John bought was the book that I had been trying to obtain for Susan. On reflection, it is clear that both sentences are ambiguous; sentence 9 can mean that what John bought was the book for Susan and sentence 10 can mean that it was for Susan that John bought the book that I had been trying to obtain. These possibilities of interpretation, however, are clearly not those that suggest themselves in the first instance. This observation of the relative availability of these two interpretations of the sentence suggests that the parsing of these sentences, in the immediate, obligatory, rapid, unconscious manner that characterizes on-line psychological processes, arrives at the preferred interpretation before the second. This, in turn, suggests certain features of the human parser. Again, to see what these features are, we need a system of representation that will allow us to capture the differences between the two possible intrepretations of these sentences.

Figures 1.1 and 1.2 indicate a variety of differences between the two structures that underline the two possible interpretations of sentences

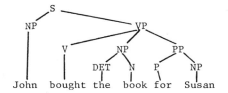

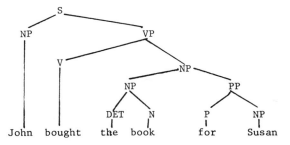

Figure 1.1

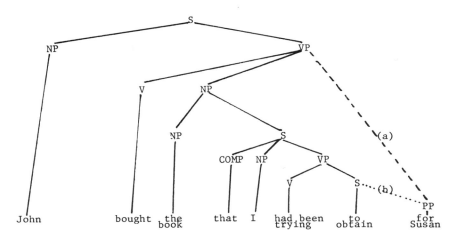

Figure 1.2

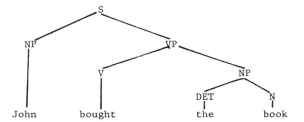

Figure 1.3

9 and 10. Following Frazier and Fodor, we see that the relevant aspect
of these structures is the structural configuration into which the node
PP is inserted in both these sentences. In the upper diagram in figure
1.1 the prepositional phrase *for Susan* is attached directly to the verb
phrase, whereas in the lower diagram it is attached to an NP that is
itself attached to the VP. In figure 1.2 the VP *for Susan* is attached
directly to a VP in both case a and case b and does not in either case
make this attachment by an additional node. This structural difference,
in this model, is critically important in deciding which is the preferred
interpretation of these two sentences. One aspect of this model postulates
that, when a structure (such as an NP) is identified in a sentence, it is
attached to the phrase marker that has previously been constructed
using the smallest number of nonterminal nodes—a principle called
minimal attachment (MA). Minimal attachment comes into play not
with the prepositional phrase *for Susan* in figure 1.1, but rather with
the previous noun phrase *the book*. Having identified *the book* as an
NP, the parser, following minimal attachment, will attach it directly
under the VP and will not postulate the second NP node intermediate
between the VP dominating *bought* and the NP dominating *the book*.
At the point where *Susan* is recognized as a PP, the phrase marker
constructed by a parser will have the form seen in figure 1.3. At that
point, there is only one place for the PP *for Susan* to be attached: as
a "sister" to the NP *the book*. Attaching the PP *for Susan* as in the
lower diagram in figure 1.1 would entail both adding an additional
node and revising the previously constructed phrase marker—something
a very natural principle of parsing would seek to avoid. Therefore, the
parser proceeds to attach the PP *for Susan* to the VP, with the result
that the sentence is interpreted to mean that it was for Susan that John
bought the book.

In the case of sentence 10, the partial structure constructed by the parser at the point where the PP *for Susan* is recognized is illustrated in figure 1.4. As can be seen, attachment of the PP *for Susan* to the VP dominating *bought* and to the VP dominating *obtain* are equally possible according to the principles thus far mentioned; neither attachment requires more nodes than the other and neither requires the revision of a previously constructed structure. What needs to be explained in sentence 10 is why the preferred and immediate interpretation of the sentence is the one in which the PP *for Susan* is attached to the VP dominating *obtain*, rather than there being a natural ambiguity which is immediately appreciated in these sentences. Here the answer lies in a third principle, which Frazier and Fodor term right attachment (RA): Terminal symbols optimally are attached to the lowest nonterminal node.

The phenomena can become considerably more complex. These principles are embedded within a two-stage model of parsing in the Frazier and Fodor proposals that leads to more detailed predictions about preferred and possible interpretations of other sentences. It should be noted that there are other models [in particular, augmented transitional network models (Wanner 1980)] that provide alternate analyses of these and other phenomena related to parsing. Nonetheless, this simplified example will serve to illustrate several important features of the human parser, itself taken as one aspect of the psychological processes utilizing linguistic representations.

As in the case of linguistic representations (and their correlates, the knowledge structures contained in a postulated mental organ for language), this analysis of the process of human parsing raises the issue of autonomy.

The first issue for autonomy is uniqueness. In the case of psychological procedures, we are interested in whether the substantive elements to which procedures apply, and the formal operations and organization of the procedures in this domain of human psychology, contain unique elements. Even this cursory and superficial presentation of some of the features of the human parser strongly suggests that the answer to these questions is affirmative; this analysis suggests that the parser makes use of substantive elements (certain types of linguistic representations) and consists of operations and constraints that, so far at least, have not found parallels in other areas of human psychology.

The second aspect of the autonomy issue with respect to psychological procedures utilizing linguistic information is whether they utilize portions

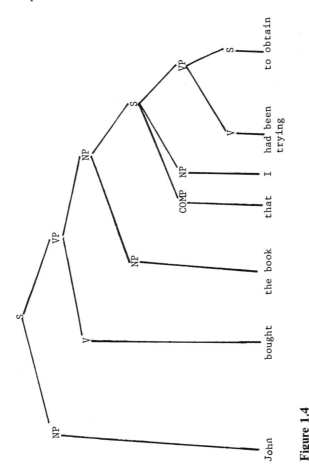

Figure 1.4

of mechanisms and procedures that are applicable to other domains of the intellect. This is largely an unexplored area. There are certainly many claims in the psychological literature, and particularly in the clinical psychological literature, that suggest that the answer to this is affirmative and that the processing of language is not autonomous in this respect. For instance, the frequent assertions that language processing makes use of the ability to sequence auditory events, or utilizes certain aspects of short-term and immediate memory, have, in many instances, been made in a spirit suggesting that these processes are applied to other domains of mental life. In fact, when one looks closely at the details of the processing of linguistic and nonlinguistic representations, the parallels are far from clear and in fact may break down. This is not to say, of course, that such a processor does not interact with other processes, such as the working memories of other cognitive systems.

The third issue that arises with respect to processes in a putative mental organ for language is their modularity. If we can accept on the basis of evidence such as that just given that the processing of human language involves a number of subsystems (such as a parser, a system of lexical access, and other such systems), we may ask whether each of these subsystems is encapsulated, such that it operates independent of each of the others, and interacts only via input and output. There are a number of suggestions in the literature that this may be the case. An oft-quoted example is a result due to Swinney (1979) that suggests that both semantic senses of a lexically ambiguous word are accessed when the word is heard in a sentential context, even in a situation where that context includes a strongly disambiguating prior segment. A variety of results, however, indicate that processes such as the accessing of lexical items are heavily influenced by ongoing analyses of other aspects of linguistic form, such as syntactic coherence of utterances, semantic expectations, and so on (Marslen-Wilson and Tyler 1980). Some models of language processing, both in the psychological literature (Morton 1979) and in the artificial-intelligence literature (Shank 1980), conceive of the process of identifying and producing linguistic representations as one that utilizes knowledge sources from, if not the entirety of human conceptual endowment, then at least a large part thereof. Here again, though the issues are not resolved, theoretical and experimental questions are becoming increasingly clear.

I have presented several simplified examples of the features of language structure and processing that have led to specific hypotheses about autonomous features of human mental life related to language. If the analyses I have sketched are at all convincing, they provide a preliminary and tentative characterization of the sorts of knowledge structures and procedures that constitute the mental aspects of an organ for language, and, coupled with the types of observations with which I began this essay, they suggest specific roles for such structures and processes in human psychology. Linguists, philosophers, and psychologists interested in language have presented variants of this argument for over two decades, and they have explicitly endorsed one implication of their position: that there is a neurological basis for this postulated mental organ, or that, as Klein (1977) suggests, "an (idealized) description of the language responsible cognitive structures as a neural structure is possible." In the third and final portion of this chapter I shall briefly explore this aspect of a mental organ for language, again focusing on the issue of autonomy.

The proposition that there is a mental organ for language that shows autonomy at the levels of knowledge structures and procedures does not entail or require the proposition that the neural basis for such a mental organ show the same properties (defined appropriately for neural structures). On the other hand, in the search for neural structures and operations underlying language, the existence of plausible arguments for the autonomy of representations and procedures in a mental organ for language surely encourages the search for a neural basis for language that contains structures and operations that are autonomous in neural terms (that is, either in the sense of being unique in terms of their physical elements or their organization or in the sense of being functionally committed only to language). The discovery of such autonomous neural structures would provide a strong and unified sense to the notion of an autonomous mental organ for language.

Before we can approach the issue of autonomy, however, we must settle on the nature of the neurobiological descriptions we shall use to characterize those neural structures related to language. I have claimed in reviews of this subject (Arbib and Caplan 1979; Caplan 1980) that the vast majority of descriptions of the neural basis for language are couched in terms of gross neuroanatomical descriptions, and I have argued that this level is empirically and conceptually inadequate. It is so critical to define and justify the terms of neurological analysis that it seems warranted to embark on a short digression from the theme of autonomy to present a characterization of this level of analysis and its shortcomings.

With exceptions, which are becoming more common and convincing, most of our knowledge about the relationship between language and the brain comes from inferences made from the study of pathological linguistic behavior in brain-injured subjects. The argument most often advanced relates a pathological performance to a deficit in the normal functional capacity of an organism, and correlates the performance (seen as a deficit) with a lesion in the brain. Since Broca's 1861 paper, the theoretically relevant level of neural description has almost universally been taken to be a macroscopic charcterization of the lesion site. The argument is that the function whose deficiency results in the pathological performance is the result of the operation of the area of the brain that is lesioned. The logic behind this argument has been criticized repeatedly, but, as has been argued more recently (Klein 1977; Caplan 1981), the argument is warranted if a number of additional assumptions are granted. Since virtually all disorders that are clinically recognizable as possible isolated disorders of language functioning follow left-hemisphere lesions, at least in the vast majority of right-handed adults, one particular application of this argument has been to assign to the left hemisphere (specifically, the "language zone" of the left hemisphere) the capacity for language in general. This claim, in one form or another, constitutes the theory of "cerebral dominance." Though the vast majority of papers in the aphasiological and neurolinguistic literature deal with these two concepts (functional localization in terms of gross neuroanatomical structures and the special case of functional localization, cerebral dominance), it would certainly be an error to allow these two concepts to exhaust the conceptual armamentarium with which we characterize the neural basis for a hypothesized mental organ for language.

This would be an error for both empirical and theoretical reasons. On the empirical side, the facts simply do not support the claim that, even in a stable population of mature right-handed speakers, the sole determinant of an aphasic syndrome is the locus of cerebral lesion. Different pathogenetic insults in grossly similar areas of the language zone produce syndromes that, even when described in the relatively coarse and inhomogeneous linguistic terms that characterize existing aphasia batteries, have been shown to be diverse. As more detailed linguistic analyses become available, we can anticipate that this diversity of behavioral manifestations of different types of lesions occupying the same grossly defined cerebral locus will increase. At the same time, the converse is also the uncomfortable truth: Specifying a syndrome does not specify a locus. At best, given a particular type of insult to the brain (such as stroke), specifying a syndrome (for instance, agrammatism) can carry the reasonable implication of a combination of a locus and a period in the temporal evolution of a lesion, as Mohr (1976) and Kertesz (1979) have shown. Thus, it would appear that, even with the investigation of the neural basis of the mental organ for language limited to what we believe to be a homogeneous population of right-handed adults, the empirical facts simply do not countenance the hypothesis, first enunciated by Broca with absolute clarity in his 1861 paper, that the sole determinant of an aphasic syndrome is the locus of lesion. What the observations minimally suggest is that, within a particular cerebral region, which varies in different populations, there are elements that can be disturbed by different pathogenetic processes at particular stages of their evolution such that language functions are specifically disturbed.

The level of macroscopic neuroanatomy is also inadequate conceptually. As Marshall (1980) has pointed out, simply gluing the components of a language processor onto the surface of the brain does not allow us to address questions regarding mechanism and, *a fortiori*, explanation. There is simply no known physiological function of grossly defined macroscopic areas of the brain that could serve as a physiology of language. Just as we are developing an understanding of the cellular and molecular biological correlates of some behaviors (Ullman 1979), we should be prepared to move from correlations of gross anatomical areas with functions—correlations that may express true generalizations—to a level of description of neural elements that offers the possibility of providing mechanistic accounts and explanations of the particular computational functions that underlie language.

The nature of these elements, their organization, and their physiology are almost unknown, and investigation at this level is an important present and future area of research. It is perhaps reasonable to speculate, as did Arbib and Caplan (1979), that these elements will be of the same electrochemical type as those that charcterize "information processing" in other species (as opposed to those neural elements and events that are related to other functions, such as endocrine functions of the brain). But it should be clearly recognized that this is no more than a speculation—indeed, one that prejudges the question of autonomy of these elements. Chomsky has, in fact, suggested the contrary, even to the point of suggesting that new forms of energy might yet be discovered to be relevant to brain processes, including those involved in language.

These empirical and theoretical considerations, then, argue for the existence of a yet undescribed level of neural structure and function at which a variety of factors—gross localization as a function of genetic load, early injury, etc.; age; possibly sex and literacy; type and evolution of pathological insult; possibly bilingualism—can combine to produce particular language pathologies. This level, by a logic identical to that leading to the conclusion of a functional localization, could be related to the representation and processing of linguistic structures in the normal state. We could, if we knew more about this level, pose the question of structural and functional autonomy about it, but it is clear that serious investigation of these questions is only in its infancy.

Despite these caveats, what we know about the relationship of language functions to macroscopically defined areas of the brain does inform us in rather interesting ways about some of the properties that these yet to be discovered elements have. For instance, as Mehler and Morton have pointed out in an unpublished paper, if we think of the dominant hemisphere as the neural basis of a postulated mental organ for language, such an organ is different from other organs in the body insofar as its removal at early stages of life leads to disturbances in its function that are extremely subtle rather than to a failure to exercise the function at all. The "plasticity" of the nervous system, in this respect, is, as Mehler and Morton point out, quite mysterious, and we are just beginning to explore it at the cellular level. At the functional level, questions relating to plasticity involve the exact role of the nondominant hemisphere in child language, the point at which it can no longer take over certain linguistic functions after insult to the left hemisphere, the nature of linguistic representations in the isolated right hemisphere, and other related matters. These questions are also in the early stages of investigation, but it seems clear that the neural basis of language does show a degree of flexibility that is not present in the neural basis of many aspects of sensory-motor function and possibly other various aspects of cognition, and is certainly quite unlike the dependence of most biological functions on a specific organic substrate.

A second important aspect of the behavior of these unknown neural elements is that their spatial localization in the brain is influenced to a considerable degree by normal genetic and environmental factors, such as those mentioned above. Some of the most significant neuropsychological work of the mid twentieth century (Zangwill 1960; Luria 1947) has explored the systematic variation in the neural basis of psychological function in various populations, and it is now abundantly clear that the interhemispheric and intrahemispheric organization of cognitive functions (including language), as determined by clinical neuropsychological testing, varies significantly in these various populations. If the knowledge structures and procedures of a mental organ for language are the same in these various human populations, this determination of different locales for these functions by such normal factors is biologically rare and poses a challenge for neurobiological theory.

Despite the variations in the locales of the elements responsible for language in various populations, and despite the fairly frequent exceptions to standard localizations in right-handed adults, there does seem to be a remarkable degree of specificity of the functional disabilities in the sphere of language that are occasioned by brain lesions in different sites, and this fact does suggest that the neural basis of language may prove to be modular and perhaps autonomous in terms of its physical elements and organization.

Thus, analyses at the macroscopic level, though best taken as a shorthand for descriptions at an empirically and conceptually more adequate level, have revealed a number of important features of the neural basis of a postulated mental organ for language. I will conclude by considering what such analyses might indicate about the autonomy of such an organ at the neural level.

There are two senses to the term autonomy at the neural level, to which I alluded earlier: the sense of uniqueness of physical structures and physiological processes and the sense of a unique functional role (language representation and processing) which a neural structure may play. With respect to the first of these senses of autonomy, it has been argued that parts of the language zone (such as the inferior parietal lobule) show sulcal complexity, patterns of connectivity, histological characteristics (which, since their function is unknown, can be taken as markers of a macroscopic area and not as a theoretically relevant description in cellular terms), and other features not found in nonhumans (Geschwind 1965). The work on lateral asymmetries in the language areas of the brain suggests that macroscopically defined areas vary in size from side to side in ways that are roughly consistent with the claims that larger areas support language (Galaburda et al. 1978). There are many important questions of fact and interpretation in this work. If these claims prove correct, they strongly suggest uniqueness of the neural structures relevant to language, both across phyla and within the human brain.

The second aspect of autonomy with respect to neural structures is whether those areas and elements in the brain that subserve language are also involved in other functions. It has classically been argued that there are disturbances that can be interpreted as (at least relatively) pure linguistic deficits, and that these disturbances follow neocortical lesions in the perisylvian area in one hemisphere. Increasingly, the aphasic syndromes identified in traditional clinical neurology and subsets of those syndromes determined by more careful and detailed descriptive taxonomic efforts (Tissot, Mounin, and Lhermitte 1973) are being analyzed in terms related to abnormal representation and utilization of particular classes of linguistic structures. Recent analyses have argued that stable syndromes of functional disability following focal cerebral lesions might be best interpreted as disorders of one or another autonomous representational or processing unit of a mental organ for language. Such suggestions have been made in the analysis of agrammatism, by Kean (1980) and Caplan (1983) with respect to linguistic representations, and by Bradley, Garrett, and Zurif (1980) and Berndt and Caramazza (1980) with respect to particular components of a language processor. Such analyses constitute the basis of a functional localization.

Demonstrating functional autonomy of the language-related areas of the brain on the basis of pathological material requires two additional steps: showing that no other lesion produces the same functional impairment and showing that no other functional impairment results from the lesion in question. The first of these steps has been addressed in considerable detail in recent work. Berndt and Caramazza (1980) have been at pains to demonstrate that superficially similar abnormal performances of Broca's aphasics and conduction aphasics on comprehension tasks involving syntactic structure reflect different mechanisms—failure of a parsing mechanism in the Broca's aphasics and failure of a short-term memory disturbance in the conduction aphasics. The second requirement is obviously false; the lesions causing aphasia obviously occasion other intellectual, as well as sensory-motor, disturbances. The issue here, in the first instance, is whether such disturbances are simply secondary effects of the language disturbance. In the second instance, we must reconsider the issue of levels of descriptions. It could well turn out that an area of the brain serves two functions because it contains two different structural elements and/or physiological operations. What we need to know is whether the elements of the brain that are relevant to linguistic function are the same ones that are relevant to some other function under their theoretically appropriate descriptions, not simply because they happen to be in the same gyrus. Here again, we simply need to know more about what brain structures and operations are involved in language to answer these questions.

In summary: The concept of a species-specific and domain-specific cognitive capacity for language, itself composed of a number of independent modules of representations and procedures, has emerged from contemporary linguistic studies and lies at the conceptual center of much of modern theoretical cognitive psychology (Pylyshyn 1980). I have reviewed some of the evidence for such a concept, presented a few examples of some tentative suggestions as to the nature of such a mental organ, and attempted to raise several questions about our current knowledge regarding this concept at three relatively accessible levels of human biology: mental representations, psychological procedures, and neural structures. This general framework and the many specific hypotheses articulated within it in contemporary linguistic, psycholinguistic, and aphasiological work can serve to integrate and stimulate investigations of human language and its biological basis.

References

Arbib, M. A., and D. Caplan. 1979. Neurolinguistics must be computational. *Behavior and Brain Science* 2: 449–483.

Berndt, R., and A. Caramazza. 1980. A redefinition of Broca's aphasia. *Applied Psycholinguistics* 1: 1–73.

Bradley, D. C., M. Garrett, and E. Zurif. 1980. Syntactic deficits in Broca's aphasia. In D. Caplan, ed., *Biological Studies of Mental Processes*. Cambridge, Mass.: MIT Press.

Bresnan, J. 1972. Theory of Complementation in English. Ph.D. diss., Massachusetts Institute of Technology.

Bresnan, J., ed. 1981. *The Mental Representation of Grammatical Relations*. Cambridge, Mass.: MIT Press.

Caplan, D. 1980. Reconciling the categories: Representation in neurology and in linguistics. In M. A. Arbib, D. Caplan, and J. C. Marshall, eds., *Neural Models of Language Processes*. New York: Academic.

Caplan, D. 1981. On the cerebral localization of linguistic functions. *Brain and Language* 14: 120–137.

Caplan, D. 1983. Syntactic competence in agrammatism: A lexical hypothesis. In M. Studdert-Kennedy, ed., *The Psychobiology of Language*. Cambridge, Mass.: MIT Press.

Chomsky, N. 1977. *Essays on Form and Interpretation*. Amsterdam: North-Holland.

Chomsky, N. 1982. *Lectures on Government and Binding*. Dordrecht: Foris.

Fodor, J. D., and L. Frazier. 1980. Is the human sentence parsing mechanism an ATN? *Cognition* 8: 417–459.

Frazier, L., and J. D. Fodor. 1978. The sausage-machine: A new two-stage parsing model. *Cognition* 6: 291–325.

Galaburda, A. M., M. LeMay, T. L. Kemper, and N. Geschwind. 1978. Right-left asymmetries in the brain. *Science* 199: 852–856.

Geschwind, N. 1965. Disconnection syndromes in animal and man. *Brain* 88: 237–294.

Kean, M. L. 1980. Grammatical representations and the description of language processing. In D. Caplan, ed., *Biological Studies of Mental Processes*. Cambridge, Mass.: MIT Press.

Kertesz, A. 1979. *Aphasia and Associated Disorders*. New York: Grune and Stratton.

Klein, B. 1977. What is the biology of language? In E. Walker, ed., *Explorations in the Biology of Language*. Montgomery, Vt.: Bradford.

Liberman, M., and A. Prince. 1977. On stress and linguistic rhythm. *Linguistic Inquiry* 8: 249–336.

Lieber, R. 1980. The Organization of the Lexicon. Ph.D. diss., Massachusetts Institute of Technology.

Luria, A. R. 1947. *Traumatic Aphasia*. Translated. The Hague: Mouton, 1971.

Marshall, J. C. 1980. On the biology of language acquisition. In D. Caplan, ed., *Biological Studies of Mental Processes*. Cambridge, Mass.: MIT Press.

Marslen-Wilson, W., and L. Tyler. 1980. The temporal structure of spoken language understanding. *Cognition* 8: 1–71.

Mohr, J. P. 1976. Broca's area and Broca's aphasia. In H. Whitaker and H. A. Whitaker, eds., *Studies in Neurolinguistics*, vol. 1. New York: Academic.

Morton, J. 1979. Word recognition. In J. Morton and J. C. Marshall, eds., *Psycholinguistics 2*. Cambridge, Mass.: MIT Press.

Morton, J., and J. Mehler. On the biology of language. Unpublished.

Pylyshyn, Z. 1980. Computation and cognition: Issues in the foundations of cognitive science. *Behavior and Brain Science* 3: 111–132.

Pylyshyn, Z. 1981. The imagery debate: Analogue media versus tactic knowledge. *Psychological Review* 87: 16–45.

Rizzi, L. 1978. Violations of the WH-island constraint in Italian and the subjacency principle. In C. Dickinson, D. Lightfoot, and Y. C. Morin, eds., Montreal Working Papers in Linguistics, vol. 11.

Shank, R. C. 1978. What makes something *ad hoc*? *Theoretical Issues in Natural Language Processing* 2: 8–13.

Swinney, D. A. 1979. Lexical access during sentence comprehension: (Re) consideration of context effects. *Journal of Verbal Learning and Verbal Behavior* 18: 645–659.

Tissot, R. J., G. Mounin, and F. Lhermitte. 1973. *L'Agrammatisme*. Brussels: Dessart.

Ullman, S. 1979. *The Interpretation of Visual Motion*. Cambridge, Mass.: MIT Press.

Wanner, E. 1980. The ATN and the sausage machine: Which one is baloney? *Cognition* 8: 209–225.

Zangwill, O. 1960. Cerebral dominance and its relation to psychological function. Edinburgh: Oliver and Boyd.

Chapter 2

Neural Mechanisms,
Aphasia, and Theories of
Language

Norman Geschwind

We are living in a golden age in the field of the neurology of behavior in general and in the field of the neurology of language in particular. Startling advances have occurred on every front. During the years since World War II, fields previously barely developed have flourished. Neurolinguistics, neuropsychology, and the anatomical study of language systems have all advanced in concert to create a level of sophistication previously unknown and unexpected.

There is, however, a jarring note that still pervades the field, a hangover from the unfortunate period between the wars when this field not only advanced very little but even suffered from a phenomenon that we scientists would like to believe does not occur: an actual regression of knowledge. Important discoveries of the first golden age, the pioneer period of our field, were forgotten or misrepresented and had to be rediscovered so that our current fruitful advances could be achieved. I refer in particular to the tendency to continue reviving the hackneyed half-truths and sophistries that were produced in the form of an endless set of apparently witty aphorisms that were supposed, erroneously, to be profound philosophical insights, and that contributed the main stock in trade of those who methodically attempted to destroy what was useful in the past while failing to add significantly to the corpus of information or useful theory. This is not a unique phenomenon in the history of science. Nothing was so effective as Darwin's theory in leading to the production of endless witticisms. One set of these appealed to the distaste of most people for the idea that their ancestors were hairy, grunting primates, lizards, and slime molds. Another set attempted to make a mockery of those who had the supreme effrontery to attempt by empirical observation to disprove what had been established beyond question by great thinkers of the past. It is, therefore, particularly odd to discover that Bishop Wilberforce is not dead, which is in itself a striking confirmation of his own belief in immortality. The advance of science having apparently given to Thomas Huxley the laurels of victory in their famous debates on evolution, Wilberforce has now returned, only thinly disguised, espousing pure reason, railing against messy empirical observations, rejecting evolution, and producing a comparable set of empty witticisms. These appeal again to the arrogance of humans in their desire for uniqueness in the order of the universe.

In an earlier era it was proved beyond question that no empirical methods could reach to the core of the soul that set us off from the beasts; the vitalists, somewhat broader in their scope, asserted that the essence of life was forever to be inaccessible to those who naively thought that puttering around with measurements of blood gases or pH could ever capture the true essence of the *élan vital*.

A war of aphorisms is a waste of time. Aphorisms encourage the substitution of shallow wit for real thought, and in the end their success is only a reflection of what has amply been pointed out repeatedly: that even sophisticated wit is hostile. The occasional witticism can hardly be objected to, but in this field the same tired phrases reappear, as if it is necessary constantly to avoid hard thought. Thus, it is almost impossible to attend a conference on aphasia without being told that "the localization of symptoms should not be confused with the localization of function," that "what language is is more important than where it is," that "the noise produced by a missing tooth in a flywheel does not prove that the tooth was there in the first place to prevent noise." Indeed, any one of these might be approriate in a discussion of some specific point, but they are instead produced in large numbers by many speakers.

There is no need for us to relive the early debates on evolution in our own field. In the end, biology is a science that depends on detail, and since we are all in a biological field we will have to look repeatedly at the validity of particular explanations of phenomena. No amount of wit or broad philosophical thinking can free us from that often onerous method of progression.

Since this volume is entitled *Biological Perspectives on Language*, I will discuss primarily some contributions that illustrate the point of my opening comments. Many advances in this field will be held up if the unfortunate tendency to downgrade the role of neurology is not checked. It was exactly this tendency that led to the unfortunate decline in the study of aphasia in the dark ages between the two world wars, a decline that harmed psychology, linguistics, and neurology alike.

The chapter by Zurif [chapter 8] is an excellent summary of many of the exciting discoveries in the linguistic study of aphasia in recent years. It is unfortunate, however, that so excellent a document should be marred by errors in scholarship. The section on Broca's aphasia begins with a historical review that is at best incorrect. Zurif notes that there have been observations that run "counter to the clinical dictum that comprehension is relatively preserved in Broca's aphasia." This is a careless use of words. Many authors have pointed out problems in comprehension in Broca's aphasia. Furthermore, the fact that there was special difficulty in grammatical comprehension appeared in the 1920s in the writings of Salomon and Bonhoeffer. Indeed, the presence of such disorders was repeatedly pointed out in the aphasia conferences of the Boston Veterans Administration Hospital Aphasia Unit from the earliest days of that organization, and it was these clinical observations that provided a major impetus to much of the recent organized study of this type of comprehension deficit. Furthermore, the very data advanced later in Zurif's chapter support the dictum that comprehension is indeed relatively preserved in Broca's aphasia, since even the most casual observation shows that the type of defect so beautifully characterized in recent studies has much less impact on the ability to comprehend the sense of a message than the type of comprehension disorder present in Wernicke's aphasia.

The statement that "the linguistic activities of speaking and listening were taken whole and the differences between them drawn in terms of contrast between sensory and motor systems" also misrepresents the authors cited, such as Wernicke and myself. It is clear that what hampered the study of aphasia for many years was precisely the widespread existence of a classification system that was supposedly based on the notion that if a patient spoke incorrectly he had a lesion in Broca's area and if he did not understand he had a lesion in Wernicke's area. If he suffered from both deficits, he was presumed to have a lesion in both areas. Yet this classification was not present in the writings of Wernicke and is not supported by my writings. I have repeatedly objected to this sensory-motor classification precisely because it is an inaccurate representation of aphasia. It is clear, for example, that Wernicke showed that the types of abnormal language produced by Broca's and Wernicke's aphasics were dramatically different. Furthermore, Wernicke made very clear that a lesion in the superior temporal gyrus produced difficulty in all forms of language. It is surprising that these facts could be presented so inaccurately.

It should also be pointed out that the studies of agrammatism at the time of this writing have avoided facing some extremely important observations. The Broca's aphasic will tend to show agrammatism in speaking and in repeating, and will have difficulties in "grammatical comprehension." One must, however, also account for the observation that, although many conduction aphasics produce fluently abnormal speech that typically lacks the markedly telegrahic quality of Broca's aphasia, many conduction aphasics will, in repetition, produce agrammatic utterances. It is obviously necessary to incorporate these data into any theory of the handling of closed-class words. The clinical. observations clearly point again to an important area for research. It seems likely that any adequate theory will have to take account of both linguistic facts and the anatomical organization of the language systems.

The chapter by David Caplan again contains a considerable amount of useful information and conceptual organization. Unfortunately, however, it presents as generally accepted a number of assumptions that at the present time are simply unproven.

At one level, it may be useful as a research strategy to work with the notion that language is an "autonomous" property of the human. This kind of assumption of autonomy is often useful as a research strategy even in such fields as immunology and endocrinology. It is a serious mistake, however, to confuse a research strategy designed to permit the construction of experiments with ultimate biological truth. The assumption of autonomy of language seems to sidestep almost completely the question of the biological advantage to humans of the high development of this capacity. Language must serve many functions. Among these must be included its value as an instrument for transmitting information about the world to other members of the species. We know so little about how this particular goal is achieved that we can hardly be entitled to make the notion of autonomy as absolute as this paper does.

Can we really argue that language is unique to humans? The history of biology contains too many examples of unique human features that have lost this exclusive position. One of the major lessons of the Darwinian theory of evolution is that precursors of any biological trait may well appear superficially to be different from the form achieved in some later evolving organism. Curiously enough, it is extremely difficult to give a definition of language. I suspect that we are far from knowing all of the underlying mechanisms necessary for this function, some of which may well be present in lower species. Perhaps language is totally unique as an evolutionary trait. We know, however, that the human brain has marked similarities to those of other mammals, in particular the higher primates. Until we know much more about the functions of the brain in language, we are at great risk in assuming something that runs counter to all biological experience in other areas. I do not wish to deny absolutely the uniqueness of human language, but I would argue that agnosticism is at this moment more appropriate than absolute faith.

Caplan states that aphasic syndromes are being increasingly analyzed as functional syndromes related to particular classes of linguistic structures, or that these disorders can be interpreted as malfunctions of one or another autonomous representation within the "mental organ for language." This statement is, however, a somewhat misleading representation of the facts. It is not clear that the agrammatic repetition of the conduction aphasic represents damage within such a modular unit. Indeed, the assumption of modularity, which is implicit in Caplan's statement, is simply a restatement of an untenable theory of aphasia: that each syndrome represents the destruction of a particular center. This view depends on an implicit idea of the neurological organization of the brain that is inconsistent with the facts, since it fails to consider the connections within the system. Caplan's statement rests on another untenable assumption: that the result of damage in any particular place leads to the ablation of some particular function. Yet there is ample evidence to show that in many syndromes the errors made depend on the use of anatomical pathways and groups of nerve cells that would not normally be used for the function under consideration. Finally, we must not neglect certain phenomena that have as yet evaded explanation on purely linguistic grounds. As I have pointed out many times, patients with Wernicke's aphasia, who fail to carry out verbal commands with the individual limbs or with the face, will correctly carry out "axial" movements (movements with the eyes or the trunk) even when the commands are syntactically identical to limb commands that they fail to execute. As I have pointed out, the only reasonable theory now existing for the curious preservation of this class of verbal commands is one that relies on knowledge of the different types of motor organization within the nervous system.

Curiously enough, when Caplan rejects the idea that the sole determinant of an aphasic syndrome is the locus of damage, he only repeats something that has been present in the literature on aphasia at least since Wernicke, who suggested in his classic monograph that there could be several reasons why damage in one particular location did not always produce the same syndrome. Yet Caplan does not follow through to its logical conclusion the implications of this well-known fact. It is certainly true that lesions in childhood in the left hemisphere are less likely to leave a permanent aphasia than lesions in the same location in adult life. Is it, in fact, true that the lesion is the same in the two cases in which one gets different effects? Is it not the case that only detailed knowledge of the neurology of the system at different points in development will enable us to explain these variations? Every textbook of anatomy contains pages and pages on variations in the structure of bones and the distribution of blood vessels, which make it possible to explain many variations in clinical symptomatology. Is it not likely to be the same in the case of the nervous system? Indeed, the very study of the neurological basis of such variations is one of the most important research frontiers of the future. Furthermore, the brain of the child is changing structurally, and thus a lesion in the early years destroys part of a system that is neurologically different from that of the adult.

Caplan goes on to cite Mehler and Morton to the effect that the language-dominant hemisphere is "different from other organs in the body insofar as its removal at early stages of life leads to disturbances in its function that are extremely subtle rather than to a failure to exercise the function at all." Caplan goes on to support the assertion by these authors that the plasticity of the nervous system is "quite mysterious." All of the assertions above are incorrect and reveal unawareness of the facts. The language-dominant hemisphere is indeed not unique. Consider even so ancient an organ as the cerebellum, which is certainly present throughout the vertebrate scale. It is well known that removal of the cerebellum in infancy or early childhood leaves the child with remarkably little impairment in comparison with what a similar removal does in adult life. Indeed, the literature on neurology is replete with examples of this type of plasticity, and the biological literature shows it both for the nervous system and for non-neurological systems.

The assertion that the plasticity of the nervous system is just another mystery reveals an unawareness of the rich literature on this topic, much of it going back many years. The mechanism of increased sensitivity to neurotransmitters after removal of the nerve supply of a particular structure has been known since the classic studies of Cannon and Rosenblueth. In recent years there have been many documentations of changes in the numbers of receptors for particular transmitters under different circumstances, of the appearance and disappearance of enzymes that metabolize certain transmitters, and of many other mechanisms. Furthermore, there is a rich literature on sprouting (the growth of the ends of nerve fibers) under a wide variety of conditions. There are many volumes in the literature that can be consulted for extensive detailed discussions of the mechanisms of plasticity. There are, furthermore, studies that document the differences in type and extent of plasticity at different times, such as during intrauterine life, in childhood, and in later life.

In brief, the statement by Caplan that "extending the notion of localization to account for these phenomena does not seem to be accomplishable in any satisfactory way" is a remarkable one that can be made only by neglecting a vast and detailed literature.

A later statement is made that "a neural analysis in terms of gross neuroanatomy cannot provide, as best we know from studies of subhuman primates, a vocabulary appropriate for the representation of the kinds of structures and processes which we have identified in language." I have a great deal of difficulty in understanding this assertion. If it simply means to say that gross anatomy in the strict sense of the term (that is, anatomy without reference to the details of cellular organization) will be inadequate for a detailed analysis of function, then I can only say that I know of no one interested in the anatomical analysis of brain and behavior who would disagree. Obviously, any biological analysis will inevitably move toward more and more detailed information. On the other hand, one should recall that no analysis of linguistic performance based purely on human utterances is likely to carry us by itself to an understanding of the most profound levels of language. I would suggest, in fact, that as information becomes more and more precise there will be a continuous interplay between linguistics and knowledge of neural organization. The studies of anatomical organization will again and again be guided by linguistics, but it is extremely likely, as has been the case in every instance in which a function has been analyzed in terms of its structural substrate, that the structure of linguistics itself will be changed.

I believe that the examples I have quoted here amply illustrate my point. It is not my intention to single out these authors, since they have made extremely important contributions. What mystifies me is why so much effort is devoted to the construction of propositions about the nervous system that are either obvious, untrue, or unprovable at the present time. I continue to have the feeling that we are reliving the great arguments that went on during the classic debates on the theory of evolution. It is quite useful to recall that the distinguished anatomist Richard Owen argued that the human brain was unique in comparison with those of other animals—an argument that was effectively destroyed by Darwin's great protagonist, Thomas Huxley. One of the major impacts of the theory of evolution was that, like the discoveries about the solar system in an earlier period, it removed humans from a unique position in the order of the universe. The Earth is no longer the center of the universe. Man is not unique physically among the creatures on Earth, and bears striking resemblance to primitive ancestors from which he has evolved. Even his brain can clearly be shown to be the outcome of further stages in a long evolutionary process.

I have the feeling that somehow the last bastion of uniqueness of the human is, in the minds of some, his possession of language, so that one finds the events of the first part of Genesis being revived among distinguished scholars. Yet we must be able to face the fact that perhaps this last fortress of human uniqueness may also fall. Perhaps we should be cautious about assuming that language will take the place of the soul in keeping us in a special position, different from that of the beasts of the field. If there is truly to be a neurolinguistics and a neuropsychology, then these fields must live with knowledge of the brain, just as research on the brain must continue to be influenced by the great discoveries in these disciplines. The idea that somehow they are in opposition to each other is an unfortunate one.

Let me close by again repeating that it is naive to assume that neurological discoveries will only serve to show how the brain succeeds in modeling some particular current theory of language. The history of biological research shows again and again that detailed knowledge of biological mechanism, although frequently guided by functional analyses, inevitably leads to important alterations in theories of function. We have no reason to assume that this will not be true in the case of language. Furthermore, we should recall that the best studies of behavioral effects of disease occur in precisely those disorders on which there is a considerable amount of anatomical knowledge. The high sophistication of modern psychological and linguistic studies of aphasia has been attained in a period in which anatomical knowledge has grown rapidly. By contrast, the confusion about the basic disorder of function in schizophrenia is paralleled by the lack of knowledge of the neurological substrate. I would strongly suggest that functional analysis always proceeds more rapidly to higher levels of sophistication when basic biological analysis advances.

Finally, we must recall that, if aphasia is a result of disease of the brain, then surely many of the methods of treating it will eventually depend on manipulations of the brain. The attempt to exclude neurological knowledge only removes another possible means of alleviating the suffering of those afflicted with this disorder.

Acknowledgments

This work was supported in part by grant NINCDS 06209 from the National Institutes of Health to the Department of Neurology of the Boston University School of Medicine and by grants from the Orton Research Fund and the Essel Foundation to the Beth Israel Hospital, Boston.

Chapter 3

Brain-Based and Non-Brain-Based Models of Language

John Morton

We have a number of lessons to learn from history. If we are lucky we can avoid making the same mistakes as thinkers in the past. The purpose of this chapter is to show the relationship between the work of the "diagram makers" of the late nineteenth century and current information-processing theory in cognitive psychology. The lesson to be learned is that the work of the early theorists was ignored because of a confusion of objectives. Specifically, what was confused was the objective of representing the elements of processing in the brain and the objective of establishing the localizability of these elements.

Let me start with a quotation from an otherwise thoughtful (i.e., approving) review of the recent book *Deep Dyslexia* (Coltheart, Patterson, and Marshall, eds.; London: Routledge & Kegan Paul, 1980): "Briefly, with regard to the other chapters: Morton and Patterson's review paper on the logogen model provides a theoretical framework for much of the clinical data. Though I would defer comment on the model to those more knowledgeable in this area, I confess to some antipathy to its general style, the discussion of 'routes' between components which have no known neural correlates." (Brown 1981, p. 389) This comment can be countered by one in which a writer states that his objections to current theorizing are "intimately related to the idea of 'localization,' i.e. of the restriction of nervous functions to anatomically definable areas, which pervade the whole of recent neuropathology." (Freud 1906, p. 1).

I do not intend to discuss the issue of localization. This issue is irrelevant to my concerns, which are to characterize the operations of the normal brain and to describe the effects of brain damage in individual patients in terms of the resulting psychological model. These are not the only tasks that face us, however. There follows the question of how the elements of the model are implemented in the brain. There is also the question of what functions are performed in particular anatomical locations. I believe these are important questions, but that the answers do not affect the form or the nature of the psychological model. The case for such distinctions has been made recently by Caplan (1981) and Marshall (1980), and I will not argue it here in detail. Rather, I will show some relationships between the work of 100 years ago and present work in the hope of illustrating what has been lost through the conceptual confusion.

On Diagrams

The problem for someone with a complex idea is to represent the idea. The form of the representation is the key to communication. My own theories, which focus on the flow of information, have always been represented in the form of diagrams resembling flow charts. I find diagrams an aid to thinking, but I have come to learn that many people cannot work with them. One reason for this may lie in a distinction between visual thinkers and verbal thinkers. Another, deeper reason relates to whether one wishes to discuss behavior or the mechanisms underlying behavior. It is the latter that the diagrams portray, and behavior may be related to diagrams in complex ways.

In the field of neuropsychology, the equivalent dichotomy is between a psychological deficit and a functional deficit (which may be localized; see Caplan 1981). The nineteenth-century diagram makers were often confused between these two because they believed in localization of function. However, I am going to ignore this aspect of their thinking and compare a number of the diagrams as if they were simply information-processing models. Some of the models were arrived at purely *a priori* and then applied by their makers to brain-damaged patients; other models resulted from careful clinical analyses of large numbers of patients. However, in the comparisons that follow I will be concerned only with the topological aspects of the models.

What I will try to do is reduce all the models to a common form. In doing so I will make one simplification and take one liberty. The simplification is to extract those aspects of the models that are concerned with the processing of single words: reading, oral comprehension, repetition, and speech. Writing will be ignored, partly because not all the diagram makers considered it and partly because current theory has less to say about the distinction between speech and writing than about the other distinctions (but see Newcombe and Marshall 1980). The

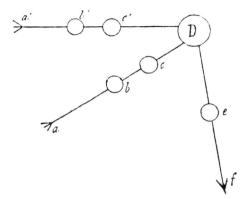

Figure 3.1

liberty I will take concerns the nature of the elements in the models. The models share the convention of diagramming "centers" of various kinds, joined together by pathways. In general, the centers are either modality-specific (that is, concerned exclusively with auditory or visual stimuli or with speech) or modality-nonspecific (such as a "center of ideas"). The different theorists ascribed varying amounts and kinds of processing to, say, the auditory center. Such distinctions will be ignored in the interest of making the comparisons among the different models easier to follow.

Many of the versions of the diagrams that follow have been taken from Moutier 1908. A few errors in this source have been spotted; no doubt many more will have been unnoticed.

The Diagram Makers

The earliest model, according to Moutier, was that of Baginski (1871). This is shown in figure 3.1. It makes a distinction between two levels of sensory analysis, "perception" and "elaboration." The model is equivalent to that illustrated in figure 3.2, in which the modality-specific functions have been collapsed into single functions. One implication of this model is that reading can proceed directly from vision. This is in contrast with other nineteenth-century (and twentieth-century) models that require that print be turned into either an acoustic or an articulatory form before it can be understood. This model is also formally equivalent to that of Langdon (1898) (figure 3.3), which distinguishes between writing, speech, and "mimicry" (that is, speech repetition without the necessity of comprehension) as output functions and also makes a variety

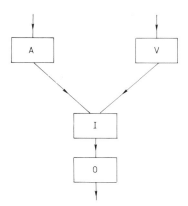

Figure 3.2

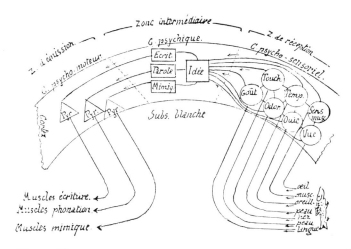

Figure 3.3

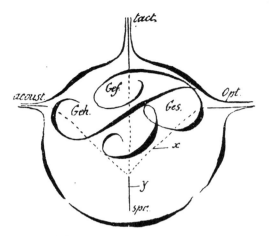

Figure 3.4

of sensory distinctions. Despite the anatomical flavor, the diagram conveys the same idea as Baginski's. This is also true of Moeli's (1891) model (figure 3.4), which further indicates the lesions leading to a couple of disconnection syndromes.

A slightly more elaborate model put forward by Ballet (1886) (figure 3.5) allowed for a direct connection between modality-specific input processes and modality-specific output processes. Thus, reading aloud from V (the visual center) to P (the speech center) could go either via the "center of the intellect" or directly via the pathway numbered 6 in the diagram. The same is true of Grasset's (1896) model (figure 3.6), in which the connections among all the modality-specific centers are direction-free.

In essence, Ballet's model is topologically identical to a simplified version of the current logogen model (Morton 1979). In the version of this model shown in figure 3.7, the direct pathways between input and output represent a lexical level. These paths could not be used to read or repeat nonwords. In the full version of the model there are also input-output paths that are nonlexical to allow for the pronunciation of new words, names, or nonwords. It is not clear whether lexical or nonlexical processes were intended by Ballet or Grasset. However, for our present purposes, this distinction is not vital.

A very different concept is represented by Kussmaul's (1877) model (figure 3.8). Here, the auditory center (*B*) is involved in speech (by the route *cbd*) and the visual (reading) center (*B'*) is involved in writing

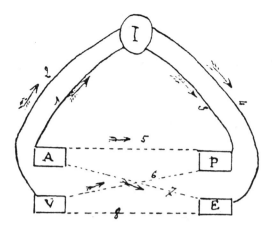

Figure 3.5

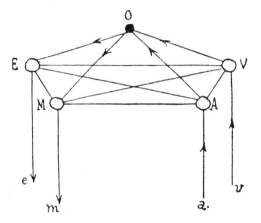

Figure 3.6

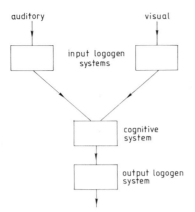

Figure 3.7

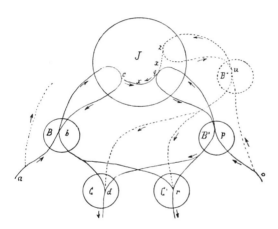

Figure 3.8

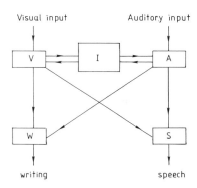

Figure 3.9

(by the pathway *qpr*). The model also allows for repetition without understanding (pathway *abd*) and reading without understanding (pathway *opd*). Kussmaul also indicated that he believed that the deaf operate in essentially the same way as the hearing. In figure 3.9 I have translated Kussmaul's model into the notation used in figure 3.2, leaving out the writing center and the special pathways for the deaf. Suggestions similar to this, in the sense of having input and output functions with shared components, have been put forward recently by Allport and Funnell (1981). Seymour's (1973) model for the processing of words and pictures (figure 3.10) is based on the same principle. As figure 3.11 shows, the same principle is a feature of the earlier versions of the logogen model (Morton 1969). One extra feature of the logogen model was that auditory and visual inputs relevant to language shared the same process. This model gave rise to predictions that were not upheld by data (Clarke and Morton 1983; Morton 1979), and it has been superseded by the model illustrated in figure 3.7.

A variant on the same theme was put forward by Charcot (1883) (figure 3.12). The difference here is that speech is possible without involving the auditory center (IC to CLA), though it is clear that the pathway between CAM and IC is bidirectional. The model thus shares features of the models illustrated in figures 3.2 and 3.9.

The two sets of models discussed above differed as to whether input and output functions were completely separate or shared some components. They shared the feature that the auditory and visual systems were independent of one another. In the three models that follow this is not the case. Rather, the recognition of print depends on mediating

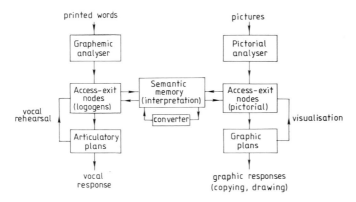

Figure 3.10

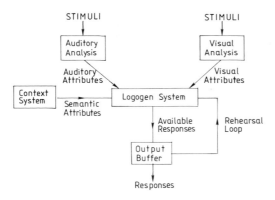

Figure 3.11

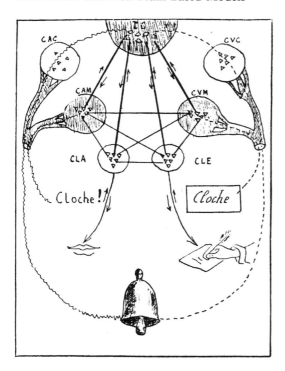

Figure 3.12

acoustic processes. The most straightforward and the best-known of these models is that of Lichtheim (1885), illustrated in figure 3.13. Here, print can be recognized only after being recoded into an auditory form via the center A. Elder (1897) and Mills (1898) produced models that included the same idea, as shown in figures 3.14 and 3.15 respectively. Both these authors added a few details on the involvement of the two hemispheres in reception, and Elder considered their involvement in production.

What Went Wrong?

From this selection of diagrams (Moutier includes ten or so more) it is clear that a great deal of effort went into the attempts to formalize a representation of brain processes. It might seem strange that proper contrasts were not made among the models and that proper tests were not made of the various alternatives, but the debate, in general, took

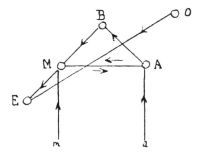

Figure 3.13

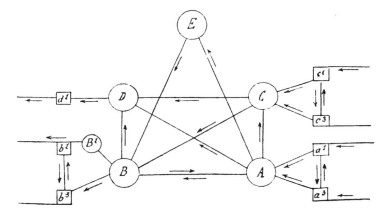

Figure 3.14

other forms. There were exceptions. Lichtheim (1885) specifically discusses how his theory might be falsified and what theoretical resources he had available (for example, postulating multiple lesions). Wernicke (1874) also had clear ideas about the way to proceed:

My conception differs from earlier ones in its consistently maintained anatomical foundation. Previous theories postulated theoretically different centers (a coordination center, a concept center, etc.), but paid no attention to anatomy in doing so, for the reason that the functions of the brain, completely unknown at that time, did not yet justify anatomical conclusions. It is a significantly different approach to undertake a thoroughgoing study of neuroanatomy and, making use of the now almost universally accepted principles of experimental psychology, to transform the anatomical data into psychological form and to construct a theory out of such material. (Wernicke 1874, p. 28; translation from Marshall 1982)

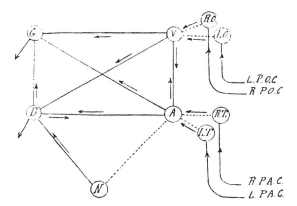

Figure 3.15

However, the purely psychological debate was minimal, and the diagrams foundered for a complex variety of reasons, most of which had to do with the relations supposed to hold between the models and the brain. The data used in the construction of the models were from patients with brain lesions. It was found that, roughly speaking, patients with lesions in the same region of the brain showed the same pattern of symptoms, or syndrome. This proper association of lesion with syndrome gradually changed until lesions were being associated either with "centers" of various kinds or with disconnections between centers. This in turn led to two major objections.

The first objection was that pathways were drawn that had no known anatomical correlates. Freud (1906, pp. 8–9) says: "Lichtheim's scheme . . . postulates new tracts, the knowledge of which is still lacking. If Lichtheim's presentation was based on new anatomical findings any further opposition would be impossible." The contrast Freud was making was with Wernicke's schema, which could be "inscribed into the brain, as the localization of the centers and fiber tracts which it contains has been anatomically verified." (Freud 1906, p. 8) In his final discussion, Freud concludes that "the apparatus of speech . . . presented itself to us as a continuous cortical area in the left hemisphere" extending between the sensory and motor areas (Freud 1906, pp. 102–103). Freud's criticism of Lichtheim, then, was effectively that, although his model was an improvement on Wernicke's, the format required localization if it was to be supported. This seems to do Lichtheim an injustice, as Lichtheim's article indicates that he was trying to separate function from anatomy. Freud also criticized Lichtheim on the grounds that patients did not conform to the predictions of the model (especially in the case of conduction aphasia). However, he failed to produce another model of the same form and failed to produce a different formalism. Head (1926) had similar problems. He comments that Kussmaul's figure (3.8 above) "lacked that definite localization or centers and paths demanded by the popular taste" (p. 64). Head goes on to discuss the "diagram makers" in general: "For every mental act there was a neural element, either identical with it or in exact correspondence. . . . They failed to appreciate that the logical formulae of the intellect do not correspond absolutely to physical events. . . ." (p. 65). The strong anti-localizationist arguments of Freud, Head, and others were sufficient to kill the notion of "centers," and with it the diagrams. Head also had fairly strong views about visual representations of ideas. He says that Kussmaul "unfortunately . . . was seduced into constructing a diagram." Because of this, and because he was interested in the effects of lesions in different parts of the brain, he missed the point that functional models (such as the "logical formulae of the intellect") could be studied independent of the brain.

The second problem with the diagrams lay in the attempt to treat each patient as representing the results of a single lesion. There were created idealized syndromes that were supposed to be related to single lesions, such as Broca's aphasia, Wernicke's aphasia, and conduction aphasia. In fact, these syndromes probably have no single clear exemplars in the whole history of neuropsychology. Attempts to circumvent the problems led to considerable contortions in the use of the models. Lichtheim (1885, p. 465) saw this problem clearly and discussed the need to consider the effects of multiple deficits. Head ignored this sophistication in his attack on Lichtheim, and Freud complained that such freedom "opens the doors to arbitrary explanations."

A third problem was that the psychological concepts were inadequate for the purpose. Some consequences of this are discussed by Arbib, Caplan, and Marshall (1982). It is a pity that there was not more interaction between experimental psychologists and neurologists at the time of the diagram makers. The legitimate rules for diagram making were often broken in the attempt to account in more detail for the symptoms of individual patients. There was no notion of testing normal people in order to get an independent justification for new "centers."

Some of these issues are discussed at greater length in Arbib et al. 1982, Caplan 1981, and Marshall 1980, 1982.

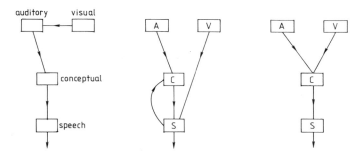

Figure 3.16

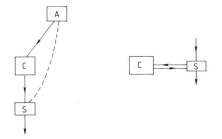

Figure 3.17

A Hundred Years Later

The major differences among the models illustrated above are easy to summarize.

• When we understand printed or written words, do we have to pass through an acoustic representation (figure 3.16, left) or an articulatory form (figure 3.16, center), or can there be "direct" access from a visual or graphemic representation to a "semantic" or "idea" representation (figure 3.16, right)? These diagrams are the simplest forms of the alternatives. None of the models considered above suggested that visual information has to pass through articulatory information, but such an idea was put forward by Jackson (1868) and others.

• Are there elements in common between the processes of auditory comprehension and speaking (and between reading and writing)? The alternatives here are illustrated in figure 3.17, with the left diagram representing independence and the right diagram representing overlapping elements.

• What are the natures of the input and output processes in the models? In fact, the answers to the two questions above are contingent on the definitions of the component processes. The main options have to do with whether a lexical representation has been reached. Thus, it is clear in some of the diagrams that, when a direct connection was indicated between input and output, the originator of the model was referring to imitation or copying (in the case of print or script). This could apply equally to linguistic and nonlinguistic stimuli. In other cases the reference was to the processing of words.

One breakthrough in contemporary work that has extended the distinctions among processes has been the careful distinction between words and nonwords. The early clinical work rarely used "nonsense" stimuli, except when patients were asked to read or repeat words from an unfamiliar foreign language. In experimental psychology, until around the early 1970s it was assumed that nonword stimuli could tell us about the way real words were processed (Morton 1979). Now it is clear that distinctions have to be made in talking about the processing of words and nonwords, including the obvious one that nonwords have (by definition) no "semantic" representation. In establishing this distinction we also go some way toward answering the first question, as we will see.

Is There Direct Comprehension of Printed Words?

There are two lines of evidence on this topic, which I will merely sketch. The first line comes from research on normal subjects. The experimental paradigm involves giving subjects a "priming" task (having the subjects make some response to a set of priming words or other stimuli) and then, between 5 and 30 minutes later, testing the effect of the priming task on the identification of these and other words. In the visual modality, the identification task involves presenting words very briefly and the measure is the amount of time the stimuli must be presented for correct identification or the percent correct at a particular exposure duration. With auditory presentation, words are presented in noise and the usual measure is the percent correct at a particular signal-to-noise ratio. It is well established that, when the same stimulus is presented in the two tasks, performance in the identification task is enhanced. The experiments then involve varying the relationship between the stimuli in the two tasks.

By looking at the resulting facilitation, we can draw conclusions about the degree of overlap in the way different stimuli are processed. First, we can examine the level of representation at which these facilitation effects take place. It is clear that it is not a sensory level. Clarke and Morton (1983) showed that when the priming task involved reading either printed words or handwritten words the facilitation of subsequent recognition of printed words was equivalent. With auditory stimuli, Jackson and Morton (1983) looked at the effects of the voice of the speaker. In this experiment the pretraining words were spoken clearly in either a male or a female voice and the test words were in the female voice. The data showed equivalent transfer in the two cases. These data show that the facilitation effect is not sensory. Other data show that it is modality-specific. Clarke and Morton (1983) could only find a small amount of facilitation from initial auditory presentation to subsequent visual recognition, whereas within-modality facilitation is great. Jackson and Morton (1983) found only a little transfer from a visual priming task to auditory recognition. Further, Gipson (in preparation) and Ellis (1982) have reported experiments in which there was significant within-modality transfer but no between-modality transfer at all. In all these cases the priming task involved some semantic processing. Thus, we deduce in answer to the first question above that there are direct routes from both the auditory and visual systems to the comprehension systems such that when one is used there is no necessary involvement of the other.

Further experiments have shown that the facilitation effect is at the level of the morpheme, not that of the word. Murrell and Morton (1974) showed that the morphemic transfer from *SEEN* in the priming task to *SEES* in the recognition task was as great as that from *SEES* to *SEES*. However, there was no transfer at all when the priming word was *SEED*. With auditory stimuli, Kempley and Morton (1982) report similar findings for pairs of words with regular inflections. However, they found no transfer at all between irregularly related pairs of words such as *man/men* and *lost/loses*. Also, they reported no effects of auditory similarity.

These experiments allow us to isolate the facilitation effect at the lexical level. The importance of this is that it helps us to begin to answer the second question, that of the common elements between input and output processes. The crucial test is whether speaking a word affects subsequent auditory recognition. On the basis of the current data we cannot decide between the options in figure 3.18. The first of these

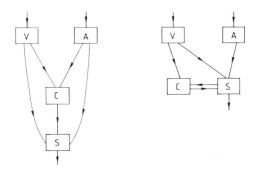

Figure 3.18

options represents the current logogen model (as in figure 3.7), whereas the second maintains the idea that there are processes in common between listening and speaking. This latter idea (though not the form of the model) is supported by Allport and Funnell (1981). Until the nature of these processes is specified more fully, it will not be possible to test between the alternatives.

Neuropsychological Evidence

Diagrams of the processes involved in the processing of verbal and pictorial stimuli are currently being used, particularly in Europe, by a number of neuropsychologists and cognitive psychologists interested in the effects of brain damage. (See, for example, Allport and Funnell 1981; Coltheart 1980a, 1980b, 1981; Marcel 1980; Morton 1981; Morton and Patterson 1980a; Marshall and Newcombe 1973; Newcombe and Marshall 1980; Ratcliff and Newcombe 1982; Patterson 1981; and Shallice 1981.) The number of scientists referring to the models or the concepts underlying the models is much greater. These authors share a belief in the utility, if not the necessity, of using single case studies and studying each patient in considerable detail with particular psychological or linguistic theories in mind. The detailed arguments in favor of single case studies are complex, and the dangers are reasonably well known (Shallice 1980). However, the dangers of using group studies, unless the groups are particularly well controlled, are even more serious if one is interested in the nature of normal psychological processes and how they can break down. The crucial distinction is between neurological and psychological defects (Beauvois 1982). It is possible to have a single neurological deficit (such as a reasonably well-localized lesion) with a number of well-defined and psychologically unrelated behavioral deficits. (See, for example, Beauvois, Derouesne, and Saillant 1980.) Notions of "typicality" of types of aphasia and dyslexia seem unimportant when the task of the neuropsychologist is no longer that of identifying the site of a lesion. Thus, it now seems essential to specify whether a patient is agrammatic, or whether receptive, productive or both, before conducting experiments on the processing of language. The mere designation of Broca's aphasia or "nonfluent" or "left anterior lesion" does not enable us to interpret the data. Marshall (1982) and Saffran (1982) have elaborated arguments along these lines. Saffran concludes (p. 333) that "if the classical syndrome categories—or worse, dichotomies such as 'anterior' vs. 'posterior'—continue to be the prevalent units of analysis in aphasia research, the contribution of neuropsychological data to the componential analysis of language will be seriously limited."

Let me say again that if the question is changed the appropriate methodology changes. If the question relates to the functions subsumed by a particular portion of the brain, then of course the location of the lesion is crucial information. If the question relates to recovery patterns, then the classical divisions, or some modification of them, will be vital for helping to characterize the patient groups.

However, the group designations, classical or modern, are of necessity fuzzy. Not all patients in a group share all the symptoms, and patients should thus be selected specifically on the basis of deficits related to the linguistic or psychological variables under investigation. (See, for example, Friederici, Schoenle, and Goodglass 1981 and Caplan, Matthei, and Gigley 1981.) In the same way, we now know that the classical symptoms of the conduction aphasic—problems in repeating single words and memory problems with lists of words or sentences—can be dissociated and so must be distinguished (Shallice and Warrington 1977).

Patients can be validly classified only when variations between patients are less than variations between groups. (See for example, Patterson 1981.) However, even this criterion is subject to restriction by the type of behavior studied. Thus, patient P. W. (Patterson 1978, 1979; Patterson and Marcel 1977; Morton and Patterson 1980a,b) is termed a deep dyslexic, but this designation is only in contrast with phonological dyslexia, letter-by-letter dyslexia, and surface dyslexia (Patterson 1981). That P. W. is agrammatic and nonfluent is irrelevant to his classification as a deep dyslexic, since other deep dyslexics exist who are neither agrammatic nor nonfluent (Coltheart 1980b; David Howard, personal

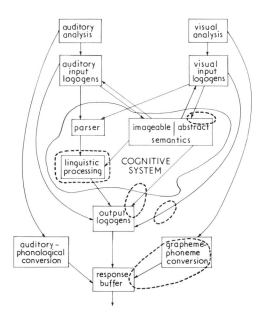

Figure 3.19

communication). Morton and Patterson (1980a) have analyzed P. W. in terms of an expanded version of the logogen model. From his reading and comprehension deficits and from his performance on a variety of other tasks we assign five distinct "lesions" on the diagram of the model (figure 3.19); however, it is arguable that only three of these processing deficits lead to the deep dyslexia symptoms. The deficits need not be disconnection deficits in the sense of Geschwind (1965). The lines in the diagram are not intended to represent well-defined fiber tracts. Some of them may correspond to fiber tracts, but that is a separate question.

I do not wish to argue the case for using such psychological models in the study of brain damage, although at least one approach to speech therapy has been based on such models (Hatfield 1982, 1983; Powell 1981). Rather, I will point to the utility of studying patients in the development of the models. In brief, the study of patients has forced changes and complications in the models, on the one hand, and provided evidence in favor of certain constructs on the other hand. To start with, the symptoms of the deep dyslexic patients demonstrate conclusively that written material can be understood without reference to phonological codes. As a second example we can take the patient described

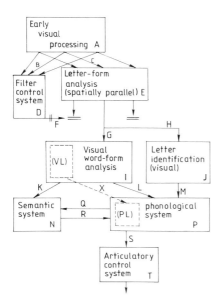

Figure 3.20
Shallice's model. VL indicates a visual logogen, PL a phonological logogen.
The components indicated by broken lines are relevant only to the approach
of Morton and Patterson (1980).

by Schwartz, Saffrin, and Marin (1980) who could read irregular words
without apparent comprehension. This patient seems to establish the
existence of the route illustrated in figure 3.17 between the visual input
logogens and the output logogens, since irregular words cannot be read
by the grapheme-phoneme conversion route and the lack of comprehension rules out the route via the cognitive system. However, the
varieties of dyslexias currently described mean that certain features of
the model in figure 3.17 are inadequate to deal with the data. One
elaboration is due to Shallice (1981), whose model is illustrated in figure
3.20. Shallice has identified nine types of dyslexia and shown how the
model can accommodate them (table 3.1).

It should be clear that the study of reading is becoming complex and
technical. It is not clear that the location in the brain of the lesions
that lead to dyslexic symptoms has anything to contribute to the study
of the psychological deficits. It is also likely that this approach does
not and cannot help to answer the questions being asked by some
people who study the brain and behavior. The diagram makers failed

Table 3.1
Acquired dyslexia syndromes.

Syndrome	Representative source[a]	Assumed impairment[b]
Neglect dyslexia	Kinsbourne and Warrington 1962	C or E or G
Attentional dyslexia	Shallice and Warrington 1977	D or F
Word-form dyslexia	Warrington and Shallice 1980	I
Surface dyslexia	Marshall and Newcombe 1973	K or N or R
Phonological alexia	Beauvois and Derouesne 1979	L or P*
Deep (or phonemic) dyslexia	Coltheart et al. 1980	(L or P*) and N
Semantic access dyslexia	Warrington and Shallice 1979	(L or P*) and N
Concrete word dyslexia	Warrington 1981	(L or P*) and N
(Nonsemantic reading	Schwartz et al. 1980	N)

Source: Shallice 1981.
a. Any empirical information about a syndrome is derived from the representative source unless otherwise stated.
b. Letters refer to figure 3.20. Asterisk indicates impairment to part of a subsystem only; in particular, P* does not involve the transformation from R to S, the articulatory output logogen of Morton and Patterson (1980).

because they, and their critics, could not separate out the different questions. The same mistake will not be made again, and brain-based and non-brain-based models of language will coexist. We can only hope that the proponents of the two types will continue to talk to each other.

Acknowledgments

I am grateful to Karalyn Patterson, John Marshall, and Tim Shallice for preserving me from some errors and oversimplifications. In spite of their efforts, my historical treatment has been idealized.

References

Note: the references marked with an asterisk have been taken from Moutier 1908.

Allport, D. A., and E. Funnell. 1981. Components of the mental lexicon. *Philosophical Transactions of the Royal Society, London, B* 295: 215–222.

Arbib, M. A., D. Caplan, and J. C. Marshall. 1982. Neurolinguistics in historical perspective. In M. A. Arbib, D. Caplan, and J. C. Marshall, eds., *Neural Models of Language Processes*. New York: Academic.

Baginsky. 1871.* Aphasie in Folge schwerer Nierener Krankungen Uraemie. *Berliner Klinische Wochenschrift* 8: 428–431, 439–443.

Ballet, G. 1886.* Le Langage Interieur et les Diverses Formes de l'Aphasie. Paris: Alcan.

Beauvois, M.-F. 1982. Optic aphasia: A process of interaction between vision and language. *Philosophical Transactions of the Royal Society, London, B298*, 35–47.

Beauvois, M.-F., and J. Derouesne. 1979. Phonological alexia; three dissociations. *Journal of Neurology, Neurosurgery and Psychiatry* 42: 1115–1124.

Beauvois, M.-F., J. Derouesne, and B. Saillant. 1980. Syndromes neuropsychologiques et psychologie cognitive. Trois examples: aphasie tactile, alexie phonologique et agraphie lexicale. *Cahiers de Psychologie* 23: 211–245.

Brown, J. 1981. Review of *Deep Dyslexia*. *Brain and Language* 14: 386–392.

Caplan, D. 1981. On the cerebral localisation of linguistics functions: Logical and empirical issues surrounding deficit analysis and functional localisation. *Brain and Language* 14: 120–137.

Caplan, D., E. Matthei, and H. Gigley. 1981. Comprehension of gerundive constructions by Broca's aphasics. *Brain and Language* 13: 145–160.

Charcot, J.-M. 1883.* *Le Differente Forme di Afasia*. Milan.

Clarke, R. G. B., and J. Morton. 1983. The effects of priming in visual word recognition. *Quarterly Journal of Experimental Psychology* 35A: 79–96.

Coltheart, M. 1980a. Reading, phonological recoding, and deep dyslexia. In Coltheart et al. 1980.

Coltheart, M. 1980b. Deep dyslexia: A review of the syndrome. In Coltheart et al. 1980.

Coltheart, M. 1981. Disorders of reading and their implications for models of normal reading. *Visible Language* 15: 245–286.

Coltheart, M., K. E. Patterson, and J. C. Marshall, eds. 1980. *Deep Dyslexia.* London: Routledge & Kegan Paul.

Elder, W. 1897.* *Aphasia and the Cerebral Speech Mechanism.* London: Lewis.

Ellis, A. 1982. Modality-specific repetition: Priming of auditory word recognition. *Current Psychological Research* 2: 123–128.

Freud, S. 1906. *On Aphasia: A Critical Study.* London: Image.

Friederici, A. D., P. W. Schoenle, and H. Goodglass. 1981. Mechanisms underlying writing and speech in aphasia. *Brain and Language* 13: 212–222.

Geschwind, N. 1965. Disconnection syndromes in animal and man. *Brain* 88: 585–644.

Gipson, P. Some auditory priming experiments inspired by the logogen model. Manuscript in preparation.

Grasset. 1896.* Des diverses varietes clinique d'aphasie. *Nouveau Montpellier med.* 121, 141, 161.

Hatfield, F. M. 1982. Diverses formes de desintegration de langage ecrit et implications pour la rééducation. In E. C. Laterre and X. Seron, eds., *Restauration fonctionnelle et Rééducation Neuropsychologique.* Brussels: Mardaga.

Hatfield, F. M. 1983. Aspects of acquired dysgraphia and implications for reeducation. In C. Code and D. J. Müller, eds., *Aphasia Therapy.* London: Arnold.

Head, M. 1926. *Aphasia and Kindred Disorders of Speech.* Volume I. Cambridge University Press.

Jackson, J. H. 1868. On the physiology of language. *Medical Times and Gazette,* pp. 275–276.

Jackson, A., and J. Morton. Facilitation of auditory word recognition. Manuscript submitted for publication.

Kempley, S. T., and J. Morton. 1982. Irregular relationships in auditory word recognition. *British Journal of Psychology* 73: 441–454.

Kinsbourne, M., and E. K. Warrington. 1962. A variety of reading disability associated with right hemisphere lesions. *Journal of Neurology, Neurosurgery and Psychiatry* 25: 339–344.

Kussmaul, A. 1877.* Die Störungen der Sprache. *Ziemssens Handbuch der speciellen Pathologie und Therapie, 12 Abhang,* 1–300.

Langdon. 1898. *The Aphasias and Their Medico-Legal Relations.* Ohio.

Lichtheim, L. 1885. On aphasia. *Brain* 7: 433–484.

Marcel, A. J. 1980. Surface dyslexia and beginning reading: A revised hypothesis of the pronunciation of print and its impairments. In Coltheart et al. 1980.

Marshall, J. C. 1980. On the biology of language acquisition. In D. Caplan, ed., *Biological Studies of Mental Processes.* Cambridge, Mass.: MIT Press.

Marshall, J. C. 1982. What is a symptom-complex? In M. A. Arbib, D. Caplan, and J. C. Marshall, eds., *Neural Models of Language Processes.* New York: Academic.

Marshall, J. C., and F. Newcombe. 1973. Patterns of paralexia: A psycholinguistic approach. *Journal of Psycholinguistic Research* 2: 175–199.

Mills, C. K. 1898.* *The Nervous System and its Diseases.* Philadelphia: Lippincott.

Moeli. 1891.* Ueber der gegenwärtigen Stand der Aphasiefrage. *Berliner Klinische Wochenschrift.*

Morton, J. 1969. The interaction of information in word recognition. *Psychological Review* 76: 165–178.

Morton, J. 1979. Facilitation in word recognition: Experiments causing change in the logogen model. In P. A. Kolers, M. E. Wrolstad, and H. Bouma, eds., *Processing of Visible Language.* New York: Plenum.

Morton, J. 1981. The status of information processing models of language. *Philosophical Transactions of the Royal Society, London* B295: 387–396.

Morton, J., and K. Patterson. 1980a. A new attempt at an interpretation. In Coltheart et al. 1980.

Morton, J., and K. Patterson. 1980b. Little words—No. In Coltheart et al. 1980.

Moutier, F. 1908. *L'Aphasie de Broca.* Paris: Steinheil.

Murrell, G. A., and J. Morton. 1974. Word recognition and morphemic structure. *Journal of Experimental Psychology* 102: 963–968.

Newcombe, F., and J. C. Marshall. 1980. Transcoding and lexical stabilisation in deep dyslexia. In Coltheart et al. 1980.

Newcombe, F., and J. C. Marshall, 1981. On psycholinguistic classifications of the acquired dyslexias. *Bulletin of the Orton Society* 31: 29–46.

Patterson, K. E. 1978. Phonemic dyslexia: Errors of meaning and the meaning of errors. *Quarterly Journal of Experimental Psychology* 30: 587–601.

Patterson, K. E. 1979. What is right with "deep" dyslexic patients? *Brain and Language* 8: 111–129.

Patterson, K. E. 1981. Neuropsychological approaches to the study of reading. *British Journal of Psychology* 72: 151–174.

Patterson, K. E., and A. J. Marcel. 1977. Aphasia, dyslexia, and the phonological coding of written words. *Quarterly Journal of Experimental Psychology* 29: 307–318.

Powell, G. E. 1981. *Brain Function Therapy*. Aldershot, England: Gower.

Ratcliff, G., and F. Newcombe. 1982. Object recognition: Some deductions from the clinical evidence. In A. Ellis, ed., *Normality and Pathology in Cognitive Functions*. London: Academic.

Saffran, E. M. 1982. Neuropsychological approaches to the study of language. *British Journal of Psychology* 73: 317–338.

Schwartz, M. F., E. M. Saffran, and O. S. M. Marin. 1980. Fractionating the reading process in dementia: Evidence for word-specific pring-to-sound associations. In Coltheart et al. 1980.

Seymour, P. H. K. 1973. A model for reading, naming, and comparison. *British Journal of Psychology* 64, no. 1: 35–49.

Shallice, T. 1980. Case study approach in neuropsychological research. *Journal of Clinical Neuropsychology* 1: 183–211.

Shallice, T. 1981. Neurological impairment of cognitive processes. *British Medical Bulletin* 37: 187–192.

Shallice, T., and E. K. Warrington. 1977. Auditory-verbal short-term memory impairment and conduction aphasia. *Brain and Language* 4: 479–491.

Warrington, E. K. 1981. Concrete word dyslexia. *British Journal of Psychology* 72: 175–196.

Warrington, E. K., and T. Shallice. 1979. Semantic access dyslexia. *Brain* 102: 43–63.

Warrington, E. K., and T. Shallice. 1980. Word-form dyslexia. *Brain* 103: 253–262.

Wernicke, C. 1874. Der aphasische Symptomen Komplex. Breslau: Cohn & Weigart.

Chapter 4

Vocal Learning and its
Possible Relation to
Replaceable Synapses and
Neurons

Fernando Nottebohm

This chapter is composed of two parts. In the first part, I review facts and theories about vocal learning in birds. I hope to give the reader a good feeling for the natural history of vocal learning and the experiments that have been done to better define this phenomenon. In the second part, I describe what is known about the anatomy of voice production in birds and the properties of the brain pathways that control this behavior (particularly in birds that learn their song).

There is strong evidence that brain pathways for vocal learning in birds provide unique opportunities for studying basic aspects of brain function. The reader with an interest in such matters may find the material useful even if he or she has no special interest in problems of animal communication. In this case birdsong becomes just a convenient example of a complex learned motor skill. But birdsong is of course used in communication, and thus it lends itself well to the study of principles of signal encoding and decoding. At this level, material presented here may suggest new approaches to the study of brain mechanisms that control human speech and language.

Vocal Learning: The Behavior

Song Imitation and Improvisation

The true, or oscine, songbirds account for about half of all the species of living birds. They are all members of the suborder *Passeres* of the order *Passeriformes*. Most of the experimental work on vocal learning in birds has been done with members of this group.

It seems likely that all oscine songbirds develop their song by reference to auditory information (Nottebohm 1972a, 1975; Kroodsma 1982). This learning process can follow either of two strategies: imitation and improvisation. In either case, vocal output is modified until it matches a preconceived or learned auditory goal.

Imitation must be preceded by identification of the correct model. Imitation is usually restricted to imitation of conspecific song. Young birds that have not yet developed their song imitate the song of older birds. It is thought that in many species recognition of the correct model depends on a genetically determined predisposition based on auditory cues. The classical work leading to this paradigm was done by Thorpe (1958) in England and Marler (Marler and Tamura 1964; Marler and Peters 1977) in the United States. However, some species that learn their song by imitation rely on social rather than auditory cues to recognize the correct model. Bullfinches (Nicolai 1959) and zebra finches (Immelmann 1969) imitate the song of their father or foster father, and will do so even if their foster father belongs to a different species.

The development of several song themes conforming to a species-specific plan, without recourse to external models, has been called "improvisation." Improvisation has been described for European blackbirds (Messmer and Messmer 1956) and Oregon juncos (Marler, Kreith, and Tamura 1962). In human terms, a set of rules says for example, "tango," and the composer improvises any of a large number of possible arrangements that will be recognized as tangos. In this manner, species that learn their song can use imitation or improvisation to enrich the complexity and enlarge the size of their song repertoires.

In canaries the same rules that guide the selective imitation of conspecific song may also guide the improvisation of a song repertoire, and either developmental strategy leads to the acquisition of correct species-specific song. Young male canaries reared in the absence of an external model can improvise their song (Metfessel 1935); however, if a young male canary can hear the song of an adult male, it will imitate it (Waser and Marler 1977). In other species, such as the chaffinch and the white-crowned sparrow, the blueprint that allows for recognition and imitation of conspecific song does not suffice for correct improvisation, and in such cases the song of young males reared in auditory isolation is markedly aberrant (Thorpe 1958; Marler and Tamura 1964).

The Importance of Subsong

Learned song does not suddenly come into being. It is preceded by months of vocal ontogeny, during which food-begging calls blend into subsong, out of which emerges "plastic song," which leads on to the stereotyped song of adults. The subsong stage may serve a crucial role in vocal learning. Subsong itself serves no communicatory function. Subsong sounds are delivered at a low amplitude and in a very variable manner—often as the young bird seems to doze, much as a child going to sleep indulges in babbling. This similarity between subsong and babbling was noticed by Charles Darwin (Thorpe and Pilcher 1958). Perhaps this is a stage during which the young bird is learning to play its musical instrument, the syrinx; later will come the serious business of learning what to play.

In some species, such as the chaffinch, subsong includes fairly stereotyped components that are, in themselves, not used in song. These components may serve as training exercises, so that the auditory consequences of well-defined motor programs are grasped early in development. Subsong experience could also bias the later choice of models, though there is as yet no evidence for this (Nottebohm 1972b). In the swamp sparrow, selective responsiveness to conspecific song can be demonstrated early in ontogeny in both males and females (Dooling and Searcy 1980); it is not clear whether this selectivity precedes the onset of subsong, which in this species can start as early as 20 days after hatching (Marler and Peters 1982a).

In swamp sparrows the subsong stage is followed by a stage of plastic song, during which the young male renders diverse improvised and copied songs. As such a bird achieves full reproductive maturity, it discards most of these learned songs, so that its adult repertoire now consists of only two songs (Marler and Peters 1982b). The purpose of a strategy that includes an early stage during which a learned repertoire exceeds the final repertoire is not clear. In canaries (Nottebohm and Nottebohm, in preparation), the adult song repertoire is built by the gradual addition of new syllable types during plastic song. The diversity of strategies for vocal learning observed in birds is sure to be of great use in the study of the neural principles that govern vocal learning.

The Female Response to Song

Female cowbirds and female song sparrows respond selectively to conspecific male song, even when cowbirds had been reared in auditory isolation from males (King and West 1977; Searcy and Marler 1981). In both cases, females in reproductive condition adopt the solicitation posture upon hearing conspecific song. This is a nice experimental situation; the female response provides a nonvocal way of determining selective responsiveness to a song model, so that the encoding and decoding functions of the vocal communication system can be tested separately. Since female cowbirds and song sparrows do not normally sing, recognition of conspecific song in them occurs in the absence of earlier experience with song production. Thus, in females, genetically coded information plays an important role in the recognition of conspecific song.

Genetically coded information used to recognize conspecific song can be supplemented by learning, as when a male learns to recognize the song of a neighbor (Emlen 1971; Kroodsma 1976a). In females, too, experience can play a role in developing song preferences. Male and female white-crowned sparrows use this combination of innate and learned information in pair-bond formation. Both males and females develop a preference for the patterns of conspecific song they hear during the first month or two after hatching, before they stray far from the hatching place. In males this leads to the formation of learned song dialects (Marler and Tamura 1964; Marler 1970). In females this early exposure leads to a preference for males that sing the home dialect. Female white-crowned sparrows without prior mating experience will adopt the solicitation posture when exposed to their home dialect, but not when exposed to an alien dialect (Baker, Spitler-Nabors, and Bradley 1981). These observations, and field studies demonstrating reduced gene flow between contiguous dialect populations (Baker and Mewaldt 1978), support the hypothesis that song dialects may serve to foster pair formation between individuals that share genetic adaptations to local environmental conditions (Nottebohm 1969a). Foreigners made miserable by their inability to seduce native beauties may console themselves with the idea that nature, in its infinite wisdom, may have chosen to restrict the exchange of genetic material between dialect nations. In this context, vocal learning in birds, as well as in humans, may serve not so much to convey new information as to hinder communication (Nottebohm 1970).

There is another possible relation between vocal learning and pair formation. The song of male birds induces ovulation in females, as first shown in the laboratory by Lott and Brody (1966). In canaries this effect is achieved only if the song is sufficiently complex (Kroodsma 1976b). In female song sparrows, who respond to conspecific song by adopting the solicitation posture, more responses are obtained to a playback with various song types than to playbacks of a single song type (Searcy and Marler 1981).

The singing prowess of male songbirds, like other acquired worldly glories (a suntan in winter, an expensive sports car), might be used as a measure of fitness. Song learning takes time and health. A large repertoire of learned songs coud be seen as an indirect measure of stamina, foraging efficiency, and past and present well-being, all of which require good genes. Other things being equal, a female choosing a good singer would be choosing a good mate.

Song-complexity information available to females is also available to males. In red-winged blackbirds, among which correlates of song-repertoire size have been studied in nature, there is a positive relation between the number of song patterns a male sings and the number of female red-winged blackbirds nesting in its territory. However, this relation is an indirect one. Large song repertoires occur among males with greater reproductive experience, who also hold larger territories. The authors of this study, Yasukawa, Blank, and Patterson (1980), concluded that among red-winged blackbirds large song repertoires confer an advantage in male-male competition for territory, and that large territories, not song per se, are advantageous in mate acquisition.

A relation between repertoire size and size of territory has been described in the European great tit. To explain it, Krebs (1977) offered the Beau Geste hypothesis: that song repertoires have evolved because males seeking territories attempt to avoid densely occupied areas, and multiple song types allow territorial males to simulate high density. It is hard to see how this hypothesis would apply to males but not females. If it applied to females, they too should interpret the larger repertoires as evidence of many males, each holding only a fraction of the territory in question. Thus, I prefer the original proposition: that song complexity is a measure of fitness, and that the same signal that is optimal for charming females may be optimal for discouraging males.

If female songbirds interpret male song as a fitness report, we may wonder how males assess the fitness of females. In monogamous species the choice of optimal partner should be equally important to either sex. "Sex appeal" and "beauty" are easy to describe within our own species, but how does a "sexy" hen look or act? Gifted songsters may know, since they may get to choose a mate from several potential ones drawn to their song. Poor songsters may have to settle for less.

Critical Periods for Vocal Learning

Up to here, the emphasis has been on *what* is learned. However, the *when* of learning can be determined just as narrowly, and here too we find tantalizing species differences. Chaffinches are able to imitate a model heard during the first 10 months after hatching (Thorpe 1958).

White-crowned sparrows imitate a model they hear during the first 50 days after hatching (Marler and Tamura 1964; Marler 1970). Zebra finches learn their song from the 40th to the 80th day after hatching (Immelmann 1969). Canaries continue to learn new song patterns after they reach full reproductive maturity (Nottebohm and Nottebohm 1978).

External factors can affect the duration of the critical period. In marsh wrens the end of the critical period for song learning depends on the hatching date. Marsh wrens hatched early in the season imitate models heard during the same season, but not those heard during the following spring. Marsh wrens hatched late in the season are still able to imitate models they first hear the following spring (Kroodsma and Pickert 1980). Hatching date may influence the pace of gonadal development and so determine the end of the critical period. A male chaffinch castrated at 6 months of age, then treated with exogenous testosterone when 2 years old, was able to imitate chaffinch song presented to him at this later time, one year after the normal end of the critical period. A surge in plasma levels of testosterone or testosterone metabolites may set an end to the critical period for song learning (Nottebohm 1969b).

The end of the critical period for song learning may reflect a loss of the motor plasticity required to imitate a model. This loss need not affect the primary ability to store new auditory memories. For example, one presumes that adult chaffinches are able to recognize the song of new neighbors. Similarly, females can presumably recognize the song of their mate for as many years as they breed, regardless of whether their mate remains the same one or changes.

The separation between sensory and motor learning is well illustrated in white-crowned sparrows. During the first 50 days after hatching, these birds acquire an auditory memory of the song they will later imitate. This temporal window must reflect a special perceptual or motivation stage. The conversion of this memory to a learned motor pattern occurs several months later (Marler 1970). An artificially induced yearly recurrence of motor plasticity could reveal the commitment of new models to the bird's memory store.

The relation between critical periods and motor plasticity is complicated by one further factor: The song of adult male chaffinches becomes quite variable each spring, at which time it is very reminiscent of the plastic song stage traversed by these birds during their first year as they learned their song (Marler 1956). However, this seeming plasticity is not used for learning new song themes, but may merely reflect changing hormone levels.

The issue of critical periods often surfaces in the context of just how many presentations are required for the learning of a song model that will later be imitated. I am not aware of a study aimed at establishing the minimal exposure that would result in imitation. Marler (1970) obtained good imitation of a model presented to young male white-crowned sparrows 4 minutes daily for a period of 21 days. Kroodsma and Pickert (1980) got good model imitation in marsh wrens exposed to a song model 90 minutes per day for a period of 5 days. One must remember that birdsong is part of a reproductive strategy. Male and female songbirds are likely to learn auditory stimuli that play a salient role in their environment, be it the song of a parent, a neighbor, or a mate or the dialect of the home area. In each case the significance of the model or models requires that it be more than a fleeting sound heard a few times and not again. In this sense, song learning in birds may not be a good system in which to determine the minimum exposure needed for learning the properties of a stimulus. Conversely, it can be argued that during exposure to a tutor model a bird is likely to keep track of the number of exposures, since this may be an important factor in deciding which of the many songs heard and memorized will be imitated. The number of songs imitated may be only a fraction of the number of songs remembered.

Why is there a critical period at all? Presumably the onset and termination of the critical period has something to do with the way in which the learned song is used. Chaffinches and marsh wrens tend to countersing in kind; as a singing male switches from theme to theme and thus works through its repertoire, a second male singing back is able to match each theme it hears with a similar theme from its own repertoire (Hinde 1958; Kroodsma 1979). Such exercises presumably relate to reproductive success by better defending the territory or better coaxing a mate. In the latter case, chorusing in kind may act as a "supernormal stimulus" (Tinbergen 1951).

The diversity of critical periods cannot but baffle even the most Darwinian of biologists. It is hard to argue that each species has found the best or unique solution for the acquisition of a learned song repertoire. What, for example, would be the disadvantages of having a critical period that recurred every year? Such flexibility would provide a chance to add to the repertoire and thus make it more complex, or to update it so as to better match that of different neighbors. Such questions are appropriate, because we know that open-ended critical periods occur (for example, in canaries). A short and sharply defined critical period is not required by brain systems for vocal learning. We are then left with a paradox. In many species, vocal learning comes to a halt at the very time when maximal vocal virtuosity has been achieved. It may be that in some species the brain circuits used for vocal learning are also used for storing the learned motor pattern of song. If so, reinstatement of the plasticity required for learning a new song repertoire could result in the loss of the acquired patterns, and in many cases this may be too great a price to pay. It may not be possible, with one set of circuits, to hold a motor memory and continue to add to it.

If the latter hypothesis is correct, how do canaries manage to add to and otherwise modify their song repertoire in successive years? A possible answer is that such birds are constantly learning; as they learn to produce new sounds, they continue to relearn sounds they had seemingly mastered. An interruption of auditory feedback, in this case, would lead to massive forgetting. This prediction is borne out. We should be prepared for the possibility that neural circuits in open-ended vocal learners differ interestingly from those in species with narrowly defined critical periods.

Critical periods for vocal learning may not be confined to birds. Adult humans can learn to produce new words but seem to find it more difficult to produce new sounds. Adults learning a foreign language often seem to use "old" sounds to generate "new" words. Not only can the French recognize foreigners speaking their language, but they also can tell if the accent is German or English: the two groups produce their French with a different set of sounds.

The Effects of Deafening

Early deafening by the removal of both cochleas affects the song of birds that acquire their repertoire by reference to auditory feedback (reviews in Konishi and Nottebohm 1969; Marler and Mundinger 1971). This is well demonstrated in the song of early-deafened white-crowned sparrows, chaffinches, and zebra finches (Konishi 1965; Nottebohm 1968; Price 1979). The song of these birds is very poorly structured and considerably more aberrant than that of "controls" reared in auditory isolation. By contrast, domestic fowl and doves, not known to acquire their vocal repertoire by improvisation or imitation, develop normal species-typical vocalizations when deafened soon after hatching (Konishi 1963; Nottebohm and Nottebohm 1971).

The effects of deafening in adulthood vary among species that learn their song. White-crowned sparrows and chaffinches retain their learned song with considerable accuracy for periods of 2 and 3 years (Konishi and Nottebohm 1969). Canaries deafened in adulthood forget all of their learned song over a period of a few months (Nottebohm, Stokes, and Leonard 1976). This difference in outcome may be related to the nature of the critical period for song learning. Birds that acquire their adult song once seem able to retain it as a long-lasting motor memory capable of persisting in the absence of auditory feedback; birds (such as the canary) that modify their song in successive years may require constant updating of their motor program by reference to auditory feedback. In the absence of such updating the pattern is forgotten, and a few months later the song of the bird deafened in adulthood is no better than that of a canary deafened one month after hatching, before the onset of song learning. This distinction between permanent (white-crowned sparrow, chaffinch) and temporary (canary) learned motor programs may be important, since it can be used to explore the neural basis for fixed versus renewable memory systems.

Deafening has been instructive in another way: By removing both cochleas at various ages during normal song ontogeny in the chaffinch, it was possible to show that the quality of adult song produced was related to the extent of the song experience before deafening. Adult song was totally lacking in structure in chaffinches deafened before they entered the subsong stage, but it was more structured in chaffinches deafened after the subsong stage (Nottebohm 1968). This experiment underscores that during song learning there is constant modification of motor programs, guided by auditory feedback.

In contrast with the cases of deafened chaffinches, white-crowned sparrows, and zebra finches, the song of canaries deafened before the onset of subsong is remarkably structured. Some of the syllables produced differ little from those of hearing birds, though their structure tends to be less stable. As in intact canaries, a recognizable syllable is repeated many times, forming a phrase; in this manner the bird can work its way through a repertoire of up to five different syllable types (Waser and Marler 1977). Quantitatively, this compares poorly with the 20–40 different syllable types of intact birds. What is worth noting is that the overall pattern of song delivery is not affected by early or late deafness, so that the motor rules governing this performance emerge independent of auditory feedback (Güttinger 1981). In canaries, vocal learning seems to be used to embellish the quality of syllable types and to add to their stereotypy and number.

Is true learning occurring in birds that "learn" their song? Is there an acquisition of new motor coordinations that otherwise would not occur in the brain? Could it be that the young bird has a large but limited number of motor programs at its disposal, and that auditory experience merely determines which of these programs are activated? The gradual and protracted acquisition of song imitations (see, for example, Nottebohm 1968 and Marler 1970) seems to argue against this idea, as does the fact that I once had an orange-winged Amazon parrot that said "Praise God!"; it is hard to imagine that this pattern had been lurking in the brains of this species since primeval times, waiting to be activated. Another, more inspired gray African parrot sang parts of an Irish song, and when it was about 20 years old, after a change of owners, it acquired a repertoire of German words. I conclude that vocal learning in birds, like speech learning in humans, constitutes a true learning of new motor coordinations that would not occur if not learned. This is not to say that learning starts from a *tabula rasa*. Motor systems, as well as sensory systems, are likely to bring their own predispositions to the learning of a motor skill (Price 1979).

Interesting questions about vocal ontogeny could be answered if it were possible to deafen birds in a reversible manner. So far, the closest thing to this is the use of white noise to mask auditory feedback. Unfortunately, to be fully effective this approach requires amplitudes of about 100 dB), and such exposure over a number of days induces partial deafness (Marler et al. 1973).

Brain Pathways for Vocal Learning

The Syrinx

In most birds, song and calls are produced during expiration (Calder 1970; Gaunt, Stein, and Gaunt 1973; Brackenbury 1979). A possible exception is the nightjar, which is thought to produce song during both inspiration and expiration (Hunter 1980). The avian larynx acts as a valve, providing for the sharp onset and termination of air flow and thus of sound pulses. The sounds themselves are produced by the syrinx (figures 4.1 and 4.2).

The songbird syrinx is at the confluence of the bronchi and the trachea. It provides each bronchus with a constriction of changeable shape and bore, determined by action of the syringeal muscles. The inner di-

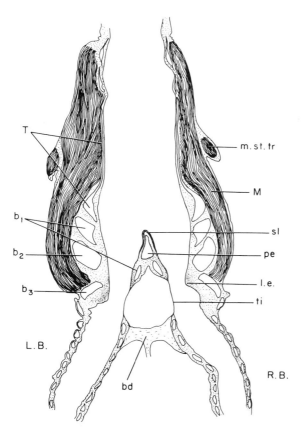

Figure 4.1
Longitudinal section of the syrinx of an adult male canary. R.B. and L.B.:
right and left bronchi; M: section through lateral mass of intrinsic syringeal
muscles; T: tympanum; b_1, b_2, and b_3: bronchial half-rings; bd: bronchidesmal membrane; pe: pessulus; sl: semilunar membrane; l.e.: labium exter-
num; ti: internal tympaniform membrane (thought to oscillate as a sound
source); m. st. tr.: sternotrachealis muscle. T, b_1, b_2, b_3 and pe are bony
components of the syrinx. The muscle mass serving the left syringeal half is
heavier than its right counterpart, which may reflect its greater use in singing. From Nottebohm and Nottebohm 1976, reproduced with permission
from *Journal of Comparative Physiology.*

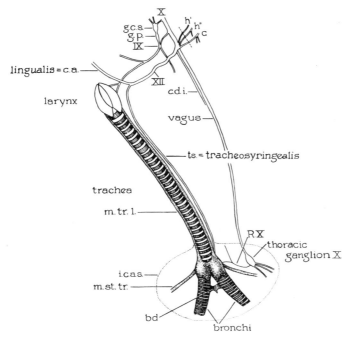

Figure 4.2
Ventral view of the syrinx of an adult male canary, surrounded by the interclavicular air sac (i.c.a.s.). The sternotrachealis muscle (m.st.tr.) anchors the
syrinx to the anterior process of the sternum (not shown); the tracheolater-
alis muscle (m.tr.l.) runs along the trachea and controls tracheal length. Innervation to the syringeal musculature is provided by the tracheosyringealis
branch (ts) of the hypoglossus (XII) nerve. The recurrens branch of the va-
gus (R.X) innervates the crop (not shown) which lies dorsal to the syrinx.
The bronchidesmal membrane anchors the syrinx to the dorsal wall of the
i.c.a.s. Bottjer and Arnold (1982) have shown that motor axons from the
caudal two-thirds of the hypoglossal nucleus (nXIIts; see figure 4.3) exit the
brain via the two lower hypoglossal roots (labeled here h″ and c), travel
along the main XII trunk and the ts nerve, and innervate the syrinx. Motor
axons from the rostral third of the hypoglossal nucleus exit the brain via the
uppermost hypoglossal root (here labeled h′), join the lingual nerve, and innervate the tongue. From Nottebohm and Nottebohm 1976, reproduced
with permission from *Journal of Comparative Physiology.*

mensions of these passages probably determine the patterns of turbulence as air flows past that point. In songbirds, the medial wall of the bronchial constriction is formed by the internal tympaniform membrane. In larger birds, such as fowl (in which the physics of voice production has been studied), the oscillation of the tympaniform membranes corresponds to the frequency of the sounds produced (Hersch 1966; Nottebohm 1975). A reasonable inference is that the oscillation of the syringeal membranes is locked in step with turbulence at that point, and that the amplitude of the membrane excursions determines the amplitude of the sounds produced.

The songbird syrinx consists of two symmetrical halves, each with its own sound source, air supply, muscular control, and innervation. The latter is provided by the tracheosyringeal branch of the hypoglossus nerve, which has both motor and sensory fibers (Nottebohm, Stokes, and Leonard 1976; Bottjer and Arnold 1982). The left and right tracheosyringeal nerves innervate only the syringeal muscles of their respective sides. In principle, then, the syrinx can be likened to two musical instruments, not just one (Greenewalt 1968; Nottebohm 1971).

The syrinx is surrounded by an airspace, the interclavicular air sac. This air sac provides a stable and predictable internal milieu within which the syrinx, as a motor appendage, can do its work. This is unlike other peripheral organs involved in motor learning. Hands and limbs must take into account environmental variables proper of manipulanda and substrate and the changing weight of the individual as it grows stouter or slimmer. Because of the stable internal milieu provided by the interclavicular air sac, motor programs for birdsong could, in principle, be learned and then persist as motor tapes (Wilson and Wyman 1965), independent of further sensory feedback. This would be the case when, after deafening, learned song persists with little change for periods of years (Konishi and Nottebohm 1969). Preliminary evidence suggests that sensory feedback from the syrinx is not important for maintenance of learned song (Bottjer and Arnold, in press).

It is most important for scientists trying to understand the brain machinery for vocal learning that the syrinx, in all likelihood, plays no role other than vocalization. The human vocal tract is far less private to this task. Our lips, tongue, and jaw are used not only in phonation, but also in facial expression and for smiling, for kissing, and for working on food; our larynx is used in respiration. With the syrinx it is different; find the part of the brain that controls the syrinx and you have the substrate for learning to sing.

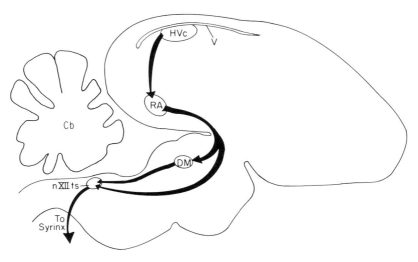

Figure 4.3
Sagittal section through the brain of an adult male canary, showing the motor pathway for song control. All connections shown are ipsilateral. HVc, RA, DM, and nXIIts: as in text; Cb: cerebellum; V: lateral forebrain ventricle. After Nottebohm, Stokes, and Leonard 1976 and Nottebohm, Kelley, and Paton 1982.

Efferent Pathways for Vocal Control

Brain pathways for syringeal control, and therefore for song control, were described in canaries several years ago (Nottebohm, Stokes, and Leonard 1976). The highest forebrain nucleus for vocal control is the hyperstriatum ventralis, pars caudalis (HVc). The HVc projects to the robust nucleus of the archistriatum (RA). The RA projects to the caudal two-thirds of the hypoglossal nucleus, which is labeled nXIIts because it is composed of the motor neurons that innervate the tracheosyringeal muscles. The RA also has indirect access to nXIIts motor neurons, via the dorsomedial (DM) nucleus of the midbrain (Gurney 1981). All these projections are ipsilateral; they are shown in figure 4.3.

   To what extent did the evolution of vocal learning in birds require the evolution of new and specialized connectivity? This question is just as important for understanding the evolution of vocal learning in our hominid ancestors, in which case, to my knowledge, it remains unanswered. In doves (an "old" group of birds in evolutionary terms) vocal learning does not occur (Nottebohm and Nottebohm 1971); however, as in canaries, there are connections between the DM and the

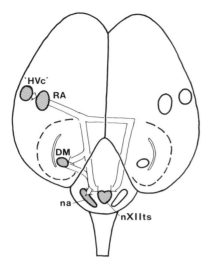

Figure 4.4
Schematized dorsal view of budgerigar brain, showing vocal-control motor
pathways. HVc, RA. DM, and nXIIts: as in text; NA: nucleus ambiguus,
which provides motor innervation to the muscles of the larynx. From Paton,
Manogue, and Nottebohm 1981, reproduced with permission of *The Journal
of Neuroscience.*

nXIIts. This suggests that the midbrain-medullary axis for vocal control
has changed little over time. There is also a very small cluster of ar-
chistriatal cells (perhaps homologous to RA) that project to the nXIIts.
We have failed to find a nucleus or connection reminiscent of the
songbird HVc (Cohen and Nottebohm, unpublished observations).

The belief that the HVc is a specialized part of the brain dealing with
vocal learning finds support in the fact that the *Psittaciformes* (including
parrots and budgerigars), which "discovered" vocal learning independent from songbirds, have a vocal-control pathway very similar to
that described in canaries, with well-developed forebrain nuclei which,
in terms of connectivity, seem homologous to the songbird HVc and
RA (Paton, Manogue, and Nottebohm 1981). There are differences,
though; in this group each RA projects to an ipsilateral and contralateral
nXIIts, and each right and left nXIIts innervates both syringeal halves
(figure 4.4). It seems likely that this arrangement reflects the special
structure of the psittaciform syrinx. The constriction of air flow occurs
in this case after the right and left bronchial bores come together,

affording a single sound source (Nottebohm 1976). In this sense the syrinx of parrots and budgerigars is more like the human larynx. Thus, in parrots, as in canaries, each HVc has access to a syringeal sound source; but whereas in parrots control of the single sound source requires simultaneous control of both syringeal halves, in canaries one-sided control suffices (Paton and Manogue 1982).

A nucleus such as the HVc, thought to preside over vocal learning, should be able to hear itself sing so that auditory feedback from the sounds produced can be compared with the auditory expectation or model. Auditory input reaches nucleas HVc (Kelley and Nottebohm 1979; Katz and Gurney 1981). However, McCasland and Konishi (1981) have shown that in adult canaries, white-crowned sparrows, and zebra finches HVc cells active just before and during song production do not respond to sound during or immediately after song. These authors refer to this as a "motor inhibition of auditory activity in HVc" and indicate that inhibition decays slowly over a period of seconds after the song terminates. This suggests that the comparison between the sounds produced and expected occurs outside the HVc, or ceases to occur in adulthood after song has been learned. The latter speculation could not apply to canaries. Adult canaries must hear themselves to maintain their learned song (Nottebohm, Stokes, and Leonard 1976). Other evidence suggests that the HVc plays an important role in the recognition of learned sounds. Margoliash (1983) reports that some cells in the white-crowned sparrow's HVc respond selectively to playbacks of the bird's own learned song. These physiological observations have not yet been related to the several afferent nuclei known to project to the HVc (Nottebohm, Kelley, and Paton 1982).

Parts of the human cortex, such as Broca's area, are involved in vocal control and partake in the phonetic decoding of speech (Ojemann and Mateer 1979). For speech control, such an area presumably projects to the motor cortex, and from there, as shown by Kuypers (1958), to motor neurons innervating the orofacial musculature. The reader will recognize analogies between this situation and what I have described in songbirds and parrots: The HVc, which responds to sound and has motor access to the syrinx, can be compared to Broca's area of the human frontal lobe; the RA can be likened to layer 5 of the motor cortex.

If nucleus HVc plays an important role in the perception as well as the production of song, as is suggested by its connections and physiology, then one would predict that the special responsiveness shown by female songbirds to conspecific or home-dialect song would disappear after lesion or removal of nucleus HVc, even if the motor response involved (solicitation) was not encoded in this nucleus. This prediction is now being tested.

Laterality

There is functional asymmetry in brain pathways controlling a behavior when the right or the left member of a symmetrical pair of organs consistently outperforms the opposite member of the pair. When this bias is systematic, so that, for example, the left side always dominates, it may point to an important strategy in brain function. Birdsong provides the best animal example of such a functional asymmetry in vertebrates, though since this was first reported (Nottebohm 1970) other examples have come to the fore. Laterality in animals may contribute important insights to our understanding of hemispheric dominance for speech and language in humans.

Left Hypoglossal Dominance

Denervation of the left but not the right syringeal half in chaffinches and canaries results in the loss of most of the components of song (Nottebohm 1970, 1971, 1972b; Nottebohm and Nottebohm 1976). Similar results have been obtained in white-crowned sparrows (Nottebohm and Nottebohm 1976) and white-throated sparrows (Lemon 1973). This phenomenon has been labeled left hypoglossal dominance for song control. Over 60 chaffinches and canaries tested in this manner have yielded not one case of right-"handedness" for song control.

Denervation of the left syringeal half before the onset of song learning is followed in chaffinches by the development of a normal song repertoire under right syringeal control (Nottebohm 1971). In canaries, section of the left ts or the left hypoglossus during the first 2 weeks after hatching also leads to the development of virtually normal song, now under right hypoglossal control. If the same operation is done during weeks 3 and 4 after hatching, the nerve regrows and the syllable repertoire is more equally divided between the two syringeal halves (Nottebohm, Manning, and Nottebohm 1979). Thus, though left hypoglossal dominance normally prevails in intact chaffinches and canaries, either hypoglossus by itself or both acting jointly can develop normal song. The question of why left hypoglossal dominance normally prevails is yet to be answered.

Hemispheric Dominance

Bilateral destruction of nucleus HVc produces a canary that is willing to sing but sings silently. The stance and dynamics of song manifest externally, but the syrinx, it seems, is not engaged. In this sense, the HVc performs as a syringeal-control nucleus. The motivation for singing, and the control of respiratory muscles to this end, must reside elsewhere (Nottebohm, Stokes, and Leonard 1976).

Destruction of the right or the left HVc nucleus produced different results. Right HVc destruction results in the loss of some syllables and some stereotypy; otherwise the pattern of delivery is not much altered, and to the ear the song sounds fairly normal. Left HVc destruction has much more drastic effects. Virtually all song syllables are lost, and the phrase structure is replaced by a wavering rendition with poorly defined, if any, phrases. The song now is so variable as to be reminiscent of the earliest stages of song development. In this sense, destruction of the left HVc seems to take away the store of learned information. Because of the marked behavioral difference after right versus left HVc destruction, the song-control system of canaries can be characterized as showing left hemispheric dominance (Nottebohm 1977). Despite this functional asymmetry, the size of the HVc and RA nuclei do not differ systematically between the right and left sides (Nottebohm, Kasparian, and Pandazis 1981).

Over weeks and months, left-HVc-lesioned canaries develop a new song repertoire, now under right-side control. In such birds, section of the right ts nerve abolishes the new repertoire. Thus, even in adulthood, male canaries can switch hemispheric dominance and develop a new song repertoire (Nottebohm 1977). This plasticity may be related to the fact that, normally, male canaries can change their song repertoire in successive years.

The right HVc seems anatomically equipped to develop normal song, yet remains subordinate. It appears to be a case of underused pathway potential. However, McCasland (1983) has reported that both left and right HVc cells are active during song production. This paradox has yet to be explained. Unlike the anatomical symmetry in canary's HVc, human left-side dominance for speech and language has been related to the larger size of the planum temporale of the left hemisphere (Geschwind and Levitsky 1968; Galaburda et al. 1978).

In humans, hemispheric specialization of function has been viewed as a clever evolutionary strategy for doubling the learning potential of the brain. If this explanation is correct, then the selective pressures that led to functional lateralization of vocal control in humans and canaries must have been different. In canaries they seem to have resulted in two parallel and potentially equivalent pathways for vocal control and learning, one of which normally becomes dominant. As long as the song sounds good, the right side exerts a relatively minor influence. Only when song is aberrant, as after left HVc destruction, does the right HVc take over. Such a reversal of dominance requires auditory feedback. After destruction of the left HVc in a deaf canary the right HVc does not become dominant for song control; the song of such a bird remains abnormally simple, with no involvement of the right syringeal half (F. Nottebohm, in preparation).

Sexual Dimorphism and Space for a Learned Skill
The HVc and the RA are several times larger in male than in female songbirds of the same species (Nottebohm and Arnold 1976). When this was reported, it was the first example of a gross sexual difference in brain structure. Now we know of other such cases in rodents (Gorski et al. 1978) and humans (Lacoste-Utamsing and Holloway 1982).

Since male songbirds sing but females usually do not, this differential apportioning of brain space seems to make good sense. The same relation between skill and space can be observed among males: Male canaries with large song repertoires have large HVc nuclei, but those with small HVc nuclei have small repertoires (Nottebohm, Kasparian, and Pandazis 1981). This relation may be a permissive one, since a male canary can have a large HVc but a small song repertoire. The same relation has been observed in a second sample of canaries and in a sample of zebra finches (F. Nottebohm, unpublished observations).

A relation between space and skill is also observed during development. Nucleus HVc first becomes recognizable at 30 days after hatching, when it has only 20% of the volume it will reach in adulthood. Subsong starts at about 40 days after hatching. The HVc continues to grow up to the 7th or the 8th month after hatching, when adult song becomes fully stable. HVc volume and the song repertoire remain stable until the end of that breeding season (F. Nottebohm, in prep.). Thus, HVc network space continues to be added as song is mastered, with no further growth after song becomes stable.

Hormonal Influences on Song-Control Pathways
Neurons in the HVc, the RA, the DM, and the nXIIts show nuclear concentration of androgen in adult male and female zebra finches (Ar-

nold, Nottebohm, and Pfaff 1976; Arnold 1980); neurons in the DM also show nuclear accumulation of estradiol (Arnold 1979). The accumulation of androgen is more marked in the male than in the female HVc (Arnold and Saltiel 1979). Androgen accumulation has also been reported for the HVc, the DM, and the nXIIts of male chaffinches (Zigmond, Nottebohm, and Pfaff 1973; Zigmond, Detrick, and Pfaff 1980). Thus, there seems to be a mechanism whereby male hormone can influence song-control pathways, possibly through genomic effects. This fits well with the observation that song in most songbirds is a seasonal phenomenon associated with territorial defense and courtship.

The adult dimorphism of the HVc and the RA seems to be attributable to early effects of hormone. Exposure to 17B-estradiol soon after hatching masculinizes the morphology of RA neurons, so that they become larger. Early exposure to androgen increases the number of neurons in the RA (Gurney and Konishi 1980; Gurney 1981). Either of these two effects, or a sexual difference in the accumulation of hormone by neurons of the vocal control system (Arnold and Saltiel 1979), could account for the observation that adult female zebra finches fail to sing when treated with testosterone. In male zebra finches, testosterone induces singing (Pröve 1974; Arnold 1975).

Adult female canaries treated with testosterone produce male-type song (Leonard 1939; Shoemaker 1939; Baldwin, Goldin, and Metfessel 1940; Herrick and Harris 1957). Adult gonadectomized females treated with physiological doses of testosterone start to sing approximately 1 week later, and by the end of their 4th week their repertoire is stable. At that time the HVc and the RA are 90% and 53% larger, respectively, than in ovariectomized control subjects treated with cholesterol. The latter birds do not come into song (Nottebohm 1980).

This remarkable plasticity in adulthood has been pursued at a microanatomical level. Earlier work had shown that the dendritic trees of RA neurons in adult female canaries are smaller than those of the same cell class in males (DeVoogd and Nottebohm 1981a). In the testosterone-treated females this difference was eliminated, which indicated that testosterone induces growth of extra dendritic material (DeVoogd and Nottebohm 1981b). Growth of the new dendrites is accompanied by the formation of new synapses (DeVoogd, Nixdorf, and Nottebohm 1982). These changes may be important for the acquisition of a learned behavior that females would not normally show.

Female canaries, like other cardueline females (Mundinger 1970), probably learn their calls by reference to auditory feedback, and this ability probably persists in adulthood. Thus, though testosterone induces adult females to sing, it is probably working on brain pathways that are already structured so as to make vocal learning possible.

A Brain for All Seasons

The next step in this story was prompted by three observations, already described in the previous pages: that song repertoires in adult male canaries change and increase from one year to the next, that there seems to be a limiting relation between size of the HVc and the size of the song repertoire, and that hormonal changes in adult females induce marked anatomical changes in song-control pathways. How do males manage, then, to develop a new song repertoire? If their available brain space is well subscribed or even replete by the end of the first singing season, how and where are the new song programs to be acquired and stored? What happens to the old programs? These questions are pertinent because the HVc and the RA undergo little if any net growth after the first breeding season (Nottebohm, Kasparian, and Pandazis 1981).

Males hatched in April were sacrificed at two ages: when 12 months old, in the middle of the spring reproductive season, after a stable song repertoire had been developed, or at 17 months, after the end of the late-summer molt, when males normally sing little if at all. Would these two stages in the yearly singing cycle be accompanied by gross changes in the HVc and the RA? To my delight, the answer was affirmative. The HVc and the RA were 99% and 76% larger in the spring group than in the fall group, though the latter group was older. Related to this, perhaps, was the observation that blood testosterone levels were high in the spring group but virtually undetectable in the fall group. My preliminary interpretation of these results is that part of the network space so diligently acquired during song learning is discarded at the end of the singing season, to make room for a new wave of growth and the acquisition of a new song repertoire (Nottebohm 1981). Combining data from the testosterone-treated females with the spring-to-fall changes observed in males, we are left with a vision of dendrites and synapses that grow and withdraw, or are shed much as trees grow leaves in the spring and shed them in the fall. These changes would be induced by seasonally rising and falling levels of testosterone or testosterone metabolites.

Throwaway Neurons

Strictly speaking, the idea of throwaway neurons is not new. Hamburger and Levi-Montalcini (1949; reviewed in Oppenheim 1981) were the first to draw attention to the widespread occurrence of this phenomenon during embryogenesis. They observed that in many parts of the nervous system excess neurons were formed and later discarded. This phenomenon has been interpreted as a way of ensuring that sufficient neurons would be available to form all the connections normally needed—a safety factor, as it were. This interpretation may or may not be correct, but the phenomenon itself, in the minds of neurobiologists, remained firmly rooted in embryogeny as one of many differences between developing and adult nervous systems. Now we know better.

In view of the remarkable growth of the female HVc under the influence of testosterone, it seemed reasonable to inquire whether some of this change could result from formation of new neurons. The standard approach to questions of neuronal birth is to use radiolabeled thymidine. (See, for example, Altman and Das 1965.) Thymidine is one of four nucleic acids that are the building blocks of DNA. New DNA formed during events leading to mitosis incorporates ^{3}H-labeled thymidine that is present at the time, and so becomes labeled.

Adult female canaries that received an implant of testosterone on day 0 were subsequently injected every 8 hours, for 2 days, with ^{3}H-labeled thymidine. When sacrificed a month later, these females had many new, labeled neurons in the HVc. To our surprise, other females that were treated with cholesterol instead of testosterone (which therefore did not come into song) showed comparable numbers of labeled neurons. This suggested the possibility that the birth of new neurons in nucleus HVc was a constantly ongoing process (Goldman and Nottebohm 1983).

Several aspects of this phenomenon are remarkable. The new neurons seem to originate from a thin layer of cells that lines the telencephalic ventrical dorsal to nucleus HVc. These cells then migrate into the HVc and presumably form connections and go to work. In this form, from 1% to 2% of all neurons in nucleus HVc are recruited every day. Since the total volume of this nucleus does not change between years 1 and 2, and density of cell packing remains similar from one year to the next, it is inferred that there must be a corresponding loss of cells. If there were no cell loss, the number of neurons in nucleus HVc would double in 49 days. Hence the term "throwaway neurons" or "replaceable neurons"; as some neurons are formed, others are discarded. Or perhaps some neurons trigger the birth of their replacements. We do not know yet if this process affects all neuronal types in nucleus HVc or only a few (for example, a local interneuron category). If the latter is the case, then the rate of turnover could be much faster than for the HVc as a whole. It is hard to imagine a more radical process of circuit rejuvenation.

Why should there be throwaway neurons at all? After all, if a neuron can grow and shed dendrites and synapses, as argued for the RA cells, this confers considerable plasticity on any network. Memories of past events, coded in synaptic changes, would be dismissed, and the neuron would start with a fresh batch of synapses. Such a neuron, one could argue, would have lost all "learned" biases. Why then kill the neuron itself? There is the possibility that forming part of a network exposed to specific patterns of use brings about changes in DNA expression, so that now some genes are turned on or off in an irreversible manner. If such changes constitute a way of storing memories, then truly fresh, uncommitted memory space can only be procured by replacing the "used" DNA with freshly minted DNA. In such a context, throwaway, replaceable neurons might be a very good thing to have.

Neuronal birth in adulthood has been documented in the brain of adult fish (Birse, Leonard, and Coggeshall 1980) and in the olfactory epithelium (Graziadei and Monti-Graziadei 1979), the olfactory lobe, and the hippocampus of rodents (Altman 1967; Kaplan and Hinds 1977; Bayer, Yackel, and Puri 1982; Bayer 1982). The significance of these other instances of adult neurogenesis remains unclear. In the olfactory epithelium, sensory neurons are replaced as in the HVc, probably to make up for the wear and eventual death they undergo in this very exposed position. In the other examples, adult neurogenesis may be related to sustained growth. It is my belief that many parts of the adult brains of different groups of animals will be found to show neurogenesis, at different rates and for different purposes. The challenge, of course, will be to find, in each case, the functional significance of this phenomenon and what controls it. In the future a novel question might be: Why are some parts of the adult brain so conservative as to show no neurogenesis?

Overview and Clinical Significance

Work on canaries has produced evidence of a very fundamental plasticity in adult nervous systems. This plasticity is demonstrated by yearly learning of new song repertoires, reversal of hemispheric dominance, forgetting after deafening, and song acquisition in hormone-treated females. Two basic phenomena were offered to explain this plasticity: the ability to form and discard synapses and the ability to form and discard neurons. Either phenomenon may provide a different approach to the "updating of software" that may be necessary for the sustained ability to acquire and discard new perceptual memories and new sensory-motor integrations. Addition and replacement of synapses and neurons, as natural processes of pathway "rejuvenation," may point to new clinical opportunities for the treatment of brain-impaired patients.

The kinds of treatment contemplated for restoring brain function may depend on assumptions about laterality and plasticity. Segregation of conflicting attentional systems to separate hemispheres may result in hemispheric dominance for different skills, as suggested by the observations of Kinsbourne and Cook (1971) and Okazaki-Smith, Chu, and Edmonston (1977). If this view is correct, there may be instances where intrahemispheric attentional conflict is avoided through unilateral underuse of networks. If a dominant side is lesioned, recruitment of a subordinate yet equipotential contralateral homologous network may reinstate the lost behavior, as I have suggested happens in canaries.

Under normal conditions, access to underused space may be limited in adulthood by loss of the plasticity needed for learning. Such plasticity could be reinstated by locally inducing replacement or addition of new synapses and neurons. However, if laterality is seen as a process of specialization, so that homologous parts of the two hemispheres discharge different functions, then there is less hope for recovery after lesion. Prognosis and treatment will be greatly influenced by a neurologist's view of hemispheric dominance.

Damage to networks often results from cell death. If dead neurons could be replaced by new neurons, networks could be repaired or reconstituted. My laboratory is presently working on the identification of conditions that induce the formation, migration, and differentiation of neurons in adulthood. If such conditions can be created locally in damaged or underused parts of the brain, it might be possible to reinstate lost functions. If neurons cannot be induced to form locally, or close enough to a desired target area, then they might be brought in. The day may come when neurology wards will have banks of fetal ventricular-zone cells ready to produce cohorts of neuroblasts which can then be brought into adult brains to busy themselves in network repair. But even if those dreams prove to be just dreams, the song system of birds promises to be uncommonly helpful for understanding how adult nervous systems learn new behaviors.

Acknowledgments

Work from my laboratory reported here was supported generously by Rockefeller University, by U.S. Public Health Service grants 5R01MH18343 and 1R01 NS17991, by National Science Foundation grant BNS7924602, by Biomedical Research Support grant 5S07 RR07065, to Rockefeller University, and by Rockefeller Foundation grant RF70095. I am deeply indebted to Marta Nottebohm for editing the manuscript and to Elizabeth Germond for typing it.

References

Altman, J. 1967. Postnatal growth and differentiation of the mammalian brain, with implications for a morphological theory of memory. In G. Quarton, T. Melnechuk, and F. O. Schmitt, eds., *The Neurosciences: A Study Program*. New York: Rockefeller University Press.

Altman, J., and G. D. Das. 1965. Autoradiographic and histological evidence of postnatal hippocampal neurogenesis in rats. *J. Comp. Neurol.* 124:319–336.

Arnold, A. P. 1975. The effect of castration and androgen replacement on song, courtship and aggression in zebra finches (*Poephila guttata*). *J. Exp. Zool.* 191:309–326.

Arnold, A. P. 1979. Hormone accumulation in the brain of the zebra finch after injection of various steroids and steroid competitors. *Soc. Neurosci. Abstr.* 5:434.

Arnold, A. P. 1980. Quantitative analysis of sex differences in hormone accumulation in the zebra finch brain: Methodological and theoretical issues. *J. Comp. Neurol.* 189:421–436.

Arnold, A. P., F. Nottebohm, and D. W. Pfaff. 1976. Hormone concentrating cells in vocal control and other areas of the brain of the zebra finch (*Poephila guttata*). *J. Comp. Neurol.* 165:487–512.

Arnold, A. P., and A. Saltiel. 1979. Sexual difference in pattern of hormone accumulation in the brain of a songbird. *Science* 205:702–705.

Baker, M., and L. R. Mewaldt. 1978. Song dialects as barriers to dispersal in white-crowned sparrows, *Zonotrichia Leucophrys Nuttalli*. *Evolution* 32:712–722.

Baker, M., K. Spitler-Nabors, and K. Bradley. 1981. Early experience determines song dialect responsiveness of female sparrows. *Science* 214:819–821.

Baldwin, F. M., H. S. Goldin, and M. Metfessel. 1940. Effects of testosterone propionate on female Roller canaries under complete song isolation. *Proc. Soc. Exp. Biol. Med.* 44:373–375.

Bayer, S. 1982. Changes in the total number of dentate granule cells in juvenile and adult rats: A correlated volumetric and ³H-thymidine autoradiographic study. *Exp. Brain Res.* 46:315–323.

Bayer, S., J. W. Yackel, and P. S. Puri. 1982. Neurons in the rat dentate gyrus granular layer substantially increase during juvenile and adult life. *Science* 216:890–892.

Birse, C. S., R. L. Leonard, and R. E. Coggeshall. 1980. Neuronal increase in various areas of the nervous system of the guppy, *Lebistes*. *J. Comp. Neurol.* 194:291–301.

Bottjer, S. W., and A. P. Arnold. 1982. Afferent neurons in the hypoglossal nerve of the zebra finch (*Poephila guttata*): Localization with horseradish peroxidase. *J. Comp. Neurol.* 210:190–197.

Bottjer, S. W., and A. P. Arnold. The role of feedback from the vocal organ: I. Maintenance of stereotypical vocalizations by adult zebra finches. *J. Comp. Physiol.*

Brackenbury, J. H. 1979. Aeroacoustics of the vocal organ of birds. *J. Theor. Biol.* 81:341–349.

Calder, W. A. 1970. Respiration during song in the canary (*Serinus canarius*). *Comp. Biochem. Physiol.* 32:251–258.

DeVoogd, T. J., B. Nixsdorf, and F. Nottebohm. 1982. Recruitment of synapses into a brain network takes extra space. *Soc. Neurosci. Abs.* 8.

DeVoogd, T. J., and F. Nottebohm. 1981a. Sex differences in dendritic morphology of a song control nucleus in the canary: A Quantitative Golgi study. *J. Comp. Neurol.* 196:309–316.

DeVoogd, T. J., and F. Nottebohm. 1981b. Gonadal hormones induce dendritic growth in the adult brain. *Science* 214:202–204.

Dooling, R., and M. Searcy. 1980. Early perceptual selectivity in the swamp sparrow. *Develop. Psychobiol.* 13:499–506.

Emlen, S. T. 1971. The role of song in individual recognition in the Indigo Bunting. *Z. Tierpsychol.* 28:241–246.

Galaburda, A. M., M. LeMay, T. L. Kemper, and N. Geschwind. 1978. Right-left asymmetries in the brain. *Science* 199:852–856.

Gaunt, A. S., R. C. Stein, and S. L. Gaunt. 1973. Pressure and air flow during distress calls of the starling *Sturnus vulgaris* (Aves; Passeriformes). *J. Exp. Zool.* 183:241–261.

Geschwind, N., and W. Levitsky. 1968. Human brain: Left-right asymmetries in temporal speech region. *Science* 161:186–187.

Goldman, S. A., and F. Nottebohm. 1983. Neuronal production, migration and differentiation in a vocal control nucleus of the adult female canary brain. *Proc. Nat. Acad. Sci.* 80:2390–2394.

Gorski, R. A., J. H. Gordon, J. E. Shryne, and A. M. Southam. 1978. Evidence for a morphological sex difference within the medial preoptic area of the rat brain. *Brain Res.* 148:333–346.

Graziadei, P. P. C., and G. A. Monti-Graziadei. 1979. Neurogenesis and neuron regeneration in the olfactory system of mammals. 1. Morphological aspects of differentiation and structural organization of the olfactory sensory neurons. *J. Neurocytol.* 8:1–18.

Greenewalt, C. H. 1968. *Bird Song: Acoustics and Physiology.* Washington, D.C.: Smithsonian Institution Press.

Gurney, M. E. 1981. Hormonal control of cell form and number in the zebra finch song system. *J. Neurosci.* 1:458–473.

Gurney, M. E., and M. Konishi. 1980. Hormone induced sexual differentiation of brain and behavior in zebra finches. *Science* 208:1380–1383.

Güttinger, H. R. 1981. Self-differentiation of song organization rules by deaf canaries. *Z. Tierpsychol.* 56:323–340.

Hamburger, V., and R. Levi-Montalcini. 1949. Proliferation, differentiation and degeneration in the spinal ganglia of the chick embryo under normal and experimental conditions. *J. Exp. Zool.* 111:457–502.

Herrick, E. H., and J. O. Harris. 1957. Singing female canaries. *Science* 125:1299–1300.

Hersch, G. L. 1966. Bird voices and resonant tuning in helium-air mixtures. Ph.D. Diss., University of California, Berkeley.

Hinde, R. A. 1958. Alternative motor patterns in chaffinch song. *Anim. Behav.* 6:211–218.

Hunter, M. L. 1980. Vocalization during inhalation in a nightjar. *Condor* 82:101–103.

Immelman, K. 1969. Song development in the zebra finch and other estrildid finches. In R. A. Hinde, ed., *Bird Vocalizations.* Cambridge University Press.

Kaplan, M. S., and J. W. Hinds. 1977. Neurogenesis in the adult rat: Electron microscopic analysis of light autoradiographs. *Science* 197:1092–1094.

Katz, L. C., and M. E. Gurney. 1981. Auditory responses in the zebra finch's motor system for song. *Brain Res.* 211:192–197.

Kelley, D. B., and F. Nottebohm. 1979. Projections of a telencephalic auditory nucleus—field L—in the canary. *J. Comp. Neurol.* 183:455–470.

King, A. P., and M. J. West. 1977. Species identification in the North American cowbird: Appropriate responses to abnormal song. *Science* 195:1002–1004.

Kinsbourne, M., and J. Cook. 1971. Generalized and lateralized effects of concurrent verbalization on a unimanual skill. *Q. J. Exp. Psychol.* 23:341–345.

Konishi, M. 1963. The role of auditory feedback in the vocal behavior of the domestic fowl. *Z. Tierpsychol.* 20:349–367.

Konishi, M. 1965. The role of auditory feedback in the control of vocalizations in the white-crowned sparrow. *Z. Tierpsychol.* 22:770–783.

Konishi, M., and F. Nottebohm. 1969. Experimental studies in the ontogeny of avian vocalizations. In R. A. Hinde, ed., *Bird Vocalizations.* Cambridge University Press.

Krebs, J. R. 1977. The significance of song repertoires: The Beau Geste hypothesis. *Anim. Behav.* 25:475–478.

Kroodsma, D. E. 1976a. The effect of large song repertoires on neighbor "recognition" in male song sparrows. *Condor* 78:97–99.

Kroodsma, D. E. 1976b. Reproductive development in a female songbird: Differential stimulation by quality of male song. *Science* 192:574–575.

Kroodsma, D. E. 1979. Vocal dueling among male marsh wrens: Evidence for ritualized expressions of dominance/subordinance. *Auk* 96:506–515.

Kroodsma, D. E. 1982. Learning and the ontogeny of sound signals in birds. In D. E. Kroodsma and E. H. Miller, eds., *Acoustic communication in birds,* vol. 2. New York: Academic.

Kroodsma, D. E., and R. Pickert. 1980. Environmentally dependent sensitive periods for avian vocal learning. *Nature* 288:477–479.

Kuypers, H. G. J. M. 1958. Corticobulbar connections to the pons and lower brain stem in man: An anatomical study. *Brain* 81:364–389.

Lacoste-Utamsing, C. de, and R. L. Holloway. 1982. Sexual dimorphism in the human corpus callosum. *Science* 216:1431–1432.

Lemon, R. E. 1973. Nervous control of the syrinx in white-throated sparrows (*Zonotrichia albicollis*). *J. Zool. (London)* 71:131–140.

Leonard, S. L. 1939. Induction of singing in female canaries by injections of male hormone. *Proc. Soc. Exp. Biol. Med.* 41:229–302.

Lott, D., and P. N. Brady. 1966. Support of ovulation in the ring dove by auditory and visual stimuli. *J. Comp. Physiol. Psychol.* 62:311–313.

McCasland, J. S. 1983. Neuronal Control of Birdsong Production. Ph.D. diss., California Institute of Technology.

McCasland, J. S., and M. Konishi. 1981. Interaction between auditory and motor activities in an avian song control nucleus. *Proc. Nat. Acad. Sci.* 78:7815–7819.

Margoliash, D. 1983. Acoustic parameters underlying the responses of song-specific neurons in the white-crowned sparrow. *J. Neuroscience* 3:1039–1057.

Marler, P. 1956. Behaviour of the Chaffinch, *Fringilla coelebs. Behaviour* Suppl. 5.

Marler, P. 1970. A comparative approach to vocal learning: Song development in white-crowned sparrows. *J. Comp. Physiol. Psychol.* monograph 71.

Marler, P., M. Kreith, and M. Tamura. 1962. Song developments in hand raised Oregon juncos. *Auk* 79:12–30.

Marler, P., M. Konishi, A. Lutjen, and M. S. Waser. 1973. Effects of continuous noise on avian hearing and vocal development. *Proc. Nat. Acad. Sci.* 70:1393–1396.

Marler, P., and P. Mundinger. 1971. Vocal learning in birds. In H. Moltz, ed., *Ontogeny of Vertebrate Behavior.* New York: Academic.

Marler, P., and S. Peters. 1977. Selective vocal learning in a sparrow. *Science* 198:519–521.

Marler, P., and S. Peters. 1982a. Structural changes in song ontogeny in the swamp sparrow *Melospiza georgiana. Auk* 99:446–458.

Marler, P., and S. Peters. 1982b. Developmental overproduction and selective attrition: New processes in epigenesis of birdsong development. *Psychobiology* 15:369–378.

Marler, P., and M. Tamura. 1964. Culturally transmitted patterns of vocal behavior in sparrows. *Science* 146:1483–1486.

Messmer, E., and I. Messmer. 1956. Die Entwicklung der Lautäusserungen und einiger Verhaltensweisen der Amsel (*Turdus merula merula L.*) unter natürlichen Bedingungen und nach Einzelaufzucht in schalldichten Räumen. *Z. Tierpsychol.* 13:341–441.

Metfessel, M. 1935. Roller canary song produced without learning from external source. *Science* 81:470.

Mundinger, P. C. 1970. Vocal imitation and individual recognition of finch calls. *Science* 168:480–482.

Nicolai, J. 1959. Familientradition in der Gesangsentwicklung des Gimpels (*Pyrrhula pyrruhula L.*). *J. Ornithol.* 100:39–46.

Nottebohm, F. 1968. Auditory experience and song development in the Chaffinch *Friagilla coelebs. Ibis* 110:549–568.

Nottebohm, F. 1969a. The song of the chingelo, *Zonotrichia capensis,* in Argentina: Description and evaluation of a system of dialects. *Condor* 71:299–315.

Nottebohm, F. 1969b. The "critical period" for song learning. *Ibis* 111:386–387.

Nottebohm, F. 1970. Ontogeny of bird song. *Science* 167:950–956.

Nottebohm, F. 1971. Neural lateralization of vocal control in a passerine bird. I. Song. *J. Exp. Zool. 177:228–261.*

Nottebohm, F. 1972a. The origins of vocal learning. *Amer. Natur.* 106:116–140.

Nottebohm, F. 1972b. Neural lateralization of vocal control in a passerine bird. II. Subsong, calls and a theory of vocal learning. *J. Exp. Zool. 179:35–49.*

Nottebohm, F. 1975. Vocal behavior in birds. In J. R. King and D. S. Farner, eds., *Avian Biology,* vol. 5. New York: Academic.

Nottebohm, F. 1976. Phonation in the orange-winged Amazon parrot *Amazona amazonica*. *J. Comp. Physiol. A* 108:157–170.

Nottebohm, F. 1977. Asymmetries in neural control of vocalization in the canary. In S. R. Harnad et al., eds., *Lateralization in the Nervous System*. New York: Academic.

Nottebohm, F. 1980. Testosterone triggers growth of brain vocal control nuclei in adult female canaries. *Brain Res.* 189:429–436.

Nottebohm, F. 1981. A brain for all seasons: Cyclical anatomical changes in song control nuclei of the canary brain. *Science* 214:1368–1370.

Nottebohm, F., and A. P. Arnold. 1976. Sexual dimorphism in vocal control areas of the songbird brain. *Science* 194:211–213.

Nottebohm, F., S. Kasparian, and C. Pandazis. 1981. Brain space for a learned task. *Brain Res.* 213:99–109.

Nottebohm, F., D. B. Kelley, and J. A. Paton. 1982. Connections of vocal control nuclei in the canary telencephalon. *J. Comp. Neurol.* 207:344–357.

Nottebohm, F., E. Manning, and M. Nottebohm. 1979. Reversal of hypoglossal dominance in canaries following syringeal denervation. *J. Comp. Physiol. A* 134:227–240.

Nottebohm, F., and M. Nottebohm. 1971. Vocalizations and breeding behavior of surgically deafened ring doves, *Streptopelia risoria*. *Anim. Behav.* 19:313–327.

Nottebohm, F., and M. Nottebohm. 1976. Left hypoglossal dominance in the control of canary and white-crowned sparrow song. *J. Comp. Physiol. A* 108:171–192.

Nottebohm, F., and M. Nottebohm. 1978. Relationship between song repertoire and age in the canary *Serinus canarius*. *Z. Tierpsychol.* 46:298–305.

Nottebohm, F., T. Stokes, and C. Leonard. 1976. Central control of song in the canary, *Serinus canarius*. *J. Comp. Neurol.* 165:457–486.

Ojemann, G., and C. Mateer. 1979. Human language cortex: Localization of memory, syntax and sequential motor-phoneme identification systems. *Science* 205:1401–1403.

Okazaki-Smith, M., J. Chu, and W. E. Edmonston. 1977. Cerebral lateralization of haptic perception: Interaction of responses to Braille and music reveals a functional basis. *Science* 197:689–690.

Oppenheim, R. W. 1981. Neuronal cell death and some related regressive phenomena during neurogenesis: A selective historical review and progress report. In W. C. Cowan, ed., *Studies in Developmental Neurobiology*. New York: Oxford University Press.

Paton, J. A., and K. R. Manogue. 1982. Bilateral interactions within the vocal control pathway of birds: Two evolutionary alternatives. *J. Comp. Neurol.* 212:329–335.

Paton, J. A., K. R. Manogue, and F. Nottebohm. 1981. Bilateral organization of the vocal control pathway in the budgerigar, *Melopsitacus undulatus*. *J. Neurosci.* 1:1279–1288.

Price, P. 1979. Developmental determinants of structure in zebra finch song. *J. Comp. Physiol. Psychol.* 93:260–277.

Pröve, E. 1974. Der Einfluss von Kastration und Testosteronsubstitution auf das Sexualverhalten männlicher zebrafinken. *J. Ornithol.* 115:338–347.

Searcy, W. A., and P. Marler. 1981. A test for responsiveness to song structure and programming in female sparrows. *Science* 213:926–928.

Shoemaker, H. H. 1939. Effect of testosterone propionate on the behavior of the female canary. *Proc. Soc. Exp. Biol. Med.* 41:229–302.

Thorpe, W. H. 1958. The learning of song patterns by birds with special reference to the song of the chaffinch (*Fringilla coelebs*). *Ibis* 100:535–570.

Thorpe, W. H., and P. M. Pilcher. 1958. The nature and characteristics of subsong. *British Birds* 51:509–514.

Tinbergen, N. 1951. *The Study of Instinct.* Oxford: Clarendon.

Waser, M. S., and P. Marler. 1977. Song learning in canaries. *J. Comp. Physiol. Psychol.* 91:1–7.

Wilson, D. W., and J. Wyman. 1965. Motor output patterns during random and rhythmic stimulation of locust thoracic ganglia. *Biophys. J.* 5:121–143.

Yasukawa, K., J. L. Blank, and C. B. Patterson. 1980. Song repertoires and sexual selection in the red-winged blackbird. *Beh. Ecol. Sociobiol.* 7:233–238.

Zigmond, R. E., F. Nottebohm, and D. W. Pfaff. 1973. Androgen-concentrating cells in the midbrain of a songbird. *Science* 179:1005–1007.

Zigmond, R. E., R. A. Detrick, and D. W. Pfaff. 1980. An autoradiographic study of the localization of androgen concentrating cells in the chaffinch. *Brain Res.* 182:369–381.

PART II

Linguistic and
Psychological Issues

Part II contains six chapters that deal with the structure of linguistic units and the nature of the psychological processing of such units, as revealed by studies that draw on populations of particular biological interest and thus bear on aspects of the biology of these representations and operations. Two of the chapters deal with limiting cases of the systems for language representation and use: The infant, who has no overt behavior that can be clearly construed as reflecting parts of the linguistic system he will very shortly master (such as word meaning), and the polygot or bilingual, who masters more than one linguistic system. The remaining four chapters deal with the breakdown of language after cerebral lesion.

Eimas's review of work on the abilities of infants to appreciate and/or form categories show that, at the earliest ages at which infants have been tested, complex and abstract aspects of the physical properties of stimuli constitute the basis for the equivalence classes infants respond to. This is true both for physical properties of acoustic stimuli that are part of the language code the child will eventually master and for features of nonlinguistic events that presumably constitute the basis for the conceptual system onto which the formal linguistic code eventually maps. In the linguistic sphere, such features as distinguish the high and low unrounded front vowels /i/ and /a/, once discriminated, are generalized to different pitches and different speakers by infants 6 months of age. In the nonlinguistic sphere, the parameters that can be abstracted intermodally by infants the same age and younger include the correlation of the tempo of occurrence of auditory events with the tempo of visual events, even when the actual occurrence of the events is not simultaneous in the two modalities. Eimas suggests a number of mental structures and psychological processes the evidence supports as basic to these abilities. The achievement of an idealized prototype on the basis of exposure to distortions of a geometric form seems to be in play with respect to structures perceived by infants, and perceptual processes (such as categorial perception) seem to be among the processes underlying the psychology of these abilities. Eimas's view is that these abilities are functional prerequisites for the development of such aspects of language as the referential properties of words. These abilities are claimed to be part of the infant's initial state, upon which further neurological maturation and exposure to language build the full intricacies of the mature language system. The studies presented by Eimas certainly show that the infant's initial state is intricate and is highly tuned to abstracting stimulus features that are relevant to adult linguistic and nonlinguistic structures. In this respect they speak to the minimal functional capacities that are genetically encoded, since it is implausible that experience before testing has had any major effect on the development of these abilities. Functional abilities of this sort must then be the very last that are "innate" in the human.

Accepting this conclusion, however, we are led to a number of further considerations. What is the neural basis for these abilities? There is some evidence bearing on this question in the studies of hemispheric preference for linguistic and nonlinguistic stimulus recognition and processing in infants. To what extent does postnatal acquisition of language depend on neural maturation rather than environmental exposure? Perhaps this is better phrased as follows: Which aspects of language development depend on particular aspects of neurological development, and how does exposure to the environment exert its effect? It is well known that contemporary linguists have presented a strongly nativist and anti-behaviorist set of answers to these questions. It is worth bearing in mind that the most interesting alternate proposals to the stimulus-response theories of environmental effects—namely theories of parameter fixation in universal general linguistic structures—required the development of particular structural theories of language for their enunciation and validation, and that answers to these questions on the functional side are not likely to be forthcoming except in relation to theories of acquired language structures. This leads to a third question: What are the limits of the initial state? More broadly: What are the characteristics of this initial stage? Eimas presents evidence for an impressive range of abilities. Can we hope to characterize these abilities in terms of structure and processing in a systematic and general way? Clearly, Eimas has not presented the entirety of the infant's capacities. In some spheres, such as the mastery of the expression of emotional force via intonation contour, the infant overtly shows considerable mastery of a system that is ultimately to be used (or adapted to be used) linguistically. How far do these remarkable abilities extend?

If we accept that essential aspects of language (such as the ability to achieve reference without stringent reinforcement schedules for a far larger number of words that has been achieved by any other species) are unique to humans, we must ask to what extent the abilities described by Eimas are helpful in explicating the development of language in humans, since at least some comparable abilities are found in both the cognitive and acoustic spheres in nonhumans. These abilities are arguably prerequisites for language, but possessing them does not seem to be sufficient for language development. What else, then, is required? When does it make its ontogenetic appearance? What is its neural basis?

The four subsequent papers deal with aphasia. Kean's paper stresses three aspects of such studies. First, they must go beyond mere description of aphasic symptoms if they are to achieve explanatory adequacy. Second, in attempting to construct mechanisms that account for aphasic symptoms, one must take care not to consider as related symptoms that are, in fact, not functionally related. Kean cites the various aspects of the syndrome of deep dyslexia and the variety of impairments seen in Broca's aphasia as collections of symptoms that may not have unified functional explanations. In both instances, she suggests that an anatomical account of the co-occurrence of these symptoms may be correct: The lesions responsible for the "syndromes" may produce a number of symptoms because they interfere with a number of neural structures. Despite these caveats, Kean clearly states her belief in functional explanations of the details of symptoms and of the co-occurrence of symptoms. Kean's third concern is related to the task of discerning these functional details and functionally occasioned syndromes. Here she makes two related points. First, the choice among descriptions of a symptom is not to be made randomly over the set of descriptions afforded by different linguistic theories and by alternate formulations of the deficit within a single linguistic theory; rather, the choice is to be dictated by the explanatory value of each description, in terms of generating empirically validated descriptions, and in terms of how the linguistic description interacts with other theories (most notably, theories of language processing) to account for details of the symptom and the co-occurrence of symptoms. The second point is that a good deal of what is manifest in an aphasic symptom or syndrome is due to the variations and effects induced by the operation of normal components of the language system. Kean cites the degree of disturbance of word order in agrammatism reported by Saffran and Schwartz as an example. In both comprehension and production, she argues, the effects reported can be attributed to the effect of normal processing mechanisms on a disturbance in utilization of the clitic vocabulary and need not be attributed to a separate psychopathological mechanism.

The chapters by Nespoulous, Zurif, and Garrett provide instances of analyses of aphasic disturbances to which Kean's observations apply.

Nespoulous first provides analyses of spoken errors that indicate that a variety of different levels of linguistic structure are the loci of error formation. These range from phonological distinctive features, through phonemes and syllables, to morphemes, words, and discourse features. Nespoulous's first topic is the justification of levels of analysis of errors. The level appropriate for describing an error such as /bato/(boat)→ /mato/ is not immediately obvious; is it the distinctive feature nasality, or is it the entire phonemes /b/ and /m/? Nespoulous argues that the circumstances under which one or the other description is appropriate are discernible and, in part, correspond to aphasic subgroups. Nespoulous distinguishes intrinsic effects on determination of the appropriate descriptive level (such as the fact that monosyllabic items are more likely to be considered verbal paraphasias than are longer items, simply because of the increased chances for phoneme substitution to result in a word output in monosyllabic words) from extrinsic effects on descriptions (such as the task in which an error is made, for example when repetition and reading provide explicit targets and thus reduce the number of unassignable errors). As Nespoulous points out, a good deal of preliminary classification of errors can be accomplished without reliance on extremely abstract analyses of errors or of syndromes, and these descriptive matters interact with the construction of more abstract and potentially more general accounts of symptoms and syndromes.

Both Zurif and Garrett approach aphasic symptoms and syndromes within the sort of theoretical framework referred to by Nespoulous and discussed by Kean. Both of these writers are principally concerned with describing aphasic performances, whether individual symptoms or functionally related groups of symptoms, in terms of the linguistic representations affected (as is Nespoulous) and in terms of models of the processes of language production and comprehension that are operative in normal language users. That is to say, the framework that provides the descriptive terms for an aphasic performance includes both a stated theory of language structure—which serves to allow identification of the linguistic structures affected by the cerebral lesion—and a theory of language processing in which these structures are operated on by particular processes and "mechanisms" in the actual tasks of language use.

Zurif's domain of investigation is primarily syntactic and semantic. He reviews and extends earlier arguments supporting the contention that agrammatic Broca's aphasics show disturbances in comprehension and production of language that are related to the construction and the interpretation of syntactic form. Since these patients are defined in terms of the relative omission of function-word vocabulary items, the analysis of their problem as the failure to construct syntactic form suggests a processing link between the utilization of the function-word vocabulary and the construction of syntactic structures. This point, also made by Kean, is strengthened by analyses linking the construction of syntactic form to the use of the function-word vocabulary that have been proposed for normals, particularly on the basis of analysis of speech errors. Zurif begins to take the next step in such investigations and theories: He suggests that the speech of Wernicke's aphasics is also abnormal with respect to syntactic structures, but that the abnormality is qualitatively different from that found in agrammatism. Broca's aphasics show a pattern of omissions and simplifications of word production, whereas Wernicke's aphasics, according to Zurif, use a small stock of well-formed syntactic frames, which are disrupted by ill-formed lexical items, anomic features of language, and the like. The qualitatively different outputs would reflect disturbances at different processing levels—the "positional" level in agrammatism and the "functional" level in Wernicke's aphasia.

In the light of this hypothesis, it is worthwhile to reconsider Geschwind's comments on Zurif's chapter. Geschwind criticizes Zurif's failure to cite some of the early literature on the nature of comprehension disturbances in Broca's aphasia, and he claims that Zurif attributes to clinical studies an incorrect espousal of a sensory-motor dichotomy as the descriptive framework of the aphasias. Zurif's introductory remarks are indeed cryptic and do not constitute a full review and explication of prior clinical observations and theories. Despite the limitations of Zurif's historically oriented introduction, however, Geschwind's remarks fail to emphasize the degree to which Zurif is correct in insisting that the framework for description and explanation of aphasic symptoms and syndromes has undergone a significant revision, which is reflected by Zurif's own work and his present analysis and suggestions. The concern to identify the linguistic elements affected in a symptom and to account for the manifestations of psychopathology upon those elements in terms of specified theories of language structure and processing, justified by normative analyses, is not new; however, it has seen greater richness and deeper development as the studies and theories of modern linguistics and psycholinguistics have been applied to research into the aphasias. Much of the value of previous work seems to lie in suggesting more specific hypotheses, phrased in modern structural and processing terms, rather than in providing the details of those hypotheses or data that bear on the detailed hypotheses now being stated. Even as detailed an analysis as Luria's work on the efferent and afferent motor aphasias is formulated within a framework of linguistic description that makes the published observations useful only in formulating, not in evaluating, hypotheses such as those presented in Zurif's chapter and in many other recent studies.

At the same time, the same considerations apply to at least some of the work of the 1970s. We cannot continue to consider syntactic structures to be all of a piece, or to consider their interpretation to be limited to the establishment of thematic relations. As more detailed studies are undertaken in response to more specific hypotheses about aphasia, we can anticipate that the hypotheses and studies that began the analysis of aphasia in modern linguistic and psycholinguistic terms will be superseded by more detailed descriptions and more elaborate formulations, and that new empirical evidence—perhaps harder to acquire, since some of these analyses may be quite abstract—will be needed to settle questions pertaining to the description and explanation of symptoms and symptom complexes. One major hope is that we can move from description in terms of major components of a grammar, like a syntactic component to formulations stated in terms of basic structural components of linguistic components (such as the basic theories of binding, government, and case currently being explored by linguists).

Garrett's chapter is similar to Zurif's in its effort to relate features of aphasic performances to aspects of sentence processing suggested by analyses of normal language. It also demonstrates the degree to which the effort to analyze aphasic data in terms of models currently being formulated in linguistics and psycholinguistics both is inspired by and goes beyond classical observations and analyses. The task set in Garrett's chapter is to accommodate aphasic errors to stages in the production model developed on the basis of analysis of speech errors. Garrett reviews the evidence from normal speech errors that has led him to postulate two separate levels of sentence construction between the determination of a "message" and the determination of the phonetic values for the realization of lexical items, intonation contours, and so on. Many data from aphasia are easily accommodated to analyses in which the abnormal performance can be attributed to failures at one or another of these levels.

Consider, for instance, the process of lexical selection. The data from aphasia support a division between the ability to produce the function-word class of lexical items and the ability to produce the content-word class (definable more formally in either psychological or linguistic terms). The contrast is between agrammatism and the various forms of "word-finding difficulty" in the posterior aphasics. Within the content-word vocabulary, the distinction between meaning-based retrieval and form-based retrieval (an essential feature of the normative production model) finds support in the performances of anomic and conduction aphasics. Some quite detailed observations regarding these patients' performances support several detailed features of the production model. For instance, anomic aphasics were found by Goodglass to be able to produce the first sound of a word only in about 5% of cases. Given the view that anomic aphasics cannot achieve the phonological pattern of a word but do achieve its semantic specification (as evidenced by their ability to describe its meaning in phrases), one hypothesis is that the "linking code" between the semantic specification and the phonological speci-fication of the word is disturbed. This code would seem to include information on initial sounds of words, syllable length, and stress place-ment, as judged by the loci of speech errors of the sound-exchange type. Thus, it is expected that anomic patients would lose these aspects of phonological information. The conduction aphasics, in contrast, were found to report the first sound of a word about a third of the time. Their disturbance, thus, is unlikely to lie at this level of lexical retrieval/ production, at least in its entirety. Rather, the disturbance seems to lie in the ability of these patients to utilize the information contained in their (possibly partially degraded) linking addresses to fully specify the phonological forms of the words they seek to utter. Further studies might specify these disturbances more specifically in relation to the production model. For instance, Garrett points out that, if the problem in the generation of phonological form on the basis of a linking address is linked to the processes of phrasal construction characterizing the positional level in his model, then increases in phonemic paraphasias occasioned by sound exchanges ought to be found in conduction aphasia. Apparently this increase does not occur. Thus, we can be more precise in our hypotheses and attribute the disturbance in conduction aphasia to the stage of lexicalization in which the segmental phonological values for individual words are retrieved.

Garrett's final remarks give an outline of the general functional ar-chitecture of the production system as he conceives of it. For each of two aspects—lexical items and syntactic form—there are meaning-based and sound-based access procedures. Many of the principal aphasic symptoms can be accommodated to the model as failures in one of these four components, according to the analyses presented by Garrett and Zurif. Some suggestive lacunae that also seem to exist in aphasic

symptomatology may be accounted for by the architecture of the production system and by constraints on where aphasic symptoms arise.
Thus, the absence of syndromes consisting of stranding errors, or word
or sound exchanges, suggests that aphasic symptoms arise as disturbances in the mapping from one part of this architecture to another,
and are not internal to a component of the system.

The latter suggestion is highly speculative at the moment, but it
illustrates an important contribution that studies of aphasia can make
to the understanding of normative systems. If such a principle could
be justified in clear cases of aphasic symptoms, it could be used to
investigate aspects of the normative model that are underspecified because of a lack of normative data. The contribution that studies of
dyslexia are currently making to the understanding of normal reading
mechanisms illustrates this potential in a different sphere of language
functioning. Even in the more unidirectional way in which the chapters
of Zurif and Garrett relate language pathologies to normative models,
according to which pathologies of language are analyzed as resulting
from disturbances at levels of processing specified in normative work
and thus as supporting existing normative analyses, the value of extending the range of populations studied to aphasic patients is clear.
This widening of the perspective on what types of language performances
bear on normal language processes to the biological sphere of aphasia
is a step—yet to be taken in detail—toward specifying the pathological
determinants of the impairment and, ultimately, the neural basis for
the functions identified.

Obler's paper provides strong evidence that the factors influencing
performance in the languages of bilinguals exert different effects on the
different languages a person knows. Differential recovery of a single
language after lesions causing aphasia is sufficiently frequent to permit
us to say that recovery at least *can* show language-specific patterns.
The factors dictating which of several languages mastered will be preserved best or recovered best are not entirely clear despite a century
of observations, but this uncertainty does not vitiate the conclusion
that some factors apply to one or another language in a way that they
do not apply to others in some polygot aphasics. These observations
are part of the impetus to investigate potential differences in the organization of languages in bilinguals. Obler reviews the results of a
decade of work seeking to delineate which types of language elements
are organized differently in the languages of a bilingual and which have
similar organizational features. An important aspect of this work is the

question of where the different languages are represented. Obler was one of the first proponents of the "stage hypothesis," according to which there is more right-hemisphere processing of language in the early stages of language acquisition than in later stages in second as well as in mother tongues. She argues that this hypothesis has continued to receive support. If it is correct, and if a bilingual is not equally competent in each of his languages, we may infer that the cerebral structures supporting the two languages are different.

One line of research, in keeping with the questions suggested by Zurif and Garrett, would be to investigate whether there are differences in the functional processing of two languages in either balanced bilinguals or bilinguals with greater proficiency in one of two languages; how any differences found are determined by acquisition factors, the nature of the languages mastered, and endogenous factors such as handedness; and how any such differences are related to brain loci. Obler's final section, which deals with concepts currently under investigation, contains an interesting suggestion regarding the functional question. Studies of speech perception in bilinguals suggest that the mastery of a second language does affect the processing of both languages: Acoustic determinants of the phonological value "voicing" in stop consonants in Hebrew-English bilinguals are not those found in either Hebrew or English monolinguals but represent the extreme values of the acoustic forms relevant to stop-consonant voicing in both languages. Thus, the perceptual system does seem to be affected by bilingualism. It is not known whether more general aspects of the organization of language processing, such as the architectural features of speech production described by Garrett and Zuriff, are also affected in this way. If they are, and for more restricted areas such as the voicing case described by Obler, it would be interesting to know whether such changes are related to any aspect of the neural mechanisms for language that can currently be investigated, such as hemispheric localization.

The chapters in part II present primarily functional analyses of the nature of psychological processes and language structures. Prerequisites for language development, aspects of the organization of language in the adult, and changes in language and influences on processing associated with bilingualism are the principal issues explored in these chapters. In some cases, these chapters directly attempt to relate their analyses to organic analyses, such as the role of either hemisphere in the languages of a bilingual. In other instances, the fact that the analyses are the results of studies of brain-injured populations raises the possibility (not yet actualized) of relating these analyses to neural mechanisms. In yet other cases, such as that of infant functioning, the organic basis of the functions described will have to be investigated by means other than those used to delineate the functions themselves. Whatever the stage of our understanding of the organic mechanisms related to the processes described in these chapters, the level of specificity of functional description presented here is sufficiently justified that we are induced to ask about neural and other organic mechanisms relevant to this level of functional description. Indeed, we are in a position to expect that efforts to relate functional analyses to neural and other organic mechanisms will continue to aim at and achieve at least this level of empirical detail of psychological and linguistic description.

Chapter 5

Infant Competence and the Acquisition of Language

Peter D. Eimas

It is now widely believed that, given the limitations of time and the available linguistic data, the acquisition of human languages during infancy can occur only if the biology of the human species imposes constraints of a linguistic nature on the acquisition process. (See, for example, Chomsky 1965 and Chomsky 1979.) Of course, the assumption that our inheritance imposes linguistic limitations does not in any way deny the profound influence of experiential factors on the acquisition process and on the final form of the acquired language. Nor does it deny the possibility that biological constraints of a more general cognitive nature are powerful determinants of the process of language acquisition. (See Bruner 1974.) Indeed, there is a growing awareness that an adequate theoretical description of the acquisition of language will involve linguistic and cognitive constructs that are biological in origin and that are functional very early in the life of the infant. (See deVilliers and deVilliers 1978.) If this metatheoretical assumption is valid, then a major task of the developmental psycholinguist who wishes to understand the acquisition of language and perhaps language itself is to provide precise descriptions of the initial state of the human infant—that is, the cognitive and linguistic "innate ideas and principles" and the manner in which they interact with environmental forces to provide the knowledge that is necessary for linguistic competence.[1]

Descriptions of the initial state and its relation to language acquisition have been slowly forthcoming in recent decades from a variety of approaches to the study of language learning and cognitive development. For example, empirical studies of the earliest utterances of young children, pioneered in their modern form by Roger Brown and his students (see Brown 1973), continue to yield a listing of the linguistic structures that are available to very young children—structures that presumably reflect, at least in part, some of our innate linguistic knowledge. A quite different approach to delineating our initial state and its relation to language learning is that of the learnability theorists, who have taken as their task the construction of a theory of language from which it can be formally demonstrated that the grammar of human language is learnable if one is given the available linguistic input. From their discussions to date, it is apparent that meeting the criterion of learnability will require the assumption of a large number of cognitive and linguistic constraints. (See Wexler and Culicover 1980.) Whether the conclusions of this formal approach to the problem will ultimately be verified by empirical studies remains a matter of conjecture.

Further descriptions of the constraints that infants bring to the task of acquiring a language are to be found in studies that have sought to uncover the cognitive precursors of a variety of aspects of linguistic competence. Some of the constraints described are quite specific, for example the semantic relation between agent and recipient (Golinkoff 1975); others are general, such as the pragmatic aspects of communicative skills (Bruner 1974). Investigations of the information processing abilities of very young infants have also begun to reveal a number of capacities, undoubtedly shaped by our inheritance, that are particularly relevant to the acquisition of language. It is with two of these abilities and their relation to language learning that I shall be concerned for the remainder of the discussion.

More specifically, I shall consider the infant's capacity to integrate information across sensory modalities (that is, to perceive the multimodal nature of objects and events) and to categorize sensory information. The discussion of the latter will focus on two lines of research. The first concerns the infant's ability to categorize visual information as evidenced by the presence of equivalence classes, that is, sets of stimuli that share some category-defining property (or properties) that presumably serves as the basis for responding to the set of stimuli in a like manner, despite discriminable variation in other, non-category-defining properties. (The presence of irrelevant information that is discriminable precludes explanations of stimulus equivalence based on an observer's inability to detect the irrelevant variation, and instead must be ascribed to some additional processing that permits this variation to be ignored, at least temporarily.) The second line of research concerns the infant's ability to categorize speech. In one approach to this problem, the existence of categories is inferred from the presence of equivalence classes that are based on the properties of speech (for example, vowel quality). In a second approach, categorical representations of speech are inferred from nonmonoticities (that is, peaks) in the discriminability functions of stimuli that lie along acoustic continua selected for study because changes in stimulus value are known to be sufficient to signal changes in the phonetic percepts of adult listeners. Moreover, the perception of these stimuli by mature listeners shows considerable consistency in the assignment of stimuli to phonetic categories, greatly enhanced discrimination at the boundaries between categories, and often near-chance levels of discrimination of stimuli from within the same phonetic category. (See Liberman et al. 1967 and Pisoni 1978 for reviews and discussions of these findings.) In other words, perception is constrained by the manner in which stimuli are assigned to phonetic categories, and thus perception is termed categorical or categorical-like. Investigations of categorical-like perception in infants, however, must rely exclusively on the presence of orderly peaks and troughs in the discriminability functions for evidence of categorization, as no methodology now available would provide functions comparable to the identification functions of adult listeners. The inference that categories exist rests on the reasonable assumption that peaks in discriminability reflect the boundaries between categories, whereas regions of lowered discriminability mark the extent of the categories *per se* (as is true for adults). Of course, in the case of young infants, the categorization of speech, when it occurs, is most likely not based on phonetic categories, in that categories of this nature probably do not exist in a prelinguistic organism. However, it is reasonable to assume that the categories that are imposed on the speech signal by the infant's processing system are the forerunners of the phonetic categories of mature languages users. It is hoped that a discussion of these very basic processing capacities will not only illustrate the very sophisticated nature of the infant's perceptual systems but also add to our understanding of the cognitive and linguistic constraints that serve, and perhaps even shape, the process of language acquisition.[2]

Intermodal Perception

The environments of human beings are complex, and the specification of the objects and events about us requires the use of higher-order relational information and the possible transformations of this information. Moreover, our specifications of environmental structures require that we use information of this nature from more than one modality. For example, a bicycle is known and recognized not only by the patterns of visual stimulation but also by the patterns of tactile, kinesthetic, and perhaps even auditory information. Despite these multiple sources of specifying information, our percepts of objects are not separable, modality-specific experiences, loosely united by associations or the temporal contiguity of the various forms of stimulation. Quite the contrary is true. Our perception of objects is a unitary experience, and it is this unity, this holistic nature of perception, that motivates us to speak of the integration of information across sensory modalities.

The acquisition of language, like the acquisition of knowledge in general, involves the use of multiple sources of information for the same event. Speech, for example, is a series of acoustic events that are, in nature, accompanied by visual patterns associated with the processes of articulation. Although it is by no means necessary that both sources of information be present for veridical perception, visual cues can aid the perception of speech. (See Summerfield 1979.) Also, incongruent visual information can alter the auditory percept of simple syllables; it creates, in effect, an auditory illusion. (See McGurk and McDonald 1976.) Moreover, asynchronous auditory and visual information specifying the same speech events can be detected by infants as young as 10 weeks of age. Dodd (1979) presented nursery rhymes with the speech sounds and lip movements either in synchrony or out of synchrony by 400 milliseconds. The infants attended to the visual display reliably less in the absence of synchrony, thereby demonstrating intermodal perception as well as the existence of an expectation of synchrony between the sounds of speech and the movements of articulation at a surprisingly early age. In view of the ability to appreciate this form of information, it would not be unreasonable to assume that infants use the visual information associated with articulation (when it is available) in acquiring competence in producing and perceiving the sounds of the parental language.

The development of referential systems in human languages—that is, the mapping of ordered sequences of speech sounds onto the infant's visual percepts—requires that information be integrated across more than one modality.[3] Inasmuch as lexical learning begins quite early, intermodal perception for auditory and visual information other than that which defines speech events should likewise be demonstrable in very young infants. The recent work of Spelke (1976, 1979, 1981) confirms this expectation. In one series of studies, Spelke (1979) showed that infants 4 months of age were able to perceive a bimodally specified event by means of the temporal patterns that were common to both the visual and auditory patterns of stimulation. More specifically, infants were found to look significantly longer at one of two simultaneously presented visual displays (a bouncing yellow kangaroo) when the tempo of visual collisions matched the tempo of bursts of sound on an accompanying soundtrack. When the tempo of the soundtrack was changed to that of a second visual display (a bouncing gray donkey), the visual preference of the infants changed and they looked longer at the second display. Moreover, Spelke was able to demonstrate intermodal perception even when the auditory events were not temporally contiguous with the visual collisions, as long as their tempos of occurrence were the same. What is particularly interesting about a number of these demonstrations is the rapidity with which young infants are able to acquire knowledge of bimodally specified events. (See Spelke 1981 in particular.) Indeed, findings of this nature support the contention that the processing abilities that support the acquisition of bimodal information are innately determined and may even actively seek, from the time of early infancy, the multimodal information that defines the concepts of mature individuals.

As Spelke (1979) has noted, these data are consistent with the hypothesis—derived from a Gibsonian view of perception and its development (E. J. Gibson 1969; J. J. Gibson 1979)—that infants perceive a bimodally defined event by directly detecting the invariant information common to both modalities (the tempo of events in the example above). The detection of the common invariance would presumably occur at an abstract level of processing, one that is essentially amodal in nature. Of course, the integration of multimodal information can also occur as a result of associative processes that are biologically tuned to detect and store the evidence for correlations that might exist in the natural world. These associative processes might require, at least in some instances, considerable experience with the environment before the existence of a correlation would or could be recognized. Regardless of the mechanisms underlying intermodal perception, it is apparent that at the time of the beginning of lexical learning the infant has had extensive experience in learning about intermodal correspondences and in developing strategies for discovering such correspondences. The latter is particularly important in that the acquisition of lexical units also requires that the infant realize that visual and tactile sources of stimulation arising from objects and events may have sounds associated with them that are not inherent characteristics of the objects and events themselves but rather are the products of human articulation.

In summary: The early appearance of intermodal perception (see also Butterworth and Hicks 1977; Lee and Aronson 1974; Meltzoff and Moore 1977), as well as its critical role in the acquisition of knowledge in humans and nonhumans, is indicative of a mode of information processing that has a very deep biological determination. Although we have made considerable progress in describing the origin of this intermodal perception, much remains to be learned of the means by which it is achieved.[4]

Categorization of Information

Visual Patterns

The ability to form categorical representations, such that under some circumstances discriminably different events and objects can be responded to as if they were equivalent, is essential if sensate creatures are to be able to exist in environments that provide a potentially infinite number of different stimuli. The ability to form categories or equivalence classes provides a means of information reduction, a cognitive economy, that prevents processing systems from being overwhelmed by information and rendered unable to respond and in so doing permits the members of a species to adapt to their environment by creating one that is in a sense more manageable.

Categorization of nonlinguistic sources of stimulation is obviously necessary for information processing in general. It is likewise a necessary condition for the development of effective communication systems, including human languages.[5] And, more specifically, it is an essential component of the acquisition of a lexicon, as Brown noted in his classic description of the process:

The original word game is the operation of linguistic reference in first language learning. At least two people are required: one who knows the language (the tutor) and one who is learning (the player). In outline form the movements of the game are very simple. The tutor names things in accordance with the semantic custom of his community. The player forms hypotheses about the [categorical] nature of the things named. He tests his hypotheses by trying to name new things correctly. The tutor compares the player's utterances with his own anticipations of such utterances and, in this way, checks the accuracy of fit between his own categories and those of the player. He improves the fit by correction. (Brown 1958, p. 194)

The questions about the formation of equivalence classes that are particularly relevant to cognitive and linguistic development are these:

What is the level of abstraction at which categorization occurs?[6] What processes underlie the formation of categories? Are there any restrictions on the nature of the information that can serve as the basis for categorization?

In the past several years, a number of investigators concerned with infant visual perception have begun to extend their research efforts to categorization in infants. Cohen and Strauss (1979), using a measure of habituation and dishabituation of visual attention, showed that infants about 30 weeks of age were able to form equivalence classes based on either "a specific female face regardless of orientation" or "female faces in general." At present we are unable to define precisely the information that provided the basis for these equivalence classes, although the research of Fagan and Singer (1979) indicates that 7-month-old infants are sensitive to a number of facial features that define such characteristics as sex and that these features can serve as a basis for categorization.[7]

As for the manner in which infants form equivalence classes, Strauss (1979) and Bomba (1980) have obtained quite convincing evidence that infants are able to abstract a "prototypical" representation of a category (see Posner and Keele 1968)—a representation that is an average of the exemplars used to illustrate the category. In the Strauss study, 10-month-old infants formed a representation of a human face. In the Bomba study, infants as young as 3 months formed representations of geometric forms. In the latter study, infants were initially shown dot patterns that were very precisely defined distortions of a perfect geometric form (a square, for example). After this familiarization period, a perfect square was presented together with a perfect triangle and the infant's preference was measured. The logic of these test trials was that, although neither of the test stimuli had actually been seen in the experiment, if the infants had formed a prototypic representation of the square during the test trials then the actual perfect square would have been effectively a familiar stimulus. Consequently, the infants would be expected to prefer to observe the novel triangle, as indeed they did. In an important control condition, Bomba also showed that the square and its distortions were readily discriminable and thus that the preference for the triangle was not simply a result of the infants' failing to discriminate the square from the previously viewed distortions.[8]

It is apparent from these studies that infants are capable of forming equivalence classes at a relatively young age. Moreover, they are able to do so with a rather diverse variety of visual information, such as a specific face or a class of faces or geometric forms, and the manner in which these classes are formed is in accord with an averaging process such as has been assumed to underlie the formation of prototypic representations. It would seem that by the time lexical learning begins the infant undoubtedly has a large number of categorical representations and the means to create new ones onto which the sounds of speech can be mapped.

Speech Patterns
The Formation of Equivalence Classes

The basic sounds of speech, those sounds that roughly correspond to the consonants and vowels of any language, are signaled by different values along a large number of acoustic dimensions. It has been evident since the earliest days of modern research on the production and perception of speech that the particular values of the temporal and spectral characteristics of any speech sound are not constant across changes in the conditions of production. (See Liberman et al. 1967.) Moreover, the variation in the acoustic parameters of a phonetic segment is considerable. This is true whether the variation results from the effects of coarticulation (that is, the differences in production of a particular segment associated with differences in the surrounding phonetic units), from differences in syllabic position, from individual variation in the vocal apparatus itself, or from differences between and within speakers with respect to rate of speech, stress, and physiological state. The variability in the acoustic parameters is often such that the distributions of the acoustic values for different segmental units overlap. Consequently, at some levels of analysis of the acoustic signal it is possible to find situations in which the same values are associated with differences to phonetic perception. There are also situations in which the same phonetic percept is associated with quite different values along some acoustic continuum, as well as more extreme instances in which the same percept actually has a number of apparently very different acoustic cues associated with it. Lisker (1978) has noted that voicing distinction in medial syllabic position may have as many as 16 different cues.

What is of course required for a theoretical description of perception in any modality is that the constancies of perception be mirrored by constancies in antecedent conditions. Conversely, differences in perception must reflect differences in prior states. There are two general classes of solutions to these demands. In the first, it is assumed that the apparent lack of invariance in the visual or auditory signal is the result of an erroneous or incomplete analysis of the signal and that a more appropriate analysis (which usually requires higher-level relational and time-dependent descriptors) would reveal the necessary physical invariants for the perceived perceptual constancies. In the domain of speech this view is held by a number of investigators, of whom most recently Stevens and Blumstein (1981) and Searle, Jacobson, and Rayment (1979) have offered the most detailed analyses of some possible invariants for phonetic distinctions based on place of articulation, and in the visual domain it has long been advocated by Gibson and his associates (see Gibson 1979) in their ecological approach to perception. The second solution to the apparent absence of physical invariance in the signal is to assume that the perceptual system imposes the necessary transformations on the incoming signal such that the isomorphism between perception and signal is achieved, albeit at a neurological level. Solutions of this nature can be found in a variety of models of speech perception, some of which assume that perception is based on listeners' tacit knowledge of articulatory mechanisms (Liberman et al. 1967; Stevens and House 1972) whereas others assume the existence of detector mechanisms that may operate in a variety of ways to achieve the desired isomorphism (Miller and Eimas 1982).

Developmental studies of perceptual constancy are probably not, in and of themselves, the means by which we will come to understand perceptual constancy in speech perception; however, such studies can begin to explicate some of the boundary conditions on the phenomenon and thereby limit the theoretical possibilities. Obviously, studies of perceptual constancy, *per se*, cannot be performed with very young, prearticulate infants; there is simply no procedure by which the necessary judgments of perceptual identity for speech or any other sensory information can be obtained from infants at this level of development. What can be obtained with speech stimuli (as with visual patterns) are assessments of the extent to which infants are able to form equivalence classes based on the categories of speech (or vision)—an obvious and necessary precondition for perception constancy.

The first investigation of the formation of equivalence classes was performed by Fodor, Garrett, and Brill (1975). They showed that infants 3–4 months old were able to learn to group together syllables of English with fewer errors when the syllable shared an initial consonant than when they did not share this consonant. In view of previous research (Liberman, Delattre, and Cooper 1952) on the particular syllables used by Fodor et al., it was most likely the case that relatively simple acoustic information (such as the spectral properties of the release burst) did not serve as the acoustic basis for the classification. Moreover, as Fodor et al. have noted, what is particularly significant is that the acoustic parameters that the infant apparently selects and finds easier to use in forming equivalence classes are those that form the basis for adult categories of speech.

More recently, Kuhl (1979, 1980) has extensively studied the classificatory behavior of infants approximately 6 months of age. She has used a variant of the visually reinforced head turning procedure, wherein an infant is reinforced if a head turn is made when there is a change in the background stimulation. For example, if the background stimulus is the vowel [a] in any of a number of acoustic variations, then a correct response occurs and is visually reinforced if a head turn is made when the vowel [i] is presented in any of a number of similar acoustic variations. In one study (Kuhl 1979), infants were trained initially on single instances of the synthetically produced vowels [a] and [i]. After acquisition of the discrimination (nine of ten consecutive correct responses), variation in the training stimuli was introduced in a slow stepwise fashion. First, the pitch of the stimuli was varied, alternating randomly between rising and falling. Next, the speaker, male or female, was randomly varied, after which both sources of variation were presented concurrently. Finally, a third speaker was introduced, a child, with a rising or falling intonation pattern. All of the infants who adapted to the apparatus completed the experiment, and the infants who were trained with the vowels [a] and [i] performed at a level not very different from that which would indicate perfect sorting of the stimuli. Kuhl (1980) has also reported the formation of equivalence classes based on the fricative contrasts [s] versus [ʃ] and [f] versus [θ] in initial and final syllabic positions. The variation in the syllables, which was again introduced gradually, was in the form of vowel quality and speaker. Although the formation of equivalence classes was more difficult with fricatives than with vowels, this may have simply been a result of procedural differences.

Katz and Jusczyk (1980) have presented evidence of a limited ability to form equivalence classes based on stop consonants, which varied with respect to the syllable-final vowel. The limitations on this ability may reflect the greater effects of coarticulation with stops as compared with fricatives, for example, and thus may indicate the necessity for greater experience with some speech sounds than with others before equivalence classes are fully formed and perceptual constancy is achieved. In sum, there is ample evidence for the ability to form equivalence classes based on the categories of speech in infants. Moreover, this ability exists well before it is necessary for the infant to know that some forms of acoustic variation are or are not essential for linguistic processing.

Kuhl would prefer to account for her results by assuming that "the infant demonstrates a proclivity to try to discover a criterial attribute which separates the two categories." (Kuhl 1979, p. 1674) Of course, as Kuhl has also noted, it is possible that the experimental procedures have at least partially shaped the formation of equivalence categories.[9] The assumption of a natural proclivity for discovering a means of classification would be better supported if, after initial training with the single exemplars of each category, a series of transfer trials were given in which all of the variation was introduced immediately and the level of transfer was high immediately after the start of the transfer trials. In two experiments designed in exactly this manner, Kuhl (1979; 1980) found at least partial support for her hypothesis. When the categories were [a] and [i] the mean percentage of correct transfer trials (of which there were 90) was 79, and when the vowel categories were [a] and [o] the mean percentage was 67. Seven of eight infants performed above chance in the former experiment, although only four of eight infants did so in the latter experiment. Part of the poorer performance in the second experiment may have stemmed from the greater acoustic difference between [a] and [i] than between [a] and [o]. Another portion of the difference may be attributable to individual variation; the second study may simply have had a greater proportion of infants who found this type of task difficult. Evidence for the latter is obtained from a comparison of infants who performed at above chance levels of performance. Their mean percentages of correct responses was 82.6 and 80.6 for the vowels [a] and [i] and the vowels [a] and [o], respectively— a difference that is not statistically reliable. What might underlie these individual differences and what their consequences might be for attaining future cognitive and linguistic competence are not known at present.

In summary: The evidence clearly shows that a major prerequisite for perceptual constancy exists early in the life of infants, certainly at least several months before it is required for the acquisition of other forms of linguistic knowledge (such as the mapping of concatenations of sounds onto the infant's conceptual categories). Without at least this rudimentary precondition for perceptual constancy, the development of referential systems would require acquisition of a potentially infinite number of associations, something that is clearly impossible. Similarly, the acquisition of grammatical rules would necessitate separate learning processes for each and every speaker that the child encounters, again an impossibility. Since the acquisition of language and the efficient processing of nonlinguistic sources of knowledge relies heavily on the ability to form equivalence classes, it is not surprising to find this process available for the encoding of a wide variety of environmental information very early in life.

Categorical Perception

There is an additional source of variation in the speech signal, not discussed in the preceding section, that results from our inherent inability to produce exactly the same value of a number of acoustic parameters relevant to phonetic distinctions at each and every instance of production. For example, Lisker and Abramson (1964, 1967) have shown that speakers from different language communities produce the voicing contrasts of their native languages with considerable variation in voice onset time, a complex acoustic continuum that is a sufficient basis for describing voicing contrasts. This variation occurs even when there is a constant phonetic environment as well as procedural circumstances that should operate to reduce acoustic differences that might arise if rate of speech and stress, for example, were completely free to vary. This form of variation is also found with other acoustic parameters, namely the formant structure signaling vowel quality (Peterson and Barney 1952) and the transition durations signaling the distinction between the stop consonant [b] and the semivowel [w] (J. L. Miller, personal communication). However, this acoustic variation does not disturb our efforts to abstract the phonetic message and ultimately the semantic intent of the speaker.

What is required to accommodate the failure of this variation to hinder the processing of speech is the assumption that the perceptual system must initially derive a phonetic categorization and on achieving this categorization make the analogic acoustic information relatively unavailable to the listener. It certainly seems to be the case that in the course of everyday conversations we are not aware of this form of variation, that is, the moment-to-moment variation in the particular values of an acoustic parameter. Of course, our casual introspections are not sufficient evidence to confirm assumptions about the operating characteristics of the speech-processing system. However, appropriate evidence does exist; indeed, it constitutes one of the best-established findings in the speech-perception literature of the past 25 years. Listeners have been found to perceive speech in a categorical manner. [Liberman et al. (1967) and Pisoni (1978) have reviewed this literature.] That is to say, listeners were able to assign synthetically produced variations of a particular phonetic contrast into rather sharply defined categories. Moreover, and more important, they were better able to discriminate equivalent acoustic differences when the difference signaled a phonetic distinction than when it signaled acoustic variants of the same phonetic category. In other words, the discriminability functions were nonmonotonic, reflecting the process of categorization and the relative unavailability of the specific acoustic information. In many instances the ability to detect within-category differences is only slightly better than chance, although it must be noted that perception is almost never perfectly categorical. Furthermore, even this approximation to categorical perception does not occur under all circumstances (Pisoni 1978); the listener's access to analogue acoustic information is more pronounced in some experimental situations than others. Be that as it may, this research supports the contention that the primary results of the initial analyses of speech are a categorization of the speech signal and a reduction in the availability of the analogue representation of the signal itself. Perception of a categorical nature has been found for the acoustic continua that underlie many of the perceived distinctions based on differences in voicing, place of articulation, manner of articulation, and to a lesser extent vowel category, if the vowel durations are quite brief.

Investigations into the ontogeny of the categorical perception of speech were begun in the late 1960s. Eimas et al. (1971) tested whether the infant's discriminability function for voice onset time was discontinuous, as it was for adult listeners whose native languages were English, Spanish, and Thai (Abramson and Lisker 1970, 1973; Lisker and Abramson 1970). These adult listeners, in showing categorical perception, were considerably better at discriminating a given difference in voice onset time when the two stimuli crossed a phonetic boundary in their native languages than when it did not. Although different languages often have a different number of categories (sometimes with substantially different boundary locations), there appear to be three major voicing distinctions, with boundary locations approximately between -20 and -60 milliseconds and between $+20$ and $+60$ milliseconds of voice onset time (Lisker and Abramson 1964). Using a high-amplitude sucking procedure that is analogous to many habituation-dishabituation procedures for assessing discriminative abilities in infants, Eimas et al. compared the infant's ability to discriminate a 20-millisecond difference in voice onset time under two conditions. In the first the two synthetic speech patterns signaled different voicing categories (the English voiced and voiceless labial stops, [b] and [pʰ], respectively), whereas in the second the two sounds signaled members of the same voicing category, voiced or voiceless. There was also a control condition against which the performance of the infants could be evaluated. The infants were either 1 or 4 months of age. The data were clear. At both age levels, the discriminability functions were nonmonotonic. The infants discriminated the stimuli that crossed the voiced-voiceless boundary in English and other languages and showed no evidence of discriminating the different acoustic tokens of the same adult voicing category. These findings have been replicated with infants from different language communities (Lasky, Syrdal-Lasky, and Klein 1975; Streeter 1976). The latter studies also demonstrated categorical-like perception at the prevoiced-voiced boundary region of the voice-onset-time continuum, which is rather strong evidence for the fact that infants come into the world with the ability to process voicing information in terms of three auditory categories that will in time be modified in varying degrees to form the voicing categories of their native languages. The evidence also indicates that, in view of the differences among languages with respect to the number and location of voicing categories, the initial dispositions to process voice onset time must be amenable to modification. Exactly how the demands of processing speech in terms of its linguistic functions determine these modifications is not yet known.

Evidence for categorical-like perception in the young infant is not limited to the continuum of voice onset time. As has been found with adult listeners, infants perceive one form of information for place of articulation (the direction and starting frequencies of the second- and third-formant transitions) in terms of categories that correspond very closely to the three phonetic categories of adult listeners (Eimas 1974; Jusczyk 1981). They also perceive two manner distinctions (a stop-semivowel contrast and a stop-nasal contrast) categorically, although by no means in an absolutely categorical manner, especially in the case of the stop-nasal contrast (Eimas and Miller 1981a,b).[10] In addition, the acoustic correlates of the distinction between [r] and [l] are perceived very nearly in a categorical manner (Eimas 1975). Finally, the differences in formant frequencies that define vowel quality in adult listeners show a tendency toward categorical perception if the duration of the sounds is very brief (Swoboda et al. 1978).

The mapping of acoustic information along a number of acoustic continua onto phonetic categories is, we are beginning to learn, a very complicated process. There does not appear to be a single range of values that define a particular category; the set of values changes quite markedly with a number of contextual factors. For example, Miller and Liberman (1979) have shown that the transition duration values that are necessary to signal the semivowel [w] as opposed to the stop consonant [b] are reliably increased when the duration of the entire syllable is lengthened. The boundary value of transition duration separating the two categories was 32 milliseconds when the syllables were 80 milliseconds in duration and 47 milliseconds when the syllables were 296 milliseconds long. As Miller and Liberman have noted, the ability to use the after-occurring context of syllable duration may provide listeners with a means of effectively normalizing the signal for changes in rate of speech.

Eimas and Miller (1980a) investigated whether infants use this same contextual information in the categorization of transition durations and found that infants who heard the 80-millisecond speech patterns discriminated transition duration values of 16 and 40 milliseconds but showed no evidence of discriminating values of 40 and 64 milliseconds. However, exactly the opposite pattern of results was obtained when the speech patterns were 296 milliseconds in duration: The infants showed evidence of discriminating the second but not the first pair of transition duration values. We have since replicated these findings, and thus there is little doubt that infants categorize this form of contextual information in very much the same way as adult listeners. Perhaps infants are also able to capitalize on this ability to use contextual information in categorizing speech in order to normalize the speech signal for variation in rate of speech. If so, the task of learning to associate the sounds of speech with our representations of the objects and events in the world about us must be markedly facilitated.

Summary and Conclusions

This chapter has focused on the abilities of infants, before they attain even rudimentary linguistic competence, to categorize and integrate information from their visual and auditory environments. The available evidence leaves little doubt that infants have highly developed, sophisticated systems for the categorization and integration of information that provide at least some of the means by which they are able to acquire linguistic and nonlinguistic knowledge and adapt to their environment. Indeed, without these abilities being largely determined by our biology it is difficult to imagine how the acquisition of knowledge would commence, let alone arrive at a state that permits abstract thought and linguistic communication.

There are of course other constraints on processing that operate relatively efficiently during infancy to promote the acquisition of knowledge. These include the ability to appreciate the organization that is inherent in visual patterns as well as in the segmental and syllabic units of speech (see Milewski 1979; Miller and Eimas 1979) and the ability to develop "frames of reference" (Pick, Yonas, and Rieser 1979). Indeed, it is likely that the conceptual categories that are potentially available to use are deeply constrained by our biology (Fodor 1975). Certainly other constraints will be discovered in our quest to describe the initial state of the human species, a quest that will ultimately further our understanding of human cognition and language.

Acknowledgments

The preparation of this chapter, and my research reported therein, were supported by grant HD 05331 from the National Institute of Child Health and Human Development. I thank Joanne L. Miller for her helpful comments on an earlier version of the chapter.

Notes

1. The term *initial state* is used quite broadly here; it should be taken to mean not only those abilities and dispositions that are present at birth but also those that arise by processes of maturation during the first year of life. In essence, a description of the initial state would include all of the innate factors (that is, the processing systems, rules, and representations) that the infant brings to the task of acquiring a human language.

2. Of course, a complete description of the initial state will include designating whether our biological constraints are cognitive or linguistic in nature. Such designations are often obvious; however, in the case of the initial dispositions for the processing of speech this is certainly not the case. It is a matter of considerable debate whether the mechanisms that underlie the processing of speech at a phonetic level are part of a species-specific system that evolved for the processing of human language. Though I believe that the processing of speech does involve such specialized mechanisms (Eimas 1982; see also Lib-

erman 1970), and consequently that the innate components of this system constitute a subset of our linguistic constraints, the discussion that follows is not predicated on this belief.

3. Sounds are also mapped onto representations of events and objects in modalities other than vision, including those of a tactile and kinesthetic nature. In the case of a blind child acquiring a first language, the integration of speech and nonvisual representations of the environment must be of critical importance.

4. By the age of 4 months, infants can be "instructed" to attend to one of two superimposed visual scenes of a dynamic nature by the simultaneous playing of a soundtrack that is appropriate to (temporally synchronous with) the visual scene that the experimenter has targeted for attention (Bahrick, Walker, and Neisser 1981). Although such research clearly demonstrates the selective nature of early perception and the ability to integrate information across modalities, it also demonstrates how information in one sensory channel can be used to focus information-gathering procedures in other channels—a function well served by language once the child has gained minimal linguistic competence.

5. Lenneberg, in his discussion of the biology of human language, has theorized that "the cognitive function underlying language consists of an adaptation of a ubiquitous process (among vertebrates) of categorization and extraction of similarities." (Lenneberg 1967, p. 374)

6. I believe it is obvious that the issue is not whether or not categories exist at the time when words are first learned, but rather what is the nature or level of abstraction of these categories. Even if the infant were acquiring a label for only a single entity, a single instance of a category, and not for a whole set of related entities, the label must still signify a category. The category would be defined as the unique entity, but in a variety of circumstances and contexts. Thus, for example, the infant's use of the word *dog* to refer to, and only to, the family pet must occur with the dog in a variety of poses and circumstances that at times dramatically alter the multimodal stimulus configurations that emanate from this creature. To explain this fact in a reasonable manner, we must assume that the child has a categorical representation of this specific dog that permits the common features to be recognized despite considerable variation in nonessential features. Whether the infant also has the ability to form categorical representations based on the characteristics (defining or fuzzy) of "dogs in general" is at the heart of the current research on categorization in early infancy.

7. As noted above, the existence of equivalence classes that cannot be explained simply as the result of the observer's inability to detect the irrelevant information requires independent evidence that the variation among members of the equivalence class is discriminable. Evidence of this nature has been provided by Cohen and Strass as well as by Fagan.

8. It is also of interest to note that the less frequently observed average information (that is, the prototypic values) was better recognized than the more frequently encountered individual variants in the Strauss study. Moreover, Bomba found greater recognition of the prototype as compared to the familiar instances even though the prototype had never been experienced.

9. That the capacity to form equivalence classes is not solely a function of the training procedures used by Kuhl is evidenced by the infant's inability to learn arbitrary classifications of the stimuli (Kuhl 1979).

10. That is to say that there is evidence that the acoustic variants of phonetic categories can at times be discriminated by infants (see Eimas and Miller 1980a,b; Miller and Eimas 1983), as is true for adults (see Pisoni 1978). In a similar vein, there is also evidence that at least some of the irrelevant variation in Kuhl's original studies of equivalence-class formation (Kuhl 1979) can be used as the basis for the formation of equivalence classes in and of itself (Kuhl and Hillenbrand 1979). Such data, of course, argue against categorization being solely a function of an inability to discriminate either irrelevant information or within-category acoustic differences.

References

Abramson, A. S., and L. Lisker. 1970. Discriminability along the voicing continuum: Cross language tests. In *Proceedings of the 6th International Congress of Phonetic Sciences*, Prague, 1967. Prague: Academia.

Abramson, A. S., and L. Lisker. 1973. Voice-timing perception in Spanish word-initial stops. *Journal of Phonetics* 1: 1–8.

Bahrick, L. E., A. S. Walker, and U. Neisser. 1981. Selective looking by infants. *Cognitive Psychology* 13: 377–390.

Bomba, P. C. 1980. The Nature and Structure of Infant Categories. Master's thesis, Brown University.

Brown, R. 1958. *Words and Things*. New York: Free Press.

Brown, R. 1973. *A First Language: The Early Stages*. Cambridge, Mass.: Harvard University Press.

Bruner, J. S. 1974. From communication to language—A psychological perspective. *Cognition* 3: 255–287.

Butterworth, G., and L. Hicks. 1977. Visual proprioception and postural stability in infancy: A developmental study. *Perception* 6: 255–262.

Chomsky, N. 1965. *Aspects of the Theory of Syntax*. Cambridge, Mass.: MIT Press.

Chomsky, N. 1979. *Rules and Representations*. New York: Columbia University Press.

Cohen, L. B., and M. S. Strauss. 1979. Concept acquisition in the human infant. *Child Development* 50: 419–424.

deVilliers, J. G., and P. A. deVilliers. 1978. *Language Acquisition*. Cambridge, Mass.: Harvard University Press.

Dodd, B. 1979. Lip reading in infants: Attention to speech presented in and out of synchrony. *Cognitive Psychology* 11: 478–484.

Eimas, P. D. 1974. Auditory and linguistic processing of cues for place of articulation by infants. *Perception and Psychophysics* 16: 513–521.

Eimas, P. D. 1975. Auditory and phonetic coding of the cues for speech: Discrimination of the [r-l] distinction by young infants. *Perception and Psychophysics* 18: 341–347.

Eimas, P. D. 1982. Speech perception: A view of the initial state and perceptual mechanisms. In J. Mehler, E. C. T. Walker, and M. Garrett, eds., *Perspectives on Mental Representation*. Hillsdale, N.J.: Erlbaum.

Eimas, P. D., and J. L. Miller. 1980a. Contextual effects in infant speech perception. *Science* 209: 1140–1141.

Eimas, P. D., and J. L. Miller. 1980b. Discrimination of the information for manner of articulation. *Infant Behavior and Development* 3: 367–375.

Eimas, P. D., E. R. Siqueland, P. Jusczyk, and J. Vigorito. 1971. Speech perception in infants. *Science* 171: 303–306.

Fagan, J. F., III, and L. T. Singer. 1979. The role of simple feature differences in infants' recognition of faces. *Infant Behavior and Development* 2: 39–45.

Fodor, J. A. 1975. *The Language of Thought*. Cambridge, Mass.: Harvard University Press.

Fodor, J. A., M. F. Garrett, and S. L. Brill. 1975. Pi Ka Pu: The perception of speech sounds by prelinguistic infants. *Perception and Psychophysics* 18: 74–78.

Gibson, E. J. 1969. *Principles of Perceptual Learning and Development*. New York: Appleton-Century-Crofts.

Gibson, J. J. 1979. *The Ecological Approach to Visual Perception*. Boston: Houghton Mifflin.

Golinkoff, R. M. 1975. Semantic development in infants: The concepts of agent and recipient. *Merrill-Palmer Quarterly* 21: 181–193.

Jusczyk, P. W. 1981. Infant speech perception: A critical appraisal. In P. D. Eimas and J. L. Miller, eds., *Perspectives on the Study of Speech*. Hillsdale, N.J.: Erlbaum.

Katz, J., and P. W. Jusczyk. 1980. Do six-month-olds have perceptual constancy for phonetic segments? Paper presented at International Conference on Infant Studies, New Haven, Conn.

Kuhl, P. K. 1979. Speech perception in early infancy: Perceptual constancy for spectrally dissimilar vowel categories. *Journal of the Acoustical Society of America* 66: 1668–1679.

Kuhl, P. K. 1980. Perceptual constancy for speech-sound categories in early infancy. In G. H. Yeni-Komshian, J. F. Kavanagh, and C. A. Ferguson, eds., *Child Phonology*. Vol. 2: Perception. New York: Academic.

Kuhl, P. K., and J. Hillenbrand. 1979. Speech perception by young infants: Perceptual constancy for categories based on pitch contours. Paper presented at biennial meeting of Society for Research in Child Development, San Francisco.

Lasky, R. E., A. Syrdal-Lasky, and R. E. Klein. 1975. VOT discrimination by four-to-six-and-a-half-month-old infants from Spanish environments. *Journal of Experimental Child Psychology* 20: 215–225.

Lee, D. N., and E. Aronson. 1974. Visual proprioceptive control of standing in human infants. *Perception and Psychophysics* 15: 529–532.

Lenneberg, E. H. 1967. *Biological Foundations of Language*. New York: Wiley.

Liberman, A. M. 1970. The grammars of speech and language. *Cognitive Psychology* 1: 301–323.

Liberman, A. M., F. S. Cooper, D. S. Shankweiler, and M. Studdert-Kennedy. 1967. Perception of the speech code. *Psychological Review* 74: 431–461.

Liberman, A. M., P. Delattre, and F. S. Cooper. 1952. The role of selected stimulus-variables in the perception of the unvoiced stop consonants. *American Journal of Psychology* 65: 497–516.

Lisker, L. 1978. Rapid vs. rabid: A catalogue of acoustic features that may cue the distinction. In Status Report of Speech Research. New Haven, Conn.: Haskins Laboratories.

Lisker, L., and A. S. Abramson. 1964. A cross-language study of voicing in initial stops: Acoustical measurements. *Word* 20: 384–422.

Lisker, L., and A. S. Abramson. 1967. Some effects of context on voice onset time in English stops. *Language and Speech* 10: 1–28.

Lisker, L., and A. S. Abramson. 1970. The voicing dimension: Some experiments in comparative phonetics. In *Proceedings of the Sixth International Congress of Phonetic Sciences*, Prague, 1967. Prague: Academia.

McGurk, H., and J. MacDonald. 1976. Hearing lips and seeing voices. Nature 264: 746–748.

Meltzoff, A. N., and M. K. Moore. 1977. Imitation of facial and manual gestures by human neonates. *Science* 198: 75–78.

Milewski, A. E. 1979. Visual discrimination and detection of configurational invariance in 3-month infants. *Developmental Psychology* 15: 357–363.

Miller, J. L., and P. D. Eimas. 1979. Organization in infant speech perception. *Canadian Journal of Psychology* 33: 353–365.

Miller, J. L., and P. D. Eimas. 1982. Feature detectors and speech perception: A critical evaluation. In D. G. Albrecht, ed., *Recognition of Pattern and Form*. New York: Springer.

Miller, J. L., and P. D. Eimas. 1983. Studies on the categorization of speech by infants. *Cognition* 13: 135–165.

Miller, J. L., and A. M. Liberman. 1979. Some effects of later-occurring information on the perception of stop consonant and semivowel. *Perception and Psychophysics* 25: 457–465.

Peterson, G. E., and H. L. Barney. 1952. Control methods used in the study of vowels. *Journal of the Acoustical Society of America* 24: 175–184.

Pick, H. L., Jr., A. Yonas, J. Rieser. 1979. Spatial reference systems in perceptual development. In M. H. Bornstein and W. Kessen, eds., *Psychological development from infancy*. Hillsdale, N.J.: Erlbaum.

Pisoni, D. P. 1978. Speech perception. In W. K. Estes, ed., *Handbook of Learning and Cognitive Processes*, vol. 6. Hillsdale, N.J.: Erlbaum.

Posner, M. I., and S. W. Keele. 1968. On the genesis of abstract ideas. *Journal of Experimental Psychology* 77: 353–363.

Searle, C. L., J. Z. Jacobson, and S. G. Rayment. 1979. Stop consonant discrimination based on human audition. *Journal of the Acoustical Society of America* 65: 799–809.

Spelke, E. 1976. Infants' intermodal perception of events. *Cognitive Psychology* 8: 553–560.

Spelke, E. 1979. Perceiving bimodally specified events in infancy. *Developmental Psychology* 15: 626–636.

Spelke, E. 1981. The infant's acquisition of knowledge of bimodally specified events. *Journal of Experimental Child Psychology* 31: 279–299.

Stevens, K. N., and S. E. Blumstein. 1981. The search for invariant acoustic correlates of phonetic features. In P. D. Eimas and J. L. Miller, eds., *Perspectives on the Study of Speech*. Hillsdale, N.J.: Erlbaum.

Stevens, K. N., and A. S. House. 1972. Speech perception. In J. Tobias, ed., *Foundations of Modern Auditory Theory*. Vol. 2. New York: Academic.

Strauss, M. S. 1979. Abstraction of prototypical information by adults and 10-month-old infants. *Journal of Experimental Psychology: Human Learning and Memory* 5: 618–632.

Streeter, L. A. 1976. Language perception of 2-month-old infants shows effects of both innate mechanisms and experience. *Nature* 259: 38–41.

Summerfield, Q. A. 1979. Use of visual information for phonetic perception. *Phonetica* 36: 314–331.

Swoboda, P. J., J. Kass, P. A. Morse, and L. A. Leavitt. 1978. Memory factors in vowel discrimination of normal and at-risk infants. *Child Development* 48: 332–339.

Wexler, K., and P. W. Culicover. 1980. *Formal Principles of Language Acquisition*. Cambridge, Mass.: MIT Press.

Chapter 6

Linguistic Analysis of   Mary-Louise Kean
Aphasic Syndromes:
The Doing and Undoing of
Aphasia Research

From time to time it has been argued that one can learn nothing of interest about an intact system from studying it when it is broken. The argument often runs like this: Given a radio, if you cut off the plug no sound will come out; you would consequently conclude that the plug was the source of the sound. Certainly the village idiot might draw such an inference (and its corollary, which is that the functions of the cord, the speaker, the miscellaneous entrails, and the cabinet were purely cosmetic). It is rare that the minimally sentient individual is quite so foolish in practice. Obviously, what is and what is not cosmetic about a radio is an empirical question. As the argument runs typically, one does not get beyond the plug; however, for the sake of discussion, we might go a step further. Considering what he takes to be the functions of the radio (the production of music, news, commercials, and so on), the village idiot might be led to look for circuits associated with each of these particular functions. But the minimally sentient scientist would hardly be so rash as to assume that his *a priori* functional classification mapped in any trivial or obvious fashion onto the entrails of the radio. A third obvious problem with the radio argument is the assumption that deficit research on a single case (one radio with its plug cut off) would be taken by anyone as yielding a sufficient body of data for determining much of anything about radios. (What of portable radios that are inherently plugless, for example?) A variant of the radio argument, the car argument, holds that one cannot learn how a car works by taking an ax to the engine and then observing that the car no longer runs. This argument is similar to the radio argument, save one rather minor embellishment: In the radio argument the damage is discrete and local; in the car argument the damage is random, and the argument goes through, but only because it is restricted in form to the single case of an ax-murdered car.

One curiosity of this line of argument against deficit research is that it is made rarely against, say, research on ants, more often against work with rats, hamsters, gerbils, and the like, and so frequently against work on the relation between brain and behavior for the higher mental processes that one might be led to conclude that there is an inverse correlation between the intelligence of researchers and the place of their immediate object of study on the phylogenetic scale. Certainly consideration of higher mental processes is, in and of itself, quite vexatious. And there are, to be sure, obvious major practical problems in performing research on a population such as humans, among which physical experiments are dictated by nature and not scientists (though this view may even turn out to be hubris on our part). But such considerations do not vitiate the enterprise of trying to understand how the human brain subserves such species-typical behaviors of humans as their linguistic capacity. As with any such research, assumptions—presumably plausible ones—are made to guide the project; even the car and radio arguments grant that much in not being based on damage to the cabinet of the radio or the trunk of the car. And people are aware of the intrinsic difficulties of the domain of study. No more than any other field of research is the domain of human deficit studies a haven for village idiots.

The rule of thumb in functional-deficit research has long been that brain damage can only cause loss of function, and that the description of the behavior impaired is a description of the function of the lesioned subsystem when intact. This assumption is, of course, unjustified. In the first place, there is no *a priori* reason to assume that the brain is so structured that functional loss is a simple subtractive process. Independent of this, there is the empirically plausible assumption that extant functional systems may be disinhibited in consequence to damage to some system that in the normal course of events is dominant. Finally, the brain can "rewire" in the face of damage, possibly creating novel functional systems. These obvious problems with the rule of thumb do not, however, vitiate its utility, indeed its necessity. If deficits arising as concomitants of focal lesions are not taken to be reflections of loss of functional capacity, then one is rejecting the assumption of functional localization. A rejection of this assumption is as unsupportable as the wholesale acceptance of the rule of thumb.

A further analytic complication is that many subsystems of higher mental functions may have distributed representations, in which case there is a problem in the detection of specific local systems. Two physical loci, A' and A'', can subserve two distinct functions, F' and F'', respectively, while at the same time each is subserving the function Q. Under such circumstances, given data on damage to A', one might be led to conclude that one's functional model was in error in distinguishing F' and Q, or (if Q is not detectably impaired), to claim that A' was the sole locus of Q as well as F', or (if Q is not impaired) to draw the conclusion that A' subserves F' alone. Obviously, no such conclusions would be warranted. Hughlings Jackson notwithstanding, however, one would be warranted in making an assumption of the necessity of A' to F' (and Q); indeed, one must if there is to be any progress in the development of our understanding of the functional organization of the brain. That one cannot know for certain is not sufficient reason to be silent; that the situation is complex is all the more reason to try to be precise so as to eventually have enough details to begin the process of demystification.

Independent of disputes about localization, the neurological literature since Wernicke (1874) has consistently included analyses of deficits in terms of hypotheses of the structure and function of the intact system, including analyses of deviant aspects of performance as functional consequences of contributions of intact components of the system. In contrast, until quite recently it was the almost exceptionless practice in the psychological literature to equate deviant performance with functional impairment, that is, to overlook the obvious logical possibility that a systematically deviant aspect of performance may, even in its deviance, be a consequence of normal function. Consider the fact that on a hierarchical clustering task agrammatic Broca's aphasics show a "disregard" for determiners, a fact that has been taken as evidence of metalinguistic agrammatism (Zurif, Caramazza, and Meyerson 1972). Further study indicates that a similar "disregard" for adjectives is found in the clusterings of agrammatic subjects (Kolk 1978). From this latter work one might be led to conclude that the category of adjectives is impaired; however, such a conclusion would not be warranted. The assumption of the paradigm is that it taps the subject's tacit understanding of gross constituent structure; from that assumption it follows that the results of these studies indicate that aphasic subjects appreciate the constituent-structure similarities of determiners and adjectives in the stimuli used. That is, this pair of studies reveals something that is normal in aphasics: that they are tacitly aware of structural similarities even though the way they express this awareness is not normal (Kean 1980a, 1981).

Dysprosody in Broca's aphasia provides a particularly compelling case for the importance of always keeping in mind the potential contributions of intact systems and for the fact that such contributions may be masked by other, sometimes extrinsic, factors. The apparent dysprosody of Broca's aphasics has almost invariably been taken to be an indication of some sort of prosodic impairment. On the basis of a Gedankenexperiment, it was conjectured that dysprosody involved no prosodic impairment but rather reflected a normal constraint on the speech output system: that any normal speaker slowed down to the halting and effortful style of speech of a Broca's aphasic would be dysprosodic (Kean 1977, 1979). The production limitation precluded the normal realization of phrasal prosody, though the representational system of phrasal prosody assignment was intact. In a dramatic demonstration of the need for close scrutiny and attention to what is normal in aphasic performance, Danly, deVilliers, and Cooper (1979) conducted a phonetic analysis of the speech of Broca's aphasics and found that, in terms of two rudimentary prosodic characteristics of fundamental frequency (the terminal falling F_0 contour and the declination of F_0 peaks in the course of an utterance), the aphasic subjects' performance was in fact not deficient even when there were pauses ranging up to 5 seconds. The deviance they did observe was in segmental timing; in particular, there was failure to produce utterance final lengthening. This highly provocative result is just the kind of result we should continuously expect: one that more closely specifies questions of functional etiology, the further consideration of which, in this case, will enhance our understanding of speech production, and, furthermore, will raise in new ways significant questions about the nature of the perceptual system. (One such question is this: To what extent is dysprosody a deficit, a limitation, of the normal apparatus of speech perception?) It is also a clear demonstration of the fallacy of the assumption of the experimenter's omniscience.

Hughlings Jackson argued in 1878 that "we must not classify [in aphasia] on the basis of a mixed method of anatomy, physiology and psychology any more than we should classify plants on a mixed natural and empirical method, as exogens, kitchen-herbs, graminaceae, and shrubs." (cited in Jakobson 1972) This position goes through only if one accepts some assumptions: (a) that there is a significant scientific distinction between what is natural and what is empirical, and either (b) that anatomy, physiology, and psychology are mere arbitrary aspects of the description of linguistic capacity or (c) that the botanical categories cited are significantly distinct levels of conceptualization in botany. As assumption a makes no sense, it is hard to see why anyone would want to accept it. As for assumption b, the best evidence is that these are not arbitrary levels of conceptualization and, therefore that assumption cannot be accepted. Assumption c fares no better; the categories cited are natural-kind terms, perhaps, but they are arbitrary-kind terms of distinct levels of analysis and not analytic levels themselves of a parity with anatomy, physiology, and psychology. The position taken here is, then, a complete rejection of Jackson's position with respect to aphasiology, while it concurs with Jackson on the matter of practice in botany. Thus, it departs as well from the position taken by Jakobson (1972). The claim here is that, whereas language structure (grammar), language processors for production and comprehension, and the neuroanatomy and neurophysiology of language are all conceptually distinct, the study of the aphasias is not itself a conceptually distinct domain in the sense that there could ever be a theory of aphasia that would be a characterization of an independent level of conceptualization and a representation of human linguistic capacity. The theoretical interest of aphasia lies in the fact that in dealing with functional analyses in aphasia one is dealing at the interface of conceptually distinct characterizations of human linguistic capacity. Aphasia research provides a test for and a means for revelation of the nature of the relations among systematic and distinct characterizations of human linguistic capacity. At its best, it will require the unembarrassed exploitation of data (i.e., analyses) from various conceptually distinct domains. Because there are significant empirical questions about what might even constitute a viable linguistic and psycholinguistic analysis of an aphasic syndrome, one cannot stipulate parameters on what will or will not be a well-formed analysis from any of these perspectives for any or all aphasic syndromes if one takes the goal of characterizing human linguistic capacity with the utmost seriousness.

It should be evident from the foregoing that the approach to the linguistic analysis of aphasic deficits being advocated here stands in contrast with the attitude, if not the position, advocated by Marshall and Newcombe (1980). The goal of their paper, which is the opening chapter of a volume on the symptom called deep dyslexia (Coltheart, Patterson, and Marshall 1980), is to establish the "conceptual status of deep dyslexia." What is of relevance here is not their debatable position that deep dyslexia is a syndrome but their posture toward deep dyslexia—which I assume is their posture toward the analysis of all symptoms and syndromes: "Our own preference is that there is indeed a disruption of a single underlying mechanism which shows itself in a meaningful cluster of surface manifestations." From the pragmatic perspective of a researcher who wants to get on with business, such a preference may be understandable. However, it cannot have any status in the actual practice of deficit studies, since it cannot help but lead the researcher to prejudge his data and bias him toward setting up arbitrary functional classifications that have no scientific merit and that in the end cannot help but be seriously misleading. Taken to its extreme, when faced with symptoms and syndromes that will not yield to univocal analyses, this attitude leads to the development of a new phrenology in which boxes proliferate on flow charts, their only justification in the end being the desire of the analyst to have at least an uninterrupted sequence of boxes to whose impairment he can attribute the deficit (Kean 1980b; Marshall and Newcombe 1980).

Aside from research bias and desire and the implicit assumption of an equation of analytic elegance with schematic tidiness, there surely must be more scientifically motivated reasons for this preference. Marshall and Newcombe are not alone in having it; they are simply the most explicit and straightforward on the matter. Being so, they give an explanation for their preference: "The effects of a chance constellation of lesions, however often observed, or even of a single lesion that for purely topological reasons . . . impairs a variety of independent systems, are unlikely to be theoretically revealing." It is hard to see how this could conceivably be true. If one takes it that there are significant conceptual distinctions among linguistic theory, processing models, functional neuroanatomy, and so on, and that there can be no direct reduction from one to another (Marshall 1980), then it follows that there will be significant empirical questions as to the relations among these different characterizations. Under such assumptions, what is theoretically revealing is where there are and where there are not correspondences and (as crucial) discongruities among the systematic conceptualizations. Is Broca's aphasia less interesting because it involves what appears to be a phonetic output limitation, a lexical access/post-access impairment, and some segmental specification "confusion" (phonemic paraphasias and so on)? Since more blatantly theoretical work, including some truly productive work, has been done on that syndrome, the answer must surely be no, if any "empirical" justification for such an answer is required. What makes the position outlined by Newcombe and Marshall particularly curious is that the set of deep dyslexia patients involved have large and in some instances diffuse lesions (Marin 1980) and show a variety of noncongruent deficits (Coltheart 1980).

Nespoulous (1981) draws an important distinction between linguistic descriptions in aphasia and linguistic analyses. The new phrenology is aided and abetted by those who fail to take that distinction seriously. For any string, be it grammatical or ungrammatical, there are a variety of systematic linguistic descriptions: phonetic, phonological, morphological, syntactic, and so forth. Although each level of representation is distinct for principled reasons, there is, across the set of representations, a coherence which means that many individual phenomena will be readily amenable to a variety of linguistic descriptions. For example, the class of bound morphemes can be discussed syntactically, lexically, and in terms of sandhi for any language. It is of course necessary that detailed descriptions be provided, and these will perforce in many cases involve sets of descriptions. Agrammatism is a striking case in point. In terms of a linguistic description it can be analyzed (at least in part) logically, syntactically, morphologically, lexically, and phonologically. But just from the fact that such descriptions are available, it does not follow that they all have equal status, or even that all have some status when it comes to the linguistic analysis of the symptom. What is required of a linguistic analysis in aphasia is that, given the analysis coupled with a theory of processing, the set of linguistic descriptions will be entailed. In my work on agrammatism, for example, the linguistic analysis is restricted to the grammatical encoding of the distinction between clitics and phonological words. That there are syntactically, lexically, logically describable aspects of the deficit is argued to follow from how these two classes, broadly, are differentially processed, and from the effects of the heterogeneity of their internal constituency, which makes the differential processing not strictly uniform across each class (Kean, *ad nauseam*). It is not my purpose here to defend this position, but rather simply to take it as an illustration of the sort of linguistic analysis we should expect. The method of arriving at it would be heartily denounced by Jackson, and the analysis itself is not the sort that is the preference of Marshall and Newcombe. I am far from alone in doing research of this character; the research of Garrett (1981) and Zurif (1981) is in quite the same spirit. If one does not attempt to go beyond a mere surface description of the facts, which is all that Jackson would permit, then the linguistic analysis of aphasic syndromes will be reduced to pseudoanalytic profligacy.

Just because a variety of linguistic descriptions of aphasiological data are inevitably going to be available, it does not follow that every symptom will have a linguistic analysis in any interesting functional sense, and should such a situation arise it would in no way vitiate the importance or relevance of linguistic theory either to aphasia research or to the theory of human linguistic capacity, *ceteris paribus.* Dysprosody is an obvious case in point. One is not conceptually set aflutter when such situations arise out toward the periphery; somehow phonetics is always granted something of a *sui generis* status. However, similar situations are likely to obtain when we consider more "central" systems, such as word order.

In a recent series of studies on comprehension and production by English-speaking agrammatic Broca's aphasics, Schwartz, Saffran, and Marin (1981) and Saffran, Schwartz, and Marin (1981) provide evidence that, in addition to omitting "function" words, agrammatic aphasics have problems with word order. On the face of this observation alone, one might be led to postulate one of a variety of linguistic (that is, grammatical) analyses. For example, one might argue that the aphasics' competence has been reduced in such a way as to render them capable of maintaining only a nonconfigurational base. More "conservatively," one might argue that the word-order deficit, to the extent that it exists, represents an inability on the part of aphasics to encode various grammatical relations: "Agrammatic speech is generated without underlying structures that represent logical relations. What [the aphasic] lacks are mediating linguistic structures that correspond to the arguments of the relation: both very abstract ones, like subject/predicate, and even case-marked categories, like agent/patient." (Saffran, Schwartz, and Marin 1981). The list of possible linguistic analyses could be extended. What is at issue is that, in postulating such analyses, one is making a linguistic claim for which there is no real evidence. To make any such arguments one would first have to argue against the theories of processing that relate function words to on-line ordering (Bradley 1978; Garrett 1980, 1981), and the analysis of the word-order deficit in terms of such theories. Since we know nothing *a priori*, it is perfectly conceivable that such linguistic analyses might be necessary for some symptoms; however, in the current case, to postulate such analyses in the face of normal processing theories that predict some such deficit without directly appealing to grammatical theory is simply to violate Occam's razor.

In fact, when we look into the data on the word-order errors made by aphasics in these studies, what we find is evidence for the functioning of normal systems. First of all, word order is not random; the verb appears in its canonical position for English between two noun phrases. This might be accounted for by appeal to the idea that the aphasics are filtering their impoverished grammatical structures through some template of possible surface forms (which does not allow topicalized sentences to pass). However, this will not do. If that is all that is at issue then the order of subject and object around the verb should be arbitrary, but this is not so. There is a tendency, when there is a mismatch of animacy, for the animate noun phrase to occur as subject and the inanimate one as object. One might postulate that this reflects a special strategy on the part of the aphasics. The "strategy" of animate first is, of course, not special to aphasics; as a myriad of acquisition studies have shown, it is ontogenetically potent as a "strategy" (Hickman 1980; Tyler 1981). Furthermore, this "strategy" does relate to aspects of grammatical configuration. As Silverstein (1976) has argued, animate referents are good agents, hence probably more highly presupposed, and therefore likely to function as the topic or thematic subject of a discourse. The initial noun phrase of a sentence in English is the topic position, and for many constructions it is also the thematic subject. Therefore, the data of deviant performance reflect, as in the case of the clustering studies noted above, something that is intact about the subjects' performance. It is crucial that what is reflected in this case is a significant appreciation for the relation between thematic roles and configurational structure as grammatically encoded (Kean, forthcoming b). To be sure, the patients' performance was on the surface far from normal. However, to the limited extent that data on the word-order deficit are available, there is no necessity to invoke any linguistic analysis *qua* analysis of linguistic impairment.

In view of this situation, even in the general terms outlined here, four points should be obvious for the linguistic analysis of aphasic deficits:

• No single set of deficit studies on (for example) Broca's aphasia, much less on any individual single case, will be able to sort out all the empirical issues.
• In doing research on the functional analysis of deficits, one cannot ignore the neurology, neuroanatomy, or neurophysiology.
• Linguistic and processing analyses cannot be carried out in isolation one from another.
• In terms of function, it is essential to consider what components of the deficit performance can be accounted for in terms of the functioning of intact components of the system.

The behavioral literature in aphasiology, especially that arising from an essentially non-neurological tradition, has with considerable consistency failed to take such obvious considerations into account. To the extent that this has been so, we must acknowledge that there is after all some truth in the radio and car arguments—truth as to practice though not in principle.

References

Bradley, D. C. 1978. Computational Distinctions of Vocabulary Type. Ph.D. diss., Massachussetts Institute of Technology.

Coltheart, M. 1980. Deep dyslexia: A review of the syndrome. In Coltheart et al. 1980.

Coltheart, M., Patterson, K., and Marshall, J. C., eds. 1980. *Deep Dyslexia*. London: Routledge & Kegan Paul.

Danly, M., de Villiers, J. G., and Cooper, W. E. 1979. Control of speech prosody in Broca's aphasia. In J. J. Wolf and D. Klatt, eds., *Speech Communication Papers Presented at the 97th Meeting of the Acoustical Society of America*. New York: Acoustical Society of America.

Garrett, M. F. 1980. Levels of processing in sentence production. In B. Butterworth, ed., *Language Production*, vol. 1. New York: Academic.

Garrett, M. F. 1981. Aphasic and non-aphasic language errors. Paper presented at Third Conference of Centre de Recherche en Sciences Neurologiques, Montreal.

Hickmann, M. 1980. Creating referents in discourse: A developmental analysis of linguistic cohesion. In J. Kreigman and A. E. Ojeda, eds., *Papers from the Parasession on Pronouns and Anaphora*. Chicago Linguistic Society.

Jakobson, R. 1972. *Child Language, Aphasia and Phonological Universals*. The Hague: Mouton.

Kean, M.-L. 1977. The linguistic interpretation of aphasic syndromes. *Cognition* 5: 9–46.

Kean, M.-L. 1979. Agrammatism: A phonological deficit? *Cognition* 7: 69–83.

Kean, M.-L. 1980a. A note on Kolk's "Judgment of sentence structure in Broca's aphasia." *Neuropsychologia* 18: 356–360.

Kean, M.-L. 1980b. From phrenology to neurolinguistics. Paper presented at Boston Colloquium for the Philosophy of Science, Boston University.

Kean, M.-L. 1981. Explanation in neurolinguistics. In N. Hornstein and D. Lightfoot, eds., *Explanation in Linguistics*. Harlow: Longman.

Kean, M.-L. 1982. Three perspectives on the analysis of aphasic syndromes. In M. Arbib, D. Caplan, and J. C. Marshall, eds., *Neural Models of Language Processes*. New York: Academic.

Kolk, H. 1978. Judgment of sentence structure in Broca's aphasia. *Neuropsychologia* 16: 617–625.

Marin, O. 1980. Appendix I: CAT scans of 5 deep dyslexic patients, with comments. In Coltheart et al. 1980.

Marshall, J. C. 1980. A new organology. *Brain and Behavioral Sciences* 2.

Marshall, J. C., and F. Newcombe. 1980. The conceptual status of deep dyslexia. In Coltheart et al. 1980.

Nespoulous, J.-L. 1981. Linguistic descriptions of the aphasias. Paper presented at Third Conference of Centre du Recherche en Sciences Neurologiques, Montreal.

Newcombe, F., and J. C. Marshall. 1980. Transcoding and lexical stabilization in deep dyslexia. In Coltheart et al. 1980.

Saffran, E. M., M. F. Schwartz, and O. S. M. Marin. 1981. The word order problem in agrammatism. II. production. *Brain and Language* 10: 263–280.

Schwartz, M. F., E. M. Saffran, and O. S. M. Marin 1981. The word order problem in agrammatism. I. comprehensions. *Brain and Language* 10: 249–262.

Silverstein, M. 1976. Hierarchy of features and ergativity. In: R. M. W. Dixon, ed., *Grammatical Categories in Australian Languages*. New Jersey: Humanities Press.

Tyler, L. K. 1981. The development of discourse mapping processes. Unpublished.

Wernicke, C. 1874. *Der aphasische Symptomkomplex*. Breslau: Cohen and Weigert.

Zurif, E. B. 1981. Psycholinguistic interpretations of the aphasias. Paper presented at Third Annual Symposium of Centre du Recherche neurologiques, Montreal. (See chapter 8 of present volume.)

Zurif, E. B., A. Caramazza, and R. Meyerson. 1972. Grammatical judgments of agrammatic aphasics. *Neuropsychologia* 10: 405–417.

Chapter 7

Clinical Description of Aphasia: Linguistic Aspects

Jean-Luc Nespoulous
André Roch Lecours

The present chapter has to do with semiologic description of aphasia in linguistic terms. Linguistic and psycholinguistic interpretations of language disturbances due to brain lesions are discussed by Kean (chapter 6) and Zurif (chapter 8). Nevertheless, attributing different titles and labels to presumably different subject matters does not necessarily provide the latter with well-defined boundaries, and neither does it prevent eventual overlaps.

Description vs. Interpretation

Because this volume was edited by researchers working in the neurological sciences, we presume that, to a certain extent, the editors may have had in mind when concocting the table of contents the classical distinction betweeen anatomy and physiology. In this context, and with the obvious relations between the two approaches set aside momentarily, description might correspond to a full account of the anatomical structure of the brain through dissection and macroscopic or microscopic observation, whereas interpretation might correspond to an attempt to figure out how the different parts in the structure of the brain may contribute to its dynamic physiological functioning. Thus, neuroanatomical description appears to rely not so much on the observer's thinking or his theoretical model as on his particular techniques, whereas neurophysiological interpretation may rely on a somewhat opposite tendency. In other words, limitations come in the former case from the observer's techniques and in the latter case from the structure of the observer's theoretical model. It is thus assumed that the former approach is more naive and less committing than the latter.

Still, the brain is a physical organ whereas language is not. Thus, if such a dichotomy appears to be reasonably valid within the context of neurology, even though it must be qualified or even questioned, to what extent does it stand when one comes to study language?

Despite these restrictions, and even if description cannot be conceived without an underlying interpretative model (even in anatomy), we will attempt to present an anatomical—or, rather, anatomo-pathological—description of aphasic speech within the context of what we will call clinical neurolinguistics, as opposed to theoretical neurolinguistics and psycholinguistic modeling, which Mary-Louise Kean and Edgar Zurif deal with in chapters 6 and 8. In relation to the classical dichotomy between Bacon's and Kepler's approaches to so-called natural phenoma (the former is based mostly on observation and inductive reasoning, whereas the latter tends to postulate directly interpretative and explicative laws whose validity relies mainly on their simplicity and their extensiveness),[1] we will thus be, at least in this chapter, disciples of Bacon, trying to be more empirical than rationalistic.

Linguistic Units and Levels

Even though speech is essentially perceived as a continuum, just as is any liquid or gas, we are inevitably led to break it up into units or components, whatever their size and limits may be, if our aim is to account for its superficial and underlying structure and its (dys)function. Otherwise, we would restrict our observations to such general and trivial comments as "It is nice" or "It is not nice" and "It is correct" or "It is not correct"[2] without ever focusing on the specific components or processes that are assumed to interact within normal verbal behavior and to be at fault in aphasia. Obviously, since descriptive units and levels of observation do not emerge from verbal data in any natural way but clearly depend on the observer's somewhat arbitrary explicit (and sometimes implicit) axioms, they are bound to vary qualitatively and/or quantitatively from one author to another. Be that as it may, discrete elements are described—particularly within the general framework of structuralism—as well-differentiated parts of a relational network that is assumed to constitute the structure of the linguistic system at different levels: phonology, lexicon, and so on. As static abstract entities, these elements were not originally conceived as potential coding elements withing a dynamic neuro-psycho-linguistic model of language processing. Nevertheless, this transformation of descriptive linguistic units into psycholinguistic processing units is most common in modern days. It is quite legitimate when it takes place within the context of a psycholinguistic experimental study, but it is open to criticism when an author carries out such a shift without mentioning it explicitly.

With these preliminaries in mind, we will examine various types of descriptive units that may be useful in the characterization of different deviant phenomena at different levels in the structure of language and of different linguistic units and levels that might correspond to different units and levels of psycholinguistic processing in speech production or reception or both. In this prospect, we will limit our presentation to the analysis of oral production of French aphasic patients, momentarily setting aside the semiology of oral comprehension and that of written production and comprehension.

The Feature Level

Many linguists, whatever their theoretical framework, tend to consider that obvious production units—phonemes and lexical items—are, as it were, molecules that can be analyzed further and split up into more basic components called features. We must admit that some deviant phenomena found in aphasic speech may indeed be plausibly attributed to an impairment disrupting the synthetic binding of the different but concomitant features required for the production of the target phoneme or word. As a matter of fact, when one is observing the verbal output of aphasic patients at the phonetic level,[3] it is fairly often tempting, when one compares the strings of phonemes these patients utter with the expected target segments, to describe some production errors (particularly some substitutions) by estimating the number of features that differentiate replacing phonemes from replaced phonemes. This is the case despite the fact that it is always a whole molecular unit that is apparently replaced by another whole molecular unit. Our temptation to quantify interphonemic distance is all the greater when such a differentiation is small, that is, when both phonemes involved in the substitution process differ only by one of their subphonemic features.

Consider the following examples:

1. /bato/ (ship) → /mato/.
2. /bizõ/ (bison) → /bidõ/.
3. /sabo/ (clog) → /saβo/.

In example 1, one will conclude that the initial target phoneme (a voiced, non-nasal, labial-stop consonant) has been replaced by its nasal equivalent. In example 2, the substitution phenomenon will be thought

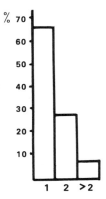

Figure 7.1
Data on interphonemic distance in pure anarthria. 66.6% of substitutions are at interphonemic distance 1 (N = 256 substitutions).

to be basically the same as the previous one, with only one feature differentiating both units involved. (It is the manner of articulation that is disturbed in this case.) In example 3, one will consider that the distance between substituted and substituting units is, again, a matter of one feature. However, in this particular case one is bound to add that such a phonetic segment as /β/ (defined as constrictive, bilabial, voiced, and oral) does not belong to the French phonological system, thus bringing in a distinction between substitutions between phonemes belonging to the same phonological system and substitutions leading to the production of what we might call neo-phonemes. Here, even a nonlinguist clinician will easily differentiate disturbances affecting distinctive features (the more so in such examples as 2, where the substitution process leads to the production of a word that happens to exist in the speaker's language) from disturbances affecting nondistinctive or redundant features, even though the underlying deficit might, at least in some cases, be the same.

The clinical adequacy of such a feature-based description seems rather obvious in cases of pure anarthria (or phonetic disintegration syndrome) and Broca's aphasia. The linguistic analysis of one patient with pure anarthria (Puel et al. 1981) allowed one of us (J.-L. N.) to emphasize the coherence of deviation types within this category of patients, with a majority of substitutions at interphonemic distance 1 (figure 7.1) and a massive production of dental phonemes (figure 7.2). In Broca's aphasia, our analysis of several thousands of phonetic and/or phonemic devia-

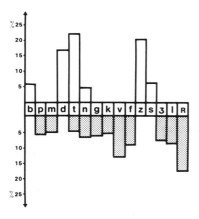

Figure 7.2
Data on replaced phonemes (shaded bars) and replacing phonemes (unshaded bars) in pure anarthria, showing preferential tendency to use dental consonants as replacing phonemes.

tions (Nespoulous and Lecours 1981b) provides us with results evidencing, again, a great coherence in deviation types (figure 7.3), even though the trends of featural misselection do not appear to be the same as the ones we mentioned in relation to pure anarthria (figure 7.4).

These observations strongly support previous studies (Blumstein 1973; Trost and Canter 1974). However, if we admit the validity for both clinical neurolinguistics and psycholinguistics of such a description, based on features, what are we to do when observing phonemic paraphasias such as that illustrated by example 4, where, within the context of the very same static analysis of the substitution process we have just mentioned, the phoneme /t/ has been replaced by the phoneme /ʃ/ ?

4. /ʃaRkytje/ (pork butcher) → /ʃaRkyʃje/

A static featural model obviously tells us at best only part of the story, on account of the fact that the substituting unit, /ʃ/, is present somewhere else in the word-structure pattern. Such a model can only emphasize that the interphonemic distance between both phonemes involved in the substitution is greater than 1.

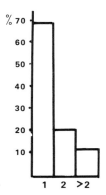

Figure 7.3
Data on interphonemic distance in Broca's aphasia. 69% of substitutions are
at interphonemic distance 1 (N = 1390 substitutions).

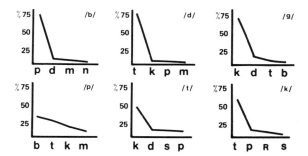

Figure 7.4
Data on preferential substitution patterns for /b/, /d/, /g/, /p/, /t/, and /k/ in
Broca's aphasia. Voicing is more affected when target phonemes are voiced;
place is more affected when target phonemes are unvoiced.

The Phonemic Level

It is our impression that both clinicians and psycholinguists are led to consider that some aphasic disturbances stand at the molecular phonemic level rather than at the atomic phonetic level whenever they cannot advocate the existence of a strong similarity (in Jakobson's terms) between substituting and substituted units and/or whenever the deviations under question appear within the context of fluent, apparently effortless speech (as opposed to the nonfluent, effortful speech of patients with anarthria and Broca's aphasia). Thus, they assume that, in such cases, misselection of phonemes does not have anything to do with the intrinsic subphonemic properties of the units involved in the substitution process. Rather, phonemes are then considered as indivisible production elements undergoing transformations within the morphemic units in which they appear. In the absence of strong similarity, such substitutions—usually identified as phonemic paraphasias (as opposed to the phonetic deviations mentioned above)—must be either random or governed by contextual factors. We tested such a hypothesis by comparing disturbances found in both Broca's aphasia and conduction aphasia.

Interphonemic Distance
In our data (Nespoulous and Lecours 1981b), interphonemic distance varies from one type of aphasia to the other, substitutions at interphonemic distance 1 being significantly less frequent among conduction aphasics (figure 7.5).

Preferential Deviation Patterns
Figure 7.6 illustrates the greater number of substitutions among conduction aphasics in which a strong similarity relationship is absent. No preferential patterns are evidenced by these patients, whereas Broca's aphasics do show preferential tendencies.

Syntagmatic Factor
Substitutions that cannot be clearly accounted for through a static featural analysis—namely substitutions between phonemes at interphonemic distance greater than 1—are very often due to a contextual contamination effect. Even though such an effect exists as well in Broca's aphasics, it is significantly less frequent (table 7.1). Be it as it may, phonemes, considered as basic obvious units in verbal behavior, seem to be more adequate concepts than features to account for some types

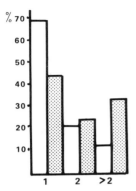

Figure 7.5
Data on interphonemic distance in Broca's aphasia (unshaded bars) and in
conduction aphasia (shaded bars).

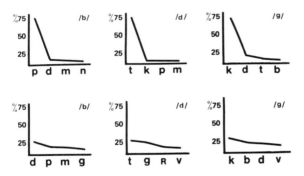

Figure 7.6
Data on preferential substitution patterns for /b/, /d/, and /g/ in Broca's
aphasia and conduction aphasia.

Table 7.1
Phoneme substitutions at interphonemic distance >1.

Aphasia type	N	$\bar{X}$	σ
Broca's	662	24.6%	8.3
Conduction	706	37.5%	3

of deviations to be found in aphasia. Substitutions with a contextual contamination effect, together with other phenomena, such as metatheses, tend to prove this adequacy. Because these phenomena are particularly frequent in conduction aphasia, it seems possible to state that the phoneme is particularly useful in describing such a type of aphasic syndrome whereas the feature would be a concept of greater adequacy to account for deviations found in patients with both anarthria and Broca's aphasia.

The Syllabic Level

At the syllabic level (which has till now been neglected by aphasiologists) the first question that arises is this: Is the syllable to constitute systematically the only basic descriptive unit to account for formal deviations? If so, the phoneme is discarded from both description and modeling of speech production and it is implied that all descriptions mentioned above have to be examined anew on the basis of a syllabic model, even in cases where errors seem to be clearly limited to a specific part of a verbal segment (as in "Bull" becoming "Pull").[4]

If such a syllable-based model is to be adopted, another question remains: How will one build up intersyllabic-distance tables in the same way as interphonemic-distance tables have been devised?

If one does not follow that somewhat revolutionary trend, is the syllable useful as a specific unit in the description of formal deviant phenomena, together with and not to the exclusion of the phoneme?

In our experience, there are indeed some deviant phenomena that cannot be accounted for through phonemic and subphonemic analysis, because they bear on segments larger than the phoneme, and cannot be interpreted as morphemic disturbances on account of the fact that in their context it seems impossible to relate them to any lexico-semantic unit (as a free form or as a bound form). Consider the following:

5. /|αpadεR/ (footlamp) → /pa|αdεR/.
6. C'est une /va/ . . . une /kravat/ (tie).

Both these examples, which are taken from a speech sample (repetition and naming task) of a patient with conduction aphasia, show the plausibility of the syllable as an important unit in both description and modeling of speech production.

The Morphemic Level (Closed-Class Items)

Derived Morphemic Paraphasias

Derived morphemic paraphasias, presented here in a slightly different manner than in earlier publications (Lecours and Rouillon 1976; Lecours and Vanier-Clément 1976), are deviations involving segments that are generally of the size of a syllable, smaller than so-called words (in traditional grammars), but semantically loaded (as may be observed in some cases of Wernicke's jargon aphasia). They lead to the production of a particular type of neologisms, such as examples 7–9, in which can be identified one or two basic morphemes of the speaker's language.

7. /silabativ/
8. /metalsjɔ̃/
9. /bɔRsjal/

When only one morpheme is identified, it is generally an affix (most of the time a suffix, at least in a language like French), as in example 9; when two elements are identified, it is because one seems to recognize, beside the affix, a lexemic base existing in the language under question (as in examples 7 and 8).

Paragrammatic Deviations

We suggest that paraphasias involving grammatical morphemes (such as prepositions and pronouns) and leading to so-called paragrammatism (or dyssyntaxia) in Wernicke's aphasia be analyzed as morphemic deviations. Such elements are in fact affixes in many languages. Whatever their status might be within a given language (free forms or bound forms), they are closed-class items, as are affixes that appear in word-formation processes. The following examples are from Ducarne and Préneron 1976:

10. Je pense très d'aller. . . . (I really think of going. . . .) [The presence of *of* is incorrect in French.]
11. L'employée de la fleuriste est apportée par ce ravissant bouquet. (The florist's assistant is brought by this lovely bouquet.)

Agrammatic Deviations

Even though agrammatic speech belongs to a clinical type of aphasia (Broca's aphasia) different from the one in which both previous types of deviations can be evidenced (Wernicke's aphasia), we mention it at this point, under the present heading, on account of the fact that for the clinician such deviations bear at least partly on grammatical morphemes, the deletion of which leads to the production of such a speech sample as the following:

12. En chambre . . . docteur . . . venu . . . embarqué . . . civière . . . ensuite . . . Montalembert . . . Purpan . . . ensuite . . . opéré. (In bedroom . . . doctor . . . come . . . taken away . . . the stretcher . . . then . . . Montalembert (school) . . . Purpan (hospital) . . . then . . . operated on.)

The Lexemic Level (Open-Class Items)

Two types of descriptive concepts appear to account at the lexemic level for deviant phenomena involving verbal items that belong to the lexical inventory of a given linguistic community.

Formal Verbal Paraphasias
In formal verbal paraphasias, substitutions occur between lexical items that show evidence of formal similarity, as in example 13.

13. Une odeur de cadavre   (a scent of cadaver)

              ↓

   Une odeur de caviar   (a scent of caviar)

Accurate borders between such a phenomenon and morphemic, syllabic, and even phonemic deviations are to be elucidated further in order to assess whether such distinctions in the observer's mind constitute really distinct impairments. It may be particularly important that formal verbal paraphasias are typically produced by conduction aphasics, patients whose main deficit is often considered to lie at the phonemic level. It may also be important that, without any particular instruction to do so, a computer programmed by one of us (A. R. L.) to generate phonemic paraphasias did produce a number of such formal verbal paraphasias (Lecours et al. 1973).

Semantic Verbal Paraphasias
In semantic verbal paraphasias, substitutions occur between conceptually related lexical items. As was suggested above for phonemes, a featural model may, here as well, be relevant to errors in word selection. But, of course, as in the case of phonemes, such a model can stand only when the atomic or semic distance between both elements involved in the substitution process is small or is organized on a generic-specific

axis and when it particularly implies nouns referring to concrete objects, as in example 14.

14. Wolf → Animal

↓

Fox

The model's adequacy to account for deviations in which a plausible contextual contamination effect can be evidenced, as in 15 and 16, still remains to be appraised.

15. Je regarde la télévision.   (I look at the television.)

Je regarde les yeux.   (I look at the eyes.)

16. La petite fille arrive chez mère-grand. Elle donne la galette à sa maman. (The little girl arrives at her granny's. She gives the cake to her mummy.)

In the latter there might be a contamination effect of the beginning of the tale "Little Red Riding Hood."

The Discourse Level

Example 16 implies that observation has here to bear on rather large speech samples, whereas most deviations evoked earlier on can be observed adequately within the context of clear-cut experimental tasks such as repetition, reading, or naming. The analysis of such samples is, in our opinion, of prominent interest for both clinical neurolinguistics and psycholinguistic modeling of language function, since it allows the observation of dynamic functional speech rather than static verbal performance in a text situation.

Without going too far into details, the analysis of samples of narrative speech produced by Wernicke's aphasics, patients with transcortical sensory aphasia, and some patients with Luria's dynamic aphasia made it possible for us to observe disturbances leading to the production of long and complex off-target semantic segments (Lecours and Rouillon 1976; Nespoulous and Lecours 1981a). We will only mention a few examples of such deviations involving speech segments longer and more complex than the microelements (phonemes, lexical items, and so on) considered above.

Example 17 illustrates narrative paraphasia with contextual contamination.

17. Dans le bois, elle rencontre le loup déguisé en grand-mère. (In the wood, she meets the wolf, dressed up as a grandmother.)

An important proposition of the narrative structure of "Little Red Riding Hood" is produced far too early in the narration.

In example 18 the patient has the wolf say something the little girl normally says later in the narration.

18. Le loup arrive chez la grand-mère et dit: Que tu as de grandes dents! (The wolf arrives at the grandmother's house and says: What big teeth you have!)

Example 19 illustrates narrative paraphasia without contextual contamination.

19. Dans le bois, elle voit le loup, l'ogre si l'on veut, qui lui demande: Et ces cailloux par terre! (In the wood, she sees the wolf, the ogre, if you like, who asks her: What about these stones on the ground!)

The patient seems to have mixed "Little Red Riding Hood" and "Tom Thumb," even though neither Tom nor the ogre says such a thing in standard versions of the latter.

In example 20, another patient evokes the moment when, in the wood, the wolf realizes that he will not be able to eat the little girl because of the presence of woodcutters. (This patient was a traveling salesman!)

20. Alors, il se dit qu'il va manquer la commande. (Then, he thinks that he is going to miss the sale.)

Such deviations as these emphasize the interest of the observation of the macroorganization of dynamic speech, an observation going far beyond that of linguistic microelements mentioned in the preceding section.

Intrinsic vs. Extrinsic Factors in Symptom Labeling

Even though one may consider the different descriptive concepts we have just enumerated fairly accurate and thus adequate for everyday clinical practice, symptom labeling in aphasia remains a most difficult task, particularly because each and every phenomenon one observes is likely to be identified in different ways. Not only can the identification vary according to the linguistic model taken as reference in the descriptive process; also—within the context of one model—variations may come from the particular clues the observer decides to consider relevant in the one specific task the patient is involved in, his decision (and thus the clues selected) tending to vary when the patient's task or speech production context changes.

Even without any discursive context, such a fictitious deviation as example 21 should be easily identified as a phonemic paraphasia, because the target word *activity* (a fairly long lexical item) is most likely to be recognized. However, the identification is easier when discursive context gives us extra clues to what the target word may be, as in example 22.

21. /aktiviki/
22. He was in his prime, in full /aktiviki/ when that accident occurred.

When observing such an item as 23, most clinicians will probably say it is a neologism on account of the fact that they cannot trace the target word. As in the previous case, discursive context may sometimes help us in our identification process; then, such a deviation as the one exemplified in 24 will probably be identified as a phonemic deviation (with deletion of the initial phoneme).

23. /ætri/
24. I must change the /ætri/ of my car.

If such an item as 25 is produced by a patient (in a naming task, for example, when the word *cat* is expected), what are we to say? Phonemic paraphasia? Verbal semantic paraphasia? And what about "The dog chases the black /bæt/"?

25. /bæt/

The first thing we want to emphasize with these examples in which the target words differ from the uttered items by only one of their phonemic constituents is that identification may vary on account of the intrinsic structure of the target word. Monosyllabic items are more likely to be identified as verbal (semantic) paraphasias than longer lexical items. Deviations on trisyllabic or tetrasyllabic items are more likely to be identified as phonemic paraphasias, unless there are too many of them for the target word to be recognized. Deviations on bisyllabic items are, in our opinion, most likely to be identified as neologisms. Because bisyllabic items are longer than monosyllables, alterations in their phonemic structure are less likely to generate items belonging to the lexical inventory of a linguistic community; being shorter than trisyllables and tetrasyllables, they do not provide the observer with sufficient clues for him to recognize the target words.

The second point we want to emphasize is that such intrinsic factors are not very often taken into account. Clinicians rely much on extrinsic clues such as the type of task in which the patient is involved,[5] contextual redundancy (particularly in large speech samples), and general semiologic context.[6]

Despite their interest, one must remain aware that extrinsic factors are extrinsic and that they might lead to erroneous descriptions and interpretations. Thus, if one adopts a basically linguistic approach to language disturbances in aphasia, one cannot but focus first on intrinsic structural factors, trying to find a common denominator to deviant phenomena that may at first sight seem rather heterogeneous and that sometimes bear different labels in everyday clinical practice (Kean, chapter 6). This does not mean, of course, that extrinsic factors should be totally discarded. The latter simply must be handled carefully, and not too early in the descriptive process.

Conclusion

Although a coherent and thorough theoretical approach to aphasia remains the aim of many a neurolinguist, it is obvious that it will be a long time before such a theory emerges. In the meantime, even though descriptive accounts of aphasic impairments may be deprived of such a powerful and unifying theoretical background, the identification of different types of verbal symptoms is a clinical necessity that must not be overlooked if one wants to avoid such general and trivial comments as the ones we mentioned above ("It's not correct") and if one's intention is to help the speech therapist build up specific programs based on specific deviation types. It is our very definite opinion that linguistics— besides the past, present, and future theoretical framework(s) it can provide to account for the different types of aphasic impairments—can help to fulfill these more pragmatic descriptive and therapeutic aims. Aphasiology and neurolinguists can wait for better models; aphasics cannot.

Acknowledgment

The work of André Roch Lecours was supported by grant MT-4210 from the Medical Research Council of Canada.

Notes

1. Although the former approach is generally assumed to be that of structuralist linguists, Hjelmslev (1953)—one of the leaders of structuralism—set aside inductive reasoning and adopted Kepler's formalism, as did Chomsky (1957).

2. In fact, in the general comment ("It is not correct"), the assumption of incorrectness already implies that one is able to analyze the verbal continuum to some extent, thus localizing specific errors.

3. The adequacy of a semantic featural model will be examined further when we present lexical substitution phenomena.

4. Morais et al. (1979) suggested that there might very well be differences among speakers as to which elements would be used as basic units in speech production (phonemes vs. syllables). According to them, illiterate subjects might preferentially use syllables whereas literate subjects would basically rely on phonemes.

5. In repetition, reading, and naming, one knows the target words the patient is to utter, whereas one very often does not know them in spontaneous speech production.

6. If a given patient is considered a conduction aphasic, we will be led to interpret as a phonemic paraphasia even such a deviation as cat → bat, despite the intrinsic ambiguity existing in such a case.

References

Blumstein, S. E. 1973. *A Phonological Investigation of Aphasic Speech*. The Hague: Mouton.

Chomsky, N. 1957. *Syntactic Structures*. The Hague: Mouton.

Ducarne, B. P., and C. Préneron. 1976. La dyssyntaxie. *La Linguistique* 2: 33–54.

Hjelmslev, L. 1953. *Prolegomena to a Theory of Language*. Bloomington: Indiana University Press.

Lecours, A. R., G. Deloche, and F. Lhermitte. 1973. Paraphasies phonémiques; description et simulation sur ordinateur. In *Colloques IRIA, Information médicale*, 311. Rocquencourt: Institut de recherche d'informatique et d'automatique.

Lecours, A. R., and F. Rouillon. 1976. Neurological analysis of jargonaphasia and jargonagraphias. In H. A. Whitaker and H. Whitaker, eds., *Studies in Neurolinguistics*, vol. 2. New York: Academic.

Lecours, A. R., and M. Vanier-Clement. 1976. Schizophasia and jargophasia. A comparative description with comments on Chaika's and Fromkin's respective looks at "schizophrenic" language. *Brain and Language* 3: 516.

Morais, J., L. Gary, J. Alegria, and P. Bertelson. 1979. Does awareness of speech as a sequence of phones arise spontaneously? *Cognition* 7: 323–331.

Nespoulous, J.-L., and A. R. Lecours. 1981a. Du trait au discours. In J.-L. Nespoulous, ed., *Etudes neurolinguistiques*. Toulouse: Service des Publications de l'Université de Toulouse-Le-Mirail.

Nespoulous, J.-L., and A. R. Lecours. 1981b. Are there qualitative differences between the phonemic deviations of aphasic patients with and without phonetic disintegration? Paper presented at Academy of Aphasia meeting, London, Ontario.

Puel, M., J.-L. Nespoulous, A. Bonafe, and A. Rascol. 1981. Etude neurolinguistique d'un cas d'anarthrie pure. In J.-L. Nespoulous, ed., *Etudes Neurolinguistiques*. Toulouse: Service des Publications de l'Université de Toulouse-Le-Mirail.

Trost, J., and G. J. Canter. 1974. Apraxia of speech in patients with Broca's aphasia: A study of phoneme production accuracy and error patterns. *Brain and Language* 1: 63.

Chapter 8

Psycholinguistic Interpretations of the Aphasias

Edgar Zurif

In this chapter I describe deficits relating to the recovery and production of features of sentence form, in both Broca's and Wernicke's aphasia. In each instance I will attempt, where possible, to connect this account to models of normal linguistic capacity, seeking to provide converging lines of evidence for a neurologically adequate characterization of syntactic processing.

Broca's Aphasia

As noted in several recent appraisals of the history of aphasia research, there have been occasional observations from the time of Broca on that have run counter to the clinical dictum that comprehension is relatively preserved in Broca's aphasia. For example, as early as 1908, Thomas noted a comprehension deficit in these patients such that, though they could understand words in isolation, they could not completely follow sentences. "Deafness for sentences" was the term applied. However, clinical insights of this general form—and there were others (see for example Goldstein 1948; for review see Caplan 1980)—remained undeveloped. Arguably, it could not have been otherwise; the linguistic activities of speaking and listening were "taken whole," and the differences between them were drawn in terms of the contrast between sensory and motor systems. (See, for example, Wernicke 1874; see Geschwind 1970 for a review and Luria 1973 for a modern form of sensory-motor decomposition.)

This situation has changed, however. Recent advances in linguistic and psycholinguistic theory (most notably with respect to the supralexical organization of language) have encouraged a reexamination of Broca's aphasia in terms of abstract language-specific operations, and the results point to the aptness of the term sentence deafness. In effect, Broca's aphasic patients seem to rely greatly on the referential values of individual words and on plausibility constraints (Caramazza and Zurif 1976; Schwartz, Saffran, and Marin 1980). When the individual word meanings in a sentence can plausibly be combined in only one way in terms of what makes factual sense, Broca's aphasics show comprehension. However, when such constraints are absent—when it "makes sense," for example, for either of two nouns in a passive sentence to serve as agent—their comprehension is reduced to chance.

The picture that emerges, therefore, is that comprehension and production skills are not so readily separated by anterior brain damage as clinical impression would have it. Indeed, there is an interesting sense in which the limitations in these two activities parallel each other: Just as Broca's aphasics speak primarily in content words, so too do they understand on this basis, inferring meaning directly from content words. In effect, they fail to appreciate grammatical pointers to meaning, and, as several studies have indicated, this limitation in comprehension (as well as, of course, in production) seems to implicate processes having to do with the use of closed-class items as syntactic place holders. (For reviews see Berndt and Caramazza 1980; Blumstein 1980; Caramazza and Berndt 1978; Marin and Gordon 1980; Zurif and Blumstein 1978.) The last point has been made, among other ways, via a paradigm in which relatedness judgments are used to elicit intuitions of syntactic structure. Specifically, and in sharp contrast with neurologically intact subjects, Broca's aphasics, when asked to judge how words in a written sentence "go best together," failed to properly integrate the closed-class items of the sentence. They tended to link contentives together, ignoring or misassigning the closed-class items (for example, by coupling two articles or an article and a verb). In short, their judgments violated the integrity of phrasal constituents (Zurif and Caramazza 1976).

Processing Accounts of Agrammatism
Accepting this broadly stated limitation, a number of investigators have lately attempted to address its interpretation in linguistic terms (by focusing on the nature of the representation of each type of linguistic information—phonological, syntactic, or semantic) or in terms of processing (by focusing on the processing systems that implement these structural representations in the act of speaking or listening). In several instances, however, investigators have assumed a direct translation between the two forms of explanation, and this has led to some confusion.

Consider in this respect Kean's (1980) claim that the relevant contrast between classes of items required for an analysis of agrammatism (that is, between the open-class items that Broca's aphasics can exploit and the items of the closed-class vocabulary that they seem relatively incapable of using) is to be found at the level of phonological representation. Kean asserts that it is only at this level that the otherwise (semantically and syntactically) heterogeneous closed-class items can be treated as one class: those items that in English are neutral with respect to stress, neither contributing to the stress of a sentence nor accepting stress except for emphatic purposes. Whatever its merit in accounting for the grammatical failure in Broca's aphasia, this is not a claim about processing such that the open-versus-closed-class vocabulary distinction is available to the processor only at the phonetic level.

In particular, although the linguistic distinction between open-class and closed-class vocabulary items is motivated by considerations of stress in English, it need not follow that the processing problem is located at the point of assigning phonetic representations to the unstressed closed-class items such that grammatical comprehension is blocked off by an initial inability to deal with the closed class at the acoustic and phonetic levels (Brown 1979; Kellar 1978). Such a direct link between grammar and processing is inconsistent with other experimental observations of the processing capacity of Broca's aphasics (Caramazza, Zurif, and Gardner 1978; Swinney, Zurif, and Cutler 1980).

Thus, as a general point, the ability of aphasics to process acoustic input at the segmental level does not reliably predict their utterance-comprehension skills (Blumstein et al. 1977). More to the point, it has been observed that closed-class items place an extra burden on the Broca's aphasic's processing capacities quite apart from acoustic and phonetic factors. In one study, although both Broca's aphasics and normal subjects responded faster to stressed words than to unstressed words when monitoring target items in sentences, only Broca's aphasics yielded a main effect of vocabulary class; this reflected the fact that they responded faster to open-class words than to closed-class words regardless of stress (Swinney, Zurif, and Cutler 1980).

How, then, ought we to characterize the Broca's aphasic's difficulty with closed-class vocabulary elements? One candidate solution has been offered by Bradley, Garrett, and Zurif (1980). (See also Bradley 1979; Zurif 1980.) Their work implicates the phonologically marked open-versus-closed-class distinction, but suggests that this distinction is exploited in processing at the stage of lexical access rather than at the stage of acoustic and phonetic processing. Specifically, they observed in a number of lexical decision tasks that, whereas neurologically intact subjects appeared to respond in systematically different ways to open-class and closed-class vocabulary items (showing, for example, a correlation between recognition time and frequency of occurrence for open-class but not for closed-class items), Broca's aphasics did not respond differently; rather, they treated closed-class items as they did open-class items, showing a frequency effect for both.

Granting for the moment the normal existence of two distinct lexical access mechanisms (one for closed-class and one for open-class items), it has seemed reasonable to attribute a parsing function to the closed-class route. That is, this route may be viewed as permitting the on-line construction of a sentence frame which in turn allows the assignment of open-class words to their form class. In this context, the Broca's aphasic's agrammatic comprehension becomes readily understandable as a reflection of the failure of the closed-class access and parsing system and an attendant failure to assign syntactic analyses to input strings. Further, it is tempting to speculate that this bifurcation of lexical retrieval mechanisms may also play some role in guiding the integration of sentence form during speech, and, correspondingly, that disruption of the closed-class route is also in some fashion implicated in the Broca's aphasic's agrammatic output.

At any rate, this work, much discussed of late, has spawned a number of attempts at replication, not all of which have been succesful (Gordon and Caramazza 1982). The evidential bases of the two-recognition-device formulation clearly remain to be more thoroughly examined, but, whether this model turns out to be correct or even dead wrong in its particulars, it has nonetheless served to indicate the sort of explicit functional decomposition required for a detailed characterization of brain-language relations. More specific, it has led, via its implication of lexical organization in on-line syntactic processing, to a crisper formulation of questions about the functional commitment of left anterior cortex.

Thus, if we impute a syntactic rationale to the organization of lexical retrieval mechanisms, should we not expect that open-class items are also, in some fashion, contacted to provide information about features of sentence form? That is, notwithstanding the fact that closed-class items are often critically involved in phrasal construction, it may still be the case that, where possible, syntactic decisions on an input string are assembled on an item-by-item basis—on open-class as well as closed-class words—and that these "open-class" decisions are also compromised in Broca's aphasia.

Although the position that highlights the contrast between form and meaning in terms of the partition between open-class and closed-class words (Bradley, Garrett, and Zurif 1980) seems to account for most of the comprehension data on Broca's aphasia, it does not account for all of it, at least not in any very transparent fashion. Specifically, there has been at least one report that Broca's aphasics have difficulty with the proper realization of word order in comprehension in linguistic contexts that ostensibly do not involve closed-class items as structural markers (Schwartz, Saffran, and Marin 1980). Clearly, should these findings prove reliable, the force of Kean's analysis and that of the parsing account of the grammatical limitation (Bradley, Garrett, and Zurif 1980) will be blunted; the failure to apply the open-versus-closed-class distinction in processing will be seen either as only part of the Broca's aphasic's problem or as the reflection of some other, more pervasive problem. However, I hasten to add that so far the findings of Schwartz, Saffran, and Marin are at odds with the general observation that Broca's aphasics do not have difficulty understanding simple verb-object sequences (Blumstein et al. 1978; Caplan, Matthei, and Gigley, in press; Goodglass 1968).

A similar concern has arisen also in relation to the output of Broca's aphasics. Although their utterances are impoverished and telegraphic, they do not seem to misorder words. Should we infer from this a capacity to provide a functional structure (of a predicate-argument nature, for example)? Or should we consider these output strings as nothing other than a series of labels planned one at a time and sequenced according to some such strategy as that of first labeling the topic of the intended preverbal message (Saffran, Schwartz, and Marin 1980)? The former alternative seems the more likely. The pertinent data are those obtained in a recent analysis of fundamental frequency (F_0) during speech production in Broca's aphasia (Danley, de Villiers, and Cooper 1979; Cooper and Zurif, in press). In normal speech, F_0 declines over the course of major constituents, permitting the inference that such constituents somehow enter as units during the course of planning speech. Surprisingly, this declination was also generally observed for Broca's aphasics (albeit over smaller domains), even though the words in their utterances were sometimes separated from each other by 5 or more seconds. In addition, continuation rises in F_0 were observed, indicating that Broca's aphasics plan for an upcoming word before completely executing a current one. In sum, these findings suggest that, even in this population, the messages sent to innervate the vocal apparatus are in chunks of more than one word, and that as a consequence these messages (or speech plans) comprise supralexical meanings. However, the form of these supralexical structures, whether they contain information on lexical categories such as noun and verb or on grammatical relations such as subject and object, remains unknown. Clearly, for both production and comprehension, many details concerning the Broca's aphasic's capacity to structure meaning relations remains to be filled in.

Syntactic-Semantic Processing Distinctions

To this point, I have sought to show that, whatever other problems Broca's aphasics have in syntactic processing, they are also (or as part of a larger problem) clearly unable to exploit closed-class items for their structure-marking function. What has not been raised, however—and this is much less certain, and difficult to confront within the perspective of linguistic theories such as Kean's—is whether Broca's aphasics can nonetheless gain the semantic or functional information associated with individual closed-class vocabulary items. It is this possibility that I now want to consider. My intention in doing so is to raise the more general issue of whether anterior brain damage honors the distinction between syntactic and semantic information on an individual-word basis.

One relevant finding stems from the study of Zurif and Caramazza (1976) in which relatedness judgments served to index the Broca's aphasic's insensitivity to syntactic constituents. Recall that in this task these patients were generally unable to integrate closed-class items in their groupings; in effect, they were unable to use these items as syntactic place holders. What deserves emphasis here, however, is that in a few instances Broca's aphasics did group closed-class items in a structurally correct fashion, namely in those instances in which closed-class items played key functional roles (as embodied, for example, in the contrasting sentences "Stories were read *to* Billy" and "Stories were read *by* Billy"). In line with the semantic-syntactic distinction, the patients misaligned the word *to* in their sortings when it did not play a semantically relevant role, when it served only the syntactic function of infinitival complementizer (as in "He likes to eat").

This sensitivity of Broca's aphasics to prepositions carrying important functional (semantic) information has recently been more fully documented via a "fill-in-the-blank" paradigm where target items having the same phonological shape could be varied in terms of whether they served only a syntactic role or whether they had semantic relevance as well. Broca's aphasics performed significantly better when faced with items of the latter sort (Friederici 1980). For example, as reported by Friederici, premorbidly German-speaking Broca's aphasics produced *auf* more often when it occurred as a free lexical element bearing some semantic information, as in "Peter steht auf dem Stuhl" ("Peter stands on the chair") than when *auf* occurred as an obligatory preposition, lexically dependent on the preceding verb, as in "Peter hofft auf den Sommer" ("Peter hopes for the summer"). In this latter instance, it may be argued that the preposition can be processed only along syntactic lines and not as an independent lexical item.

In this context, consider again the findings of Bradley, Garrett, and Zurif (1980)—in particular the finding that, although Broca's aphasics no longer had access to the special closed-class route, they nonetheless recognized the closed-class words as belonging to their language. In view of this pattern, it seems reasonable to suggest that closed-class items are normally registered or activated once via their special route and once via the frequency-sensitive route that also contacts the open-class vocabulary, that this latter route serves some sort of semantic or semantically interpretable function in complementary fashion to the parsing role hypothesized for the closed-class mechanism, and that it is via a disruption to the first-mentioned route and a sparing of the second that anterior brain damage yields a separation of linguistic form and meaning. Clearly this is speculation at this point, and it is wholly dependent on the reliability of the data of Bradley, Garrett, and Zurif (1980), but for the present it seems to me to be speculation of the sort that provides a rather directed search for the further elaboration of language-brain relations.

Wernicke's Aphasia

In contrast with Broca's aphasics, Wernicke's aphasics have standardly been described as having a relatively spared syntactic facility (Goodglass and Kaplan 1972). Some recent analyses suggest otherwise, however.

Wernicke's aphasics clearly have a problem in attaining the meaning of individual lexical items. However, it must not be overlooked that they are not completely disbarred from connecting verbal labels to their referents (Goodglass and Baker 1976); indeed, they can more than occasionally infer the content of strings of words (Caramazza and Zurif 1976). It is in this context, then, that we can seek to determine whether—and if so, how—syntactic indications of meaning might enhance their comprehension.

The evidence that bears on this issue is less decisive than one would wish. Wernicke's aphasics have been shown in a sentence-picture matching task to benefit when sentences are cast in the canonical S-V-O form (Blumstein et al. 1979). They have also been shown (the evidence here is from Germany) to be able to sort phrases to form sentences on the basis of the morphological form of articles (von Stockert and Bader 1976), and in general they appear more sensitive to form-class designations than Broca's aphasics (Friederici 1980). Yet, though they are perhaps better than Broca's aphasics in each of these instances, in none of them can their performance be viewed as entirely normal. Further, they are certainly less than normal in dealing with more complicated constructions, such as those containing relative clauses (Blumstein et al. 1979; Caramazza and Zurif 1976). On balance, then, the Wernicke's aphasic's comprehension problem seems to implicate, at least to some extent, syntactically supported language processes as well as lexical semantic processes. Despite some sensitivity to form-class distinctions and to the semantic implications of word order, they do not appear able to compute the normal range of semantically relevant structural descriptions. That is, they do not appear capable of systematically applying features of sentence form to constrain the inferential possibilities of whatever partial and imprecise lexical semantic units they have retrieved.

The findings from production also call into question the dogma of "intact syntax" in Wernicke's aphasia. Although neologistic intrusions and semantic paraphasias may often seem nothing other than "local" signs of a word-finding difficulty, they cannot always be viewed so unambiguously. As Marshall (1980) puts it, "Aphasic errors do not carry their linguistic interpretation with them." Consider the example he offers, "I went the train to London." and the question he asks of it: Is the patient's syntax intact, and the disruption to be stated in terms of lexical processing, the patient maintaining correct form-class assignments but substituting *went* for *caught* or *the* for *by*? Alternatively, is there a paragrammatism such that the arguments of the verb are somehow misaligned or mischosen, or (to state the disruption in terms of its syntactic reflection) has the patient incorrectly subcategorized *go* as $(+NP+PP)$?

The first of these alternatives—that syntax is unimpaired—is supported by Butterworth's (1980) detailed analysis of pausal and hesitation phenomena in the speech of Wernicke's aphasics. Butterworth, mostly concerned to elaborate on the processes of word finding and noting a disproportionate number of hesitations and long pauses preceding neologistic elements, has fashioned a model of production by which he frames the notion that neologistic elements represent an adaptation to a failure of lexical search. Of relevance here, however, is his report that abnormal hesitation pauses seem specifically to index a lexical search problem. Put positively, pauses at major syntactic boundaries seem normal, on which basis Butterworth suggests that the data are consistent with the notion of an intact syntactic component in Wernicke's aphasics.

It might be, however, that such aspects of syntax as appear normal in Wernicke's aphasics (the above-mentioned distribution of pauses, for example) attest to nothing other than these patients' memory for a limited number of surface forms that have the status of "prefabricated routines" as opposed to constructions planned over well-formed (and semantically informed) structural representations. There are several lines of evidence to support this possibility. First, notwithstanding past claims that the speech of Wernicke's aphasics contains complex constructions (Goodglass and Kaplan 1972), a recent detailed analysis of their output shows that they are far less likely than normals to depart from the simple active declarative form in their utterances (Gleason et al. 1980). Second, to the extent that Wernicke's aphasics do occasionally produce relative clauses, these constructions often contain striking semantic discontinuities across clausal boundaries (Delis et al. 1979). In effect, they appear to produce a syntactic frame with little regard for its semantic function. Presumably, the frame is "run off" without fully specified instructions from a higher-level syntactic system in which the semantically relevant structural facts are represented.

Finally, an analysis of F_0 contours in sentences uttered by Wernicke's aphasics also points to a disruption in the flow of syntactic information during utterance construction (Cooper, Danly, and Hamby 1979; Cooper and Zurif, in press). Specifically, whereas normal subjects take into account the overall length of a target utterance in programming their first F_0 peak (the longer the material to be uttered, the higher the initial peak), Wernicke's aphasics generally start their first F_0 peak at a level that does not vary as a function of utterance length. Again, the deficit appears to encompass constituents larger than the single word. In sum:

In production as in comprehension, Wernicke's patients seem to have a problem with syntactic processing.

Broca's and Wernicke's Aphasia Compared

I have argued that both Broca's and Wernicke's aphasics are disrupted in their ability to implement syntactic facts. Does it follow, therefore, that left-anterior and left-posterior cortex have a similar functional commitment? Not necessarily. As amplified in Garrett's (1975, 1976, 1980) analyses of speech errors (a detailed consideration of which is beyond the scope of this chapter), there appear to be at least two distinct levels of processing concerned with aspects of the syntactic structure of sentences. We are now seeking to determine whether they can be distinguished from each other by the contrasting syndromes of Broca's and Wernicke's aphasia.

The first of these levels, termed by Garrett the functional level, represents phrasal membership and grammatical functions of words. Garrett suggests as one possibility a predicate-argument form of representation at this level that requires the specification of form-class information (e.g., verb, noun), subcategorization features (e.g., that a verb is transitive, that it takes a complement), and possibly also grammatical relations (e.g., subject, object). The second level—the "positional" level—is that at which the serial order of words is specified via a so-called inflectional frame, a surface arrangement of inflectional and free grammatical morphemes.

This analysis begins to provide a framework for distinguishing Broca's from Wernicke's aphasia on the basis of syntactic impairments. In particular, the failure of Broca's aphasics to exploit closed-class elements, and their corresponding inability to mark at least some of the phrasal constituents that enter into grammatical relations, seem to implicate the positional level much more than the functional level. Thus, though they produce telegraphic, palpably "frameless" utterances, Broca's aphasics nonetheless seem to plan speech over functional arrangements encompassing at least two words, as indicated by F_0 observations. Likewise, in comprehension—though here the evidence is at present less compelling—they seem capable of gaining some form of relational meaning so long as the syntactic indications to this meaning do not critically depend on, at least, closed-class items.

Wernicke's aphasics, by contrast, seem to have access to at least some positional frames, but their selection of these frames seems rather arbitrarily related to constraints at the functional level. Either the lexical semantic bundles are insufficiently specified for assignment to their appropriate argument or predicate roles at the functional level or, if correctly accessed, they are misassigned. According to another scenario, even should an appropriate predicate-argument structure be attained, it will not be accessible to those mechanisms underlying the selection of a "planning frame." At this point, I am indulging in what some might term irresponsible speculation; nonetheless, on all accounts, the disruption in Wernicke's aphasia seems to demand ultimately a sententially based characterization, the formation of which a number of us are currently working on (Menn et al. 1981).

Acknowledgments

The research reported in this chapter was supported in part by the National Institutes of Health under grants NS 11408 and 06209 to the Aphasia Research Center of the Boston University School of Medicine. I thank Francoise Pastouriaux for comments on an earlier version of this work.

References

Berndt, R. S., and A. Caramazza. 1980. A redefinition of the syndrome of Broca's aphasia: Implications for a neuropsychological model of language. *Applied Psycholinguistics* 1: 225–278.

Blumstein, S. E. 1980. Neurolinguistics: Language-brain relationships. In S. B. Filskov and T. J. Boll, eds. *Handbook of Clinical Neuropsychology*. New York: Wiley.

Blumstein, S. E., W. E. Cooper, E. B. Zurif, and A. Caramazza. 1977. The perception and production of voice-onset time in aphasia. *Neuropsychologia* 15: 371–384.

Blumstein, S. E., S. Statlender, H. Goodglass, and C. Biber. 1979. Comprehension strategies determining reference in aphasia: A study of reflexivization. Paper presented at Academy of Aphasia, San Diego, Calif.

Bradley, D. C. 1979. Computational Distinctions of Vocabulary Type. Ph.D. thesis, Massachusetts Institute of Technology.

Bradley, D. C., M. Garrett, and E. B. Zurif. 1980. Syntactic deficits in Broca's aphasia. In D. Caplan, ed., *Biological Studies of Mental Processes*. Cambridge, Mass.: MIT Press.

Brown, J. 1979. Comments on Arbib paper. Presented at University of Massachusetts Conference on Neural Models of Language.

Butterworth, B. 1979. Hesitation and the production of verbal paraphasias and neologisms in jargon aphasia. *Brain and Language* 8: 133–161.

Caplan, D. 1980. Cerebral localization and Broca's aphasia. Paper presented at Academy of Aphasia, Dennis, Mass.

Caplan, D., E. Matthei, and H. Gigley. 1981. Comprehension of gerundive constructions by Broca's aphasics. *Brain and Language* 13: 145–160.

Caramazza, A., and R. Berndt. 1978. Semantic and syntactic processes in aphasia: A review of the literature. *Psychological Bulletin* 85: 898–918.

Caramazza, A., and E. B. Zurif. 1976. Dissociation of algorithmic and heuristic processes in language comprehension: Evidence from aphasia. *Brain and Language* 3: 572–582.

Caramazza, A., E. B. Zurif, and H. Gardner. 1978. Sentence memory in aphasia. *Neuropsychologia* 16: 661–669.

Cooper, W. E., M. Danly, and S. Hamby. 1979. Fundamental frequency (F_0) attributes in the speech of Wernicke's aphasics. In *Speech Communication Papers Presented at the 97th Meeting of the Acoustical Society of America*, J. J. Wolf and D. H. Klatt, eds. New York: Acoustical Society of America.

Cooper, W. E., and E. B. Zurif. In press. Aphasia: Information processing in language production and reception. In *Language Production*, vol. 2, B. Butterworth, ed. London: Academic.

Danly, M., J. G. deVilliers, and W. E. Cooper. 1979. Control of speech prosody in Broca's aphasia. In *Speech Communication Papers Presented at the 97th Meeting of the Acoustical Society of America*, J. J. Wolf and D. H. Klatt, eds. New York: Acoustical Society of America.

Delis, D., N. S. Foldi, S. Hamby, H. Gardner, and E. B. Zurif. 1979. A note on temporal relations between language and gestures. *Brain and Language* 8: 350–354.

Friederici, A. 1980. Structure and semantic processing in Broca's aphasia. Paper presented at Academy of Aphasia, Dennis, Mass.

Garrett, M. 1975. The analysis of sentence production. In *The Psychology of Learning and Motivation*, vol. 9, G. Bower, ed. New York: Academic.

Garrett, M. 1976. Syntactic processes in sentence production. In *New Approaches to Language Mechanisms*, E. Walker and R. Wales, eds. Amsterdam: North-Holland.

Garrett, M. 1980. Levels of processing in sentence production. In *Language Production*, vol. 1, B. Butterworth, ed. London: Academic.

Geschwind, N. 1970. The organization of language in the brain. *Science* 170: 940–944.

Gleason, J. B., H. Goodglass, L. Obler, M. Hyde, and S. Weintraub. 1980. Narrative strategies of aphasic and normal-speaking subjects. *Journal of Speech and Hearing Research* 23: 370–382.

Goldstein, K. 1948. *Language and Language Disturbances.* New York: Grune and Stratton.

Goodglass, H. 1968. Studies on the grammar of aphasics. In *Developments in Applied Psycholinguistics Research,* S. Rosenberg and J. H. Koplin, eds. New York: Macmillan.

Goodglass, H., and E. Baker. 1976. Semantic field, naming, and auditory comprehension in aphasia. *Brain and Language* 3: 359–374.

Goodglass, H., and E. Kaplan. 1972. *The Assessment of Aphasia and Related Disorders.* Philadelphia: Lea and Febiger.

Gordon, B., and A. Caramazza. 1982. Lexical decision for open- and closed-class items: Failure to replicate differential frequency sensitivity. *Brain and Language.* 15: 143–160.

Kean, M.-L. 1980. Grammatical representations and the description of language processing. In *Biological Studies of Mental Processes,* D. Caplan, ed. Cambridge, Mass.: MIT Press.

Kellar, L. 1978. Stress and syntax in aphasia. Paper presented at Academy of Aphasia, Chicago.

Luria, A. R. 1973. *The Working Brain.* New York: Basic.

Marin, O., and N. Gordon. 1980. Neuropsychological aspects of aphasia. In *Modern Neurology II,* H. Tyler and D. Dawson, eds. New York: Wiley.

Marshall, J. 1980. Review of *Studies in Neurolinguistics,* vol. 4. *Linguistics* 18: 362–368.

Menn, L., G. Miceli, J. Powelson, E. B. Zurif, and M. Garrett. 1981. Paper presented at BABBLE, Niagara Falls, Canada.

Saffran, E. M., M. F. Schwartz, and O. Marin. 1980. Evidence from aphasia: Isolating the components of a production model. In *Language Production,* vol. 1: Speech and Talk, B. Butterworth, ed. New York: Academic.

Schwartz, M. F., E. M. Saffran, and O. Marin. 1980. The word order problem in agrammatism: Comprehension. *Brain and Language* 10: 249–262.

Swinney, D., E. B. Zurif, and A. Cutler. 1980. Interactive effects of stress and form class in the comprehension of Broca's aphasics. *Brain and Language* 10: 132–144.

von Stockert, T. R., and L. Bader. 1976. Some relations of grammar and lexicon in aphasia. *Cortex* 12: 49–60.

Wernicke, C. 1874. *Die Aphasische Symptomencomplex.* Breslau.

Zurif, E. B. 1980. Language mechanisms: A neuropsychological perspective. *American Scientist* 68: 305–311.

Zurif, E. B., and S. E. Blumstein. 1978. Language and the brain. In *Linguistic Theory and Psychological Reality*, M. Halle, J. Bresnan, and G. A. Miller, eds. Cambridge, Mass.: MIT Press.

Zurif, E. B., and A. Caramazza. 1976. Psycholinguistic structures in aphasia: Studies in syntax and semantics. In *Studies in Neurolinguistics*, vol. 1, H. Whitaker and H. A. Whitaker, eds. New York: Academic.

Chapter 9

The Organization of Processing Structure for Language Production: Applications to Aphasic Speech

Merrill F. Garrett

These remarks represent a preliminary effort to associate some generally acknowledged patterns of aphasic language disorder with the structure of normal language production. The characterization of normal processes used derives primarily, though by no means exclusively, from the distributional features of word, morpheme, and sound errors made in spontaneous speech by mature, normal speakers of English. The effort is preliminary precisely because it is an *ad hoc* analysis rather than the result of direct tests of hypotheses about aphasic performance based on a theory of their deficit modulo the normal error-generation mechanisms.

There are two ways to think about the comparison of speech errors in normal speakers with errors observed in aphasias. We may ask how the aphasic symptoms or syndromes can be *assimilated* to a model of the production process that is based on normal error patterns. Sound-based errors in aphasia would be assigned to the stage of the normal production process that is responsible for the segmental interpretation of lexical and grammatical formatives, meaning-based errors to the stage responsible for semantic interpretation, and so on. Another way is to *compare directly* the error phenomena from the two populations. Normal word-substitution errors would be compared with verbal paraphasias, normal sound errors with literal paraphasias, and so on.

In the main, I will take the former tack. In so doing, I will draw on certain direct correspondences and differences of normal and aphasic error types. It seems best to do this in two steps, the first of which will acquaint the reader with a working model of the normal production process and with some of the error data that support it. Second, I will note some quite general properties of aphasic disorders in relation to the model and then contemplate more specific claims for certain aphasic errors.

A Working Model of Language Production

The model I will use is outlined in figure 9.1. I stress that its features are based on real-time language-processing data, not on *a priori* assumptions about the relation between formal grammar and processing structure. The obvious correspondences between the organization of the processing model and contemporary proposals for that of grammars is, therefore, a consequence of the phenomena examined, not of the observer's theoretical predilections. The observations leading to the model summarized are discussed in Garrett 1975, 1976, 1980a, and 1980b; see also Fromkin 1971.

Note first that the basic distinctions are quite traditional (though not wholly uncontroversial in spite of this venerable estate); they appear in virtually every published discussion of language production. Such treatments isolate a general conceptual level, a language-specific sentence level, and a motor level of articulatory control. In Figure 9.1, these are called the *message* level, for general conceptual processes, the *functional*, *positional*, and *phonetic* levels for sentence processes, and the *articulatory* level for motor control processes. The speech-error data permit one to associate by constructive argument some quite specific claims about processes of lexical selection and phrasal integration with these traditional distinctions between conceptual, linguistic, and motor processes. I will comment briefly on each of the hypothesized processing types and illustrate their function with normal error data.

Message-level processes build the real-time conceptual construct that determines sentence-level processes. They are based on the speaker's current perceptual and affective state and his general knowledge of the world. Such representations are compositional; that is, a conceptual syntax builds complex expressions from a basic (but not small) vocabulary of simple concepts. Message-level representations are the proximal cause of sentence construction; thus, language production is the development, under message-level control, of sentence-level representations that are sufficient to determine an appropriate control structure for articulation.

I do not intend the message level to be identified with the semantic level of various formal grammatical proposals. Such an identification is inappropriate because the message level is, by hypothesis, responsive to both linguistic and nonlinguistic facts. (See Fodor et al. 1980 and the references therein.) This is in no way to claim that meaning is irrelevant to the message level—quite the contrary. The message level

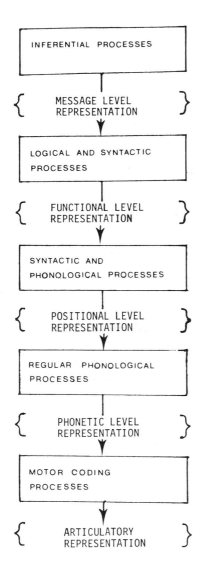

Figure 9.1

is the intended locus of the inferential processes that determine structured discourse.

Articulatory-level processes are identified with the translation from
sentence-level structures to articulatory structures (including those responsible for respiration). At this stage of language production, the
distinctive feature matrices representing the phonetic and prosodic sentence structures must be translated into instructions for control of the
respiratory and articulatory systems.

The sentence level of processing is the focus of my observations here.
It must account for real-time construction of representations of both
the logical and the phonological properties of sentences—roughly, for
what might be called inferential packaging and pronunciation packaging.
Whether speaking of the logical force of a sentence or of the constraints
imposed by its pronunciation in a given language, one must have recourse to descriptions that specify its lexical elements and their phrasal
arrangement. My analysis is dominantly concerned with processes of
lexical selection and phrasal construction. A central expectation must
be that the structures of these processes will reflect an accommodation
between the two principal sources of constraints on sentence form: the
requirements of accent-free pronunciation and of communicatively appropriate interpretation. The internal structure of the sentence level,
as described in figures 9.2 and 9.3, expresses such accommodation.

In figure 9.2, the translation from message-level to functional-level
representations is the step from the general inferential level to the first
specifically linguistic structures. The properties ascribed to the functional
level here are only those that can be inferred from speech errors and
hesitation phenomena. The specification of this level is thus most certainly incomplete, given the computational requirements imposed by
its place in the model and by the properties ascribed to the next-ordered
level of planning. This is simply to say that the logically dictated requirements of the functional processing level exceed the features currently certifiable on grounds of empirical observation of normal
spontaneous speech.

The features of planning indicated in figure 9.2 are lexical selection
on the basis of meaning relations, specification of functional structures,
and assignment of lexical elements to structural role positions. The
types of speech errors that illustrate these planning modes are meaning-
based word substitutions (see examples 1) and whole-word exchanges
between phrases (see examples 2). Recall that the assignment of error
types to processing levels is decidedly nonarbitrary, since the processing

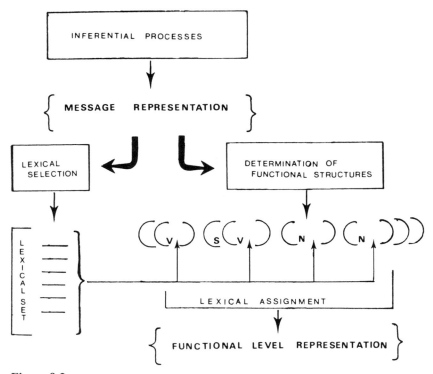

Figure 9.2

distinctions are motivated in the first instance by the error distributions. Thus, the functional level in the model is a multiphrasal level of planning, in which the assignment of major lexical-class items to phrasal roles is accomplished, precisely because errors in which two words exchange positions (as in examples 2) are predominantly those involving words of corresponding grammatical role in distinct phrases.

1. Meaning-based word substitutions:
   "He rode his bike to school tomorrow." (yesterday)
   "You don't just write a book and then send it off to an author." (publisher)
   "All I want is something for my shoulders." (elbows)
   "I'm chronically on the fringe of, uh, on the verge of making a break."

2. Word-exchange errors (exchanged elements underlined):
   "You're not allowed to put use to knowledge."

"Cat, it's too <u>hungry</u> for you to be <u>early</u>."
"I have a pinched <u>neck</u> in my <u>nerve</u>."
"They <u>left</u> it and <u>forgot</u> it behind."

There is one other claim about the functional level that I wish to call to attention even though I do not have occasion to explore its basis in detail. It is this: Observations of hesitation phenomena (Ford 1978) and some experimental results (Ford and Holmes 1978) indicate that the development of the functional level is controlled in terms of verb-dominated groups of simple phrases. That is, functional structures are selected and elaborated in an order that is reflected in the surface sequencing of verbs and their associated simple phrases. Speech-error data indicate that this may typically involve simultaneous processing of two such verb groups (Garrett 1982).

In figure 9.3, the translation from the functional to the positional level represents a transition from a logic-oriented to a pronunciation-oriented representation. Its properties include retrieval of the segmental structure of lexical items, determination of surface phrasal geometry, assignment of lexical formatives to phrasal positions, and interpretation and siting of grammatical formatives in the surface sequence of sentence elements. The following examples illustrate this with speech errors:

3. Form-based word substitutions:
   ". . . a slip which considered . . ." (consisted)
   "People don't know—people who have never sat on an envelope." (elephant)
   "It looks as if you're making considerable process." (progress)
   "Does it make tradition, uh, transition at the appropriate point?"

4. Sound exchanges:
   "A disorder of speech, spictly streaking, is. . . ." (strictly speaking)
   "She's a real rack pat." (pack rat)
   "I was towing Warren around in that little blo baut." (blue boat)
   "I was just gonna rock on the nong door." (knock/wrong)

5. Stranding exchanges
   "It waits to pay." (pays to wait)
   "You have to square it facely." (face it squarely)
   "They are turking talkish." (talking Turkish)
   ". . . by the time I got the park trucked . . ." (truck parked)

6. Word and morpheme shifts (presumed shifted element underlined)
   "You have to <u>do</u> learn that." (you do have to learn)

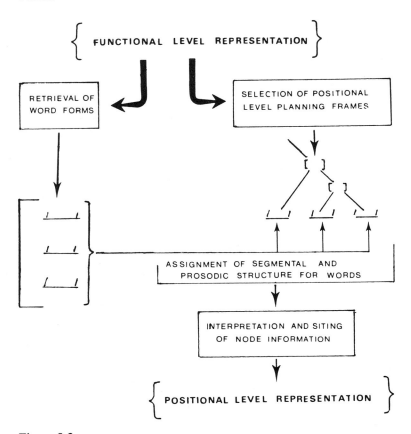

Figure 9.3

"What do you attribute to your longevity?" (your longevity to)
"That would be the same as add tenning." (adding ten)
"I'd forgot abouten that." (forgotten about)

The word-substitution errors of examples 3 contrast sharply with those of examples 1. No meaning relation is a plausible mediator for these errors, just as, in complementary fashion, no form relation suggests itself for the meaning-related pairs of examples 1. Note that the parameters of form that govern lexical retrieval operations at this stage may be inferred from the points of similarity between target and intrusion for errors like those in examples 3. These include initial phonetic elements, syllable length, and stress locus. I will return to this feature of lexical retrieval; for the moment, I simply stress that the lexical selection processes of these two levels of process are clearly disjoint.

In a similar sense, the exchange errors involving sound elements contrast with the word-exchange errors. Similarity of the exchanged sound elements and of their environments is the rule for errors like those of examples 4, but no such similarity of form is evident for word exchanges like those in examples 2. Beyond this, the sound exchanges contrast with word exchanges in another important respect: Sound exchanges are dominantly phrase-internal; hence, the positional level is construed as a single-phrase planning level rather than a multiphrase planning level (though it should be borne in mind that the relevant phrasal types may well be different—syntactic at the functional level and phonological at the positional level). Thus, sound exchanges are presumed to arise in the process of segmental interpretation of the lexical items that are assigned to positions in the phrasal frames of this planning level.

The processes just discussed are restricted to the major lexical classes because those are the classes involved in word or sound exchanges or in substitutions. Minor category elements, both bound and free, are assumed to be explicitly introduced only as features of the planning frames of figure 9.3; hence, such elements are not subject to exchange processes. This separation of types is illustrated quite clearly in errors like those in examples 5 and 6. In examples 5, for the stranding errors, we see the evidences of the association of inflectional elements with planning frames and their consequent insulation from exchange processes. Indeed, exchanges involving elements that are stranded (as in examples 5) are not to be found; stranded elements do not themselves exchange in a fashion analogous to that of the sound elements of the exchange errors in examples 4. It is, however, exchange processes in particular, not error processes in general, from which these inflectional morphemes are insulated; as examples 6 illustrate, they do undergo movement error. But these are errors of misplacement of a single item in the terminal string of sentence elements. Such errors are presumed to arise in the process of interpreting features of the phrasal planning frame and siting them in the terminal string. This process holds for both bound and free forms.

There are other aspects of error distributions that reinforce the conclusions I have summarized; some of these have to do with stress features. Put succinctly, phrasal stress, which is a feature of surface phrasal structure (and hence of the planning frames in the working model), is preserved (maintains its original locus) under exchange pressures, but is violated in shift errors.

Figure 9.3 also represents the translation from the positional level to a subsequent level of phonetic representation. Why postulate such an additional process, given that the segmental structure of words is implicated in the processes of the positional level itself? The answer lies in the existence of another class of error phenomena in which the detailed phonetic character of elements involved in an error, or in the immediate environment of an error, is altered to conform to the regular phonological constraints of the language. In examples 7, one sees the form of the indefinite article, and the shape of inflectional elements accommodating to their error-induced environment.

7. Accommodation errors:
    ". . . what that add up*s* to . . ." (add*s* up to)
    ". . . *a* money's aunt . . ." (*an* aunt's money)
    ". . . easy enoughly . . ." (easily enough)
        /izi̱/   /izali̱/

Similarly, phonetic changes induced in stems by affixation (vowel reduction) appear appropriately. Such errors show that, at the points of shift errors and sound exchanges (which processes are taken to be diagnostic of positional-level processes), the phonetic character of elements subject to regular phonological processes remains to be specified, and this is a quite general case not restricted by vocabulary type in the way that positional-level processes seem to be.

The question why one distinguishes levels might also be raised for the positional versus the functional level, or for the functional versus the message level. The answer to such "why" questions is explicit in the preceding discussion: The constraints on the errors that characterize the processing levels are disjoint. Word-exchange errors are between phrases, whereas sound exchanges are within phrases. Word exchanges are not constrained by morphological structure or by segmental and prosodic similarity, whereas sound exchanges and most standing errors are so constrained. Word-selection processes are constrained by vocabulary type, both with respect to syntactic and phonological roles and with respect to meaning and form, and in each case the effective descriptions are particular to the levels of processing and representation that inform the working model.

The nature of the argument is simple. If speech error patterns are taken to reflect normal processing structure, the properties of error types and their interactions should tell us what structures are being computed by the system at given points in the elaboration of a sentence. We may thereby assert claims for the computational vocabularies and operations that are characteristic of sentence-level processing. The working model is a summary statement of the distributional properties of sample error corpora under such assumptions.

I now turn to a comparison of some generally accepted characterizations of aphasic language disorder and the properties of the production model just outlined. The first major aspect of the comparison that I wish to emphasize is the most obvious, but nonetheless it is of fundamental import. It is the distinction between meaning and form, and I wish to address this matter in the two domains of lexical selection and phrasal construction.

Meaning and Form in Lexical Selection

Note that figure 9.1, which gives an overview of the working model, shows no connection between the message level and any level other than the functional. The properties of sentence form that determine meaning are assumed to be fixed at the functional level, including the processes of meaning-dependent lexical access. Recall the observations about normal speech errors that prompt this formulation.

• Word-substitution errors are of two quite different types: those in which the lexical relations mediating error are meaning-dependent, and those in which they are form-dependent. These two types do not intersect; meaning sets are not cross-classified by form, nor form sets by meaning.

• Word exchanges and the source words for sound exchanges are not meaning-related; that is, the assignment processes for phrasal role and phrasal position are not directly coded for meaning. Shifts of bound and free morphemes are not meaning-constrained; attachment processes at this level are form-constrained only.

Consider some analogous phenomena from aphasia. There are several well-known but enduringly relevant observations, perhaps the most compelling of which is the dissociation of meaning and form in anomic or amnesic aphasia. Word-finding difficulties are common to several classes of language disorders, but here the anomia is the principal notable dysfunction. The speech of anomic aphasics is (barring lexical hesitations, of course) fluent and well formed; comprehension processes are not impaired. In these respects they contrast with Wernicke's aphasics, who, though fluent, show comprehension deficits and syntactic, phonetic, and semantic anomalies in their speech (Goodglass and Geschwind 1977). Anomic aphasics provide a striking example of the failure to access word form in the presence of strong evidence that the meaning of the offending word is available. A common observation is their ability to provide "definitional phrases" with pantomimic gestures, such as when *pen* is the target word and the patient may say "It's for writing" and make a writing gesture (Buckingham 1979). These patients are very certain that they "know" the word they cannot utter, and in this respect they have been compared to normal subjects in a tip-of-the-tongue state (Brown and McNeil 1966). In such states normal subjects can provide some information about the form of the target word, such as its initial sound and its syllable length. Anomics are notably deficient in this ability (Goodglass et al. 1976).

A second type of relevant observation concerns the performance of conduction aphasics. They show good comprehension, but their speech contains frequent sound errors and they show great difficulties with repetition tasks. Most striking for our present purpose is their resort to a strategy of "synonymic shifts" when confronted with a pronunciation difficulty. They often make repeated attempts at a target word, showing thereby a successful word identification, which does not, even so, ensure correct execution of the target form. They are observed (Joanette, Keller, and Lecours 1980) to sometimes break such a series and shift, without hesitation, to an alternative, semantically equivalent form.

The neologistic speech of jargon aphasics (usually Wernicke's aphasics) also bears on this point. The production of neologisms has been construed by several investigators—for example, Buckingham (1979) and Lecours (1982)—as combining a failure of meaning-based word-selection processes with failures of the segmental interpretation process. The resultant nonword forms are simply not, therefore, readily associable with a contextually inferrable target. Nonetheless, many significant aspects of the processes that deal with word form are preserved in spite of grave impairment of the meaning-based system. Both Lecours and Buckingham, and more recently Butterworth (1979), have proposed that when the lexical retrieval process fails word-formation processes may be invoked to provide a "gap filler" or "place holder" in the utterance, thus preserving fluent output. The mechanisms of such a process do not distinguish clearly between the two loci of retrieval failure given in the working model. However, certain aspects of the form-based systems that are dissociated in jargon aphasia do connect with error processes at the functional and positional levels of the model— namely, those of morphological structure. Lecours and Lhermitte (1972) (discussed in Lecours 1980) reported a case that illustrates the point clearly (see also Buckingham and Kertesz 1976; Buckingham 1979). This case ("General Alex") provided numerous examples of what Lecours calls "derived morphemic deviations," in which the constituent morphemes of a neologism are licit forms of the language but the combination is not. Moreover, these are not meaningful neologisms; no principle of semantic compositionality is evident. (This is significant, for normally one thinks of such word-formation processes as in the service of and dependent on the meaning of the constituent morphemes, whether inflectional or derivational.) Nonetheless, we see that the dependence is not direct, for the computational machinery of language production represents separately the meaning constraints and the combinatorial devices that ensure the well-formedness of the morphologically complex structures. [See also Buckingham's (1979) argument for preservation of morpheme sequencing constraints in Wernicke's aphasia.] The stranding and shift errors of normal language production reflect a similar separation of morphological structure and meaning constraints.

There is a point of difference, however, between the word-movement and word-substitution errors of normal speakers and the features of jargon aphasia I have referred to. Jargon aphasics do not appear to distinguish sharply between inflectional and derivational morphemes, whereas the working model of the production process does—derivational processes are deemed complete prior to the positional level of processing, and hence the model as it stands provides no reconstruction of the derivational processes of word formation. This may constitute a language-specific failure of the current formulation of the model.

Syntax and Semantics

I now turn to the dissociation of meaning-based and form-based processes at the level of phrasal structure. As I remarked earlier, movement errors at the functional and positional levels do not show evidence of meaning-based constraint. To compare these with aphasic disorder requires a bit more detail. The mechanisms of phrasal construction suggested by the stranding and shift errors and by exchange distinguish two classes of vocabulary, which I shall refer to hereafter as the open-class and closed-class vocabularies. This terminological convention emphasizes the computational force of the distinction. It is, of course, roughly the contrast we are accustomed to referring to as that between content and function words. In the current context, it is the vocabulary contrast supported by the differential sensitivity of major-class words (N, V, Adj) and minor-class words (Conj, Quant, Comp, Rel, Demonst, etc.) to sound-exchange processes—the former contribute to sound exchange, the latter do not. Exchange and shift errors also distinguish the two classes, for the former are dominantly movements of open class elements while the latter are dominantly movements of closed class elements, and this applies with equal force to both bound and free forms. Recall that in the speech-error model the representation of closed-class vocabulary is identified with surface phrasal geometry.

The major features of anterior and posterior types of language disorders also distinguish the two vocabulary classes. In jargon aphasia, closed-class vocabulary (bound or free) is relatively preserved in spite of grave difficulties with open-class forms (Lecours 1982; Buckingham 1980). And, here too, local phrasal organization is preserved. Lecours gives a systematic catalog of neologisms from samples of the speech of two such cases; neologisms were entirely confined to nouns, verbs, adjectives, and a few lexical adverbs, and none occurred in any other category (including that of auxiliary verbs).

In sharp contrast with the prosodically and phrasally organized jargon of Wernicke's aphasia, the agrammatism of Broca's aphasia presents the complementary deficit. Here the characteristic failure is with grammatical morphemes, bound and free (what I am calling members of the closed class). The other side of the coin, of course, is that agrammatic speakers produce and comprehend open-class forms much better than closed-class forms (Saffran, Schwartz, and Marin 1980; Goodglass 1976). Open-class performance is by no means entirely spared; the range of open-class items is substantially reduced in their output. In general, however, agrammatic aphasics are quite evidently in better possession than jargon aphasics of the lexical inventory as it is represented for inferential purposes (Zurif and Caramazza 1976). What the agrammatics are notably deficient in is the language-specific devices of phrasal integration. These are, for the speech-error model, most obviously related to the mapping from functional to positional levels of representation, though it is by no means certain that the deficit should be characterized solely in terms of the devices I have identified with that level. In particular, there is no overt expression of the role of verb structure in constraints on the positional level of the speech-error model, though the natural place to express such would be at the functional level of representation in that model. In this regard, it must be remarked that agrammatic speech is notably lacking in its use of verbs (Saffran, Schwartz, and Marin 1980; Berndt and Caramazza 1981).

At this point, it is perhaps enough to note that the principal features of agrammatism in Broca's aphasia and jargon in Wernicke's aphasia and conduction aphasia reinforce the conclusions reached on the basis of speech-error analysis: Closed-class and open-class vocabularies play distinct computational roles in sentence construction. That difference implicates the processes of phrasal construction, which in turn implicates the contrast on meaning-based and form-based processes in production. It is striking that the aspects of the form-based system required for the detailed description of normal error processes should so closely approximate the principal clinical contrasts in which lexically mediated inferential processes and the vehicles of phrasal integration are dissociated. How far that detail may be pursued without being found wanting is something to which I cannot now attest. I will, however, consider some more detailed examples that seem promising. The first concerns a formal interpretation of the speech-error model and a parallel formal interpretation of Broca's aphasia.

The Positional Level as Phonological

The affinity of positional-level processes in speech errors with phonological processes in formal grammars is obvious. How strongly should we draw the correspondence? I will present one argument that supports identification of a formal level of linguistic representation with the positional processing level, and at the same time will draw out a more specific claim about correspondence of speech errors and aphasic disorders. The argument connects a feature of speech errors with agrammatism, and the speech-error case, as does the agrammatic case, concerns prepositions.

In the speech-error data one finds an apparent anomaly: Prepositions, which in ordinary parlance are classed with "function words," nevertheless appear in word-exchange errors. If not for this, one might properly say that only the major category words (N, V, Adj) that are conventionally thought of as the "content" words can appear in the word exchanges typical of the functional level of processing in the production model (i.e., exchanges between words of corresponding grammatical category from different phrases). In fact, of course, the appearance of prepositions in such errors is not anomalous, for they, like nouns, verbs, and adjectives, introduce a major phrasal class (prepositional phrases). Therefore, one might express the regularities of word exchanges in terms of phrasal heads. If one does so, however, the other horn of a hidden dilemma appears. The generalization about exchanges of words between phrases also predicts which words contribute to sound-exchange errors, except for prepositions. In sound-exchange errors, prepositions behave like other minor-category words. Sound segments from nouns, verbs, and adjectives appear in sound exchanges, but not those from prepositions. Thus, prepositions, like inflectional morphemes, quantifiers, articles, and conjunctions, seem "insulated from sound-exchange processes"—they too are closed-class words. How can an element be both closed-class (as for sound exchanges) and open-class (as for word exchanges)? An answer lies in the already-established separation of the functional and positional levels of processing. At the functional level prepositions are lexical, and at the positional level they are not. The "demotion" of prepositions to closed-class estate as a consequence of that level shift is quite *ad hoc* if one construes both planning levels solely in syntactic terms. However, if one were to construe the shift from functional to positional level as a shift from a specifically syntactic and/or logical level to the systematic phonemic or phonological level (i.e., the level of representation to which regular phonological processes apply), the prepositional shift in status would be required, for, phonologically, most prepositions behave with the minor classes.

This line of argument parallels one of a number made by Kean (1977, 1980) for the formal characterization of the symptoms of agrammatism.

Kean notes that a description of agrammatism as a specifically syntactic deficit runs afoul of this same problem. The sentence elements that agrammatics experience most difficulty with are a mixed bag syntactically. At the phonological level of analysis, this heterogeneous set is rendered uniform. Thus, Kean claims that the agrammatism of Broca's aphasia should be considered, formally, as a phonological deficit.

Whether for the positional level in speech errors or for the characterization of agrammatism, the identification of the descriptive type as phonological does not eliminate syntactic variables from a role in the processing; the phonological level of representation includes the major features of surface syntactic phrasal organization. However, though the role of syntactic variables in speech processes at that level is thus entirely supportable, our interpretation of error data and aphasic data must be affected in diverse ways if we opt for such a course. One example in particular is the domain of error interactions for sound and word exchanges—on the current argument, the former must be phonological phrasing and the latter syntactic phrasing. It will often be difficult to distinguish these two in the spontaneous speech of normal or aphasic speakers, but the theoretical issue is clear.

Lexical Access and Pronunciation

At this point I wish to shift the discussion from general features of the production model and aphasic syndromes to a more specific and hence more speculative class of observations about processes of lexical retrieval and segmental interpretation of words and phrases. These processes are multifaceted activities that include two distinct processes to determine correct word selection and several steps to ensure correct execution of the target forms. I will briefly trace the steps for a given word target:

1. meaning-based identification of an underlying word   (Derivational structure is fixed at this point.)
2. assignment of the word to a functional structure role
3. retrieval of the segmental specification of the word   (This must be accomplished via a *linking address* that is part of the information provided at step 1. The parameters of access are those derived from form-related word-substitution errors i.e., initial phonetic shape, stress locus and syllable length; see Fay and Cutler 1977 for related discussion.)
4. assignment of segmental information to a phrasal frame position
5. specification of phonetic detail (via regular phonological processes)
6. motor programming

Consider, first, two disorders that show different patterns of "word-finding" difficulty: pure anomic or amnesic aphasia and conduction aphasia. Consider pure anomics in light of the sequence listed above. They might be characterized as having lost the linking addresses in step 3. If so, and if that is assumed to be the principal deficit, we can expect several consequences on the basis of the production processes outlined here, most of which have been previously noted in the aphasia literature:

- Meaning-based access should be intact. This comports with the observations I noted earlier of pantomimic substitutions and ability to provide "definitional" information.
- Execution of form should be intact. This predicts a relative absence of literal paraphasias, which again appears to be correct.
- Unrelated verbal paraphasias should be more common than semantic verbal paraphasias. This is because the latter presumes correct entry to the form inventory; if the linking address is gone, "random" entries to the lexicon should be more likely. The facts of the matter are not clear on this point (though see Geschwind 1967). This prediction must be evaluated with attention to the incidence of literal paraphasias, for such can also produce apparent cases of unrelated verbal paraphasia by chance generation of word forms from sound errors (Lecours 1982). Such word substitutions are not at all common in normal speakers.
- There should be a specific loss of linking information about word form. There is support for this in the findings of Goodglass et al. (1976). In about 5% of the cases tested, anomics were able to provide correct information about the initial letter of the target word, as compared with about 34% for conduction aphasics; similarly, a large contrast in knowledge of syllable length was observed. Such differences were obtained in spite of about the same level of failure in the naming task for the anomics and conduction aphasics. Goodglass et al. note that the performance levels for anomics on first-letter information was at chance and their performance on syllable-length information was less than chance; conduction aphasics' performance substantially exceeded chance in both categories.

The observations about conduction aphasics made by Goodglass et al. (1976) suggest that they indeed have an intact linking address available to them, but are impaired in the use of the information so obtained. For the speech-error model, this might be understood in terms of the processes of segmental interpretation of word forms in the phrasal frame (item 4 above), in terms of the phonological specification of the details of phonetic form (step 5 above), or in terms of the construction of motor programs based on the phonetic specification of phrases (item 6 above). Depending on which of these processes are the loci of difficulty, one will entertain different expectations about the error patterns to be observed. The commonly reported failures of conduction aphasia seem best understood as a disorder of the phonological interpretation of the lexically specified positional frames (that is, of step 5 above). Thus:

• Meaning-based and form-based lexical access processes should be intact. The relative absence of verbal paraphasias, whether of meaning or form, comports with this.

• The characteristic failure of repetition for such patients is, though not logically dictated, certainly understandable. If we assume (as the results of Goodglass et al. indicate) that linking information is available, and hence that there is access to a specification of word form, it follows that having such information is not sufficient to ensure correct pronunciation. The repetition failure may simply indicate that whether the form specification is internally generated or externally provided has no material consequences; the segmental interpretation process fails.

• If the processes at step 4 were the locus of conduction aphasics' problems, we would, on the model of normal speech errors, expect to see frequent sound-exchange errors, for such errors are assumed to arise as a part of the processes that associate lexical content with surface phrasal configurations. So far as I know, no current observations indicate such an exacerbation of exchange errors for conduction aphasics. Note also that the repetition failures of conduction aphasics are manifest for isolated words as well for phrasally integrated words; i.e., the pronunciation difficulties do not seem tied to phrasal structure (in the way that sound-exchange processes are in the speech-error model of production processes).

• If processes at step 6 (motor programming of the phonetic description of phrases) were the locus of conduction aphasics' difficulty, one would expect violations of language-specific phonetic categories and perhaps of phonetic sequencing. Again, published reports indicate the contrary (Lecours and Rouillon 1976).

These speculations about conduction aphasia can be placed in the further perspective of an aphasic disorder that does seem best described in terms of the processes at step 6 (motor programming for articulation of speech)—namely, apraxia of speech. A frequently cited constraint on error mechanisms common to both aphasia and normal errors is the preservation of the sound-sequencing constraints—language-specific restrictions on the possible word-internal succession of segment types. Wells (1951) pronounced it a "law of the slip" for normal speech that such constraints were not violated in the error products in normal spoonerisms or similar sound errors, and all other observers have been in substantial agreement. Similarly, the paraphasias of conduction aphasia and Wernicke's aphasia, including severe jargon cases, and (barring dysarthric distortions) those of Broca's aphasia, all honor the specific segment-ordering conventions of the language (Buckingham 1980). What is interesting about this is the robustness in aphasia disorders of these particular language-specific conventions. Clearly, such constraints cannot be lexical (the conventions are honored in neologisms, whether normal or aphasic in origin) or articulatory (the sequencing constraints are regularly "violated" by the juxtapositions of normally fluent speech—for example, /sbɛnɑ/ = "it's been a . . ." or /vwɩgat/ = "have we got"). The locus of such constraints must reside in the part of the system that is specific to the control of the articulatory apparatus for speech. Thus, it would reside in the translation from the phonetic level of representation to motor coding for speech. There may be violations of sequencing constraints in apraxia of speech or pure anarthyria (Johns and Lapointe 1976; Lecours and Lhermitte 1976)—cases in which the motor control of the articulatory system (for nonspeech gestures) is substantially intact except for the integration of speech gestures at normal rates. The content and the precise nature of such violations in apraxia of speech are not yet documented, but preservation of the constraints in other aphasias that display frequent sound errors suggests that this contrast should be examined carefully.

Concluding Remarks

I wish finally to comment on some features of normal error patterns for which I did not find ready counterparts in aphasia reports. These concern the movement errors that justify a great deal of the internal structure of the working model of speech production that I outlined above.

Exchange errors, whether of words, of morphemes, or of sounds, do not seem common. This is a risky observation, of course, because of the difficulty in assessing patients' output for those disorders most productive of sound and word error—the jargon aphasias. However, where comprehension is not markedly impaired, as in conduction aphasia, we might hope to find indications of such error patterns if they occur with pathological incidence. The general pattern is for an error mechanism to display itself as a pronounced exacerbation of the normal error mechanism. Thus, one finds cases in which there is a great increase in segmental selection errors, cases in which there is a great increase in anticipatory and perseveratory word and sound errors, cases in which there is a great increase in lexical selection errors based on meaning or on form, and cases in which there is a great increase in lexical and morphological blending, but so far as I have been able to discern there are no cases that are characterized by a dramatic surge in the frequency of stranding errors, word exchanges, and sound exchanges. It may simply be that conduction aphasia has not been looked at with these sorts of error patterns in mind, of course. But if it is true that such errors are not more frequent in aphasia than in normal speech, what might such a circumstance mean?

One interesting hypothesis is that clear correspondences between error types and aphasic syndromes will be found only for errors that arise out of the need to integrate the products of independent subsystems, not for errors that arise out of the detail of the computational processes internal to subsystems. The principal classes of aberrant syntax are agrammatism and paragrammatism, both of which are readily understandable in terms of the structure of the working model but neither of which represents a "partial" disturbance of the mechanisms revealed by exchanges and shifts. They seem, rather, to be cases in which the specifically syntactic or abstract phonological systems are largely preserved, though distorted by word selection and sound errors (i.e., paragrammatism), or largely destroyed and rendered opaque by greatly reduced output (agrammatism). Roughly, language-construction processes can be divided into two computational tasks: vocabulary selection and phrasal integration. Each of these has two facets to its structure: one based on meaning and one based on pronunciation. The structural parallels between normal speech errors and aphasic errors seem to be readily interpretable in such terms; however, the details of the internal structure of the major components may not be readily isolable in the aphasic data, if one takes the normal error data as definitive. Perhaps the picture will change with more detailed observation based on the expectations derived from such a hypothesis about the relation between the organization of normal language production and the possibilities for stable aphasic disruption of language function.

References

Berndt, R., and A. Caramazza. 1980. A redefinition of the syndrome of Broca's aphasia: Implications for a neuropsychological model of language. *Applied Psycholinguistics* 1.

Brown, R., and D. McNeil. 1966. The tip of the tongue phenomenon. *Journal of Verbal Learning and Verbal Behavior* 5: 325–327.

Buckingham, H. 1979. Linguistic aspects of lexical retrieval disturbances in the posterior fluent aphasias. In H. Whitaker and H. A. Whitaker, eds., *Studies in Neurolinguistics*, vol. 4. New York: Academic.

Buckingham, H. 1980. On correlating aphasic errors with slips of the tongue. *Applied Psycholinguistics* 1: 199–220.

Buckingham, H., and A. Kertesz. 1976. *Neologistic Jargon Aphasia: Neurolinguistics III*. Amsterdam: Swets and Zerlinger.

Butterworth, B. 1979. Hesitation and the production of verbal paraphasias and neologisms in jargon aphasias. *Brain and Language* 8: 133–161.

Fay, D., and A. Cutler. 1977. Malapropisms and the structure of the mental lexicon. *Linguistic Inquiry* 8: 505–520.

Fodor, J. A., M. Garrett, E. Walker, and C. Parkes. 1980. Against definitions. *Cognition* 8: 263–267.

Ford, M. 1978. Planning Units and Syntax in Sentence Production. Ph.D. diss., University of Melbourne.

Ford, M., and V. Holmes. 1978. Planning units and syntax in sentence production. *Cognition* 6: 35–53.

Fromkin, V. 1971. The non-anomalous nature of anomalous utterances. *Language* 47: 27–52.

Garrett, M. 1975. The analysis of sentence production. In G. Bower, ed., *Psychology of Learning and Motivation*, vol. 9. New York: Academic.

Garrett, M. 1976. Syntactic processes in sentence production. In R. Wales and E. Walker, eds., *New Approaches to Language Mechanisms*. Amsterdam: North-Holland.

Garrett, M. 1980a. Levels of processing in sentence production. In B. Butterworth, ed., *Language Production*, vol. 1. London: Academic.

Garrett, M. 1980b. The limits of accommodation. In V. Fromkin, ed., *Errors in Linguistic Performance*. New York: Academic.

Garrett, M. 1982. Production of speech: Observations from normal and pathological language use. In A. W. Ellis, ed., *Normality and Pathology in Cognitive Functions*. London: Academic.

Geschwind, N. 1967. The varieties of naming errors. *Cortex* 3: 97–112.

Goodglass, H. 1976. Agrammatism. In H. A. Whitaker and H. Whitaker, eds., *Perspectives in Neurolinguistics and Psycholinguistics*. New York: Academic.

Goodglass, H. and N. Geschwind. 1977. Language disorders (aphasia). In E. C. Carterette and M. Friedman, eds., *Handbook of Perception*, vol. 7. New York: Academic.

Goodglass, H., E. Kaplan, S. Weintraub, and N. Ackerman. 1976. The "tip-of-the-tongue" phenomenon in aphasia. *Cortex* 12: 145–153.

Joanette, Y., E. Keller, and A. R. Lecours. 1980. Sequences of phonemic approximations in aphasia. *Brain and Language* 11: 30–44.

Johns, D., and L. LaPointe. 1976. Neurogenic disorders of output processing: Apraxis of speech. In H. A. Whitaker and H. Whitaker, eds., *Studies in Neurolinguistics*, vol. 1. London: Academic.

Kean, M.-L. 1977. The linguistic interpretation of aphasic syndromes. *Cognition* 5: 9–46.

Kean, M.-L. 1980. Grammatical representations and the description of language processing. In D. Caplan, ed., *Biological Studies of Mental Processes*. Cambridge, Mass.: MIT Press.

Lecours, A. R. 1982. On neologisms. In J. Mehler, S. Franck, E. Walker, and M. Garrett, eds., *Perspectives of Mental Representation*. Hillsdale, N.J.: Erlbaum.

Lecours, A. R., and F. Lhermitte. 1976. The "pure form" of the phonetic disintegration syndrome (pure anarthria): Anatomo-clinical report of a historical case. *Brain and Language* 3: 88–113.

Lecours, A. R., and R. Rouillon. 1976. Neurolinguistic analysis of jargonaphasia and jargonagraphia. In H. Whitaker and H. A. Whitaker, eds., *Studies in Neurolinguistics*, vol. 2. New York: Academic.

Lecours, A. R., G. Deloche, and F. Lhermitte. 1973. Paraphasies phonemiques: Description et simulation sur ordinateur. *Colloques I.R.I.A. Informatique Medicale* 1: 311–350.

Saffran, E., M. Schwartz, and O. Marin. 1980. Evidence from aphasia: Isolating the components of a production model. In B. Butterworth, ed., *Language Production*, vol. 1. London: Academic.

Wells, R. 1951. Predicting slips of the tongue. *Yale Scientific Magazine* 26, no. 3: 9–30. Reprinted in V. Fromkin, ed., *Speech Errors as Linguistic Evidence*. The Hague: Mouton, 1973.

Chapter 10

The Neuropsychology of Bilingualism

Loraine K. Obler

At first blush, the bilingual would seem to be the ideal experimental subject for the modern neuropsychologist. In the bilingual we have personified the essence of the experiment with two competing channels: one brain, two simultaneous systems, no fancy experimental technology necessary, no problems with finding controls matched for age, gender, education, or handedness. Thus it was with great enthusiasm that in 1974 Martin Albert and I found ourselves in Jerusalem, a city where virtually everyone speaks at least two languages, and we set out to examine the neurolinguistics of bilingualism.

It turns out, of course, that no two bilinguals look alike on any test you give them, although groups of them can appear to have language and neurolinguistic behaviors in common. In this chapter I will set out a number of elements of the complex system we have discovered bilingualism to be, first by briefly reviewing the questions of the first century of modern aphasiological research, as they pertained to bilingualism, and then by treating the numerous questions about bilingualism that neurolinguists began posing around the early 1970s. Among the latter I will treat such topics as a critical age for second-language acquisition, independent versus interdependent organization of the two languages, mechanisms for switching from one language to the other, the translation process, and cerebral lateralization. After a sketch of the research advances in the last decade, I will turn to the three concepts that in my opinion will govern our thinking about they neuropsychology of bilingualism in the coming decade: parsimony, potency, and attrition.

The First Eighty Years

The great bulk of the neurolinguistic literature on bilingualism of the final decade of the nineteenth century and the first seven decades of this century revolved around a single question: In the aphasic polyglot, which language returns first, and why? Much thinking about language representation in the brain during this period evolved out of studies of monolingual aphasics (and, we may assume, bilingual aphasics who were only tested in one language). With respect to the bilinguals or polyglots, most of the interesting cases were those in which at least one language was entirely inaccessible while in the other language or languages the patient displayed aphasic symptoms—a pattern termed differential recovery. More recently it has been observed (by Penfield in 1953, by Paradis in 1977, and Whitaker in 1978) and documented (by Charlton in 1964, by Lhermitte et al. in 1966, and at my laboratory with Martin Albert since 1976) that the 200-odd cases in the literature of differential recovery from aphasia in polyglots by no means represent the clinical norm; the norm is that both or all languages are impaired and recover in like manner, in proportion to how well they were known before the aphasia-producing accident. Thus, what becomes interesting is the range of individual differences represented. Which polyglot aphasic will show differential recovery, and which patient will show parallel recovery? It was only as recently as 1977 that Paradis articulated the numerous possibilities for order and type of selective impairment in the polyglot aphasic, giving us further evidence of individual differences.

In the early part of this century, however, authors focused on those patients they saw with differential recovery and served to encourage investigators' speculations on the reasons why one language or another recovered first. The answers they accepted, I maintain, had to do with a notion of potency; the language learned first returned first (Ribot 1882), or the language most familiar to the aphasic returned first (Pitres 1895), or the language bearing most emotional weight returned first (Minkowski 1963; Krapf 1954, 1957). These explanations were not intended to be mutually exclusive, of course; the underlying implication for any of these answers was that a more potently represented language would return first. More recently, the discussion has centered on an explanation of inhibition (specifically, selective inhibition of the more inaccessible language), but I will argue below that the potency approach is still viable, particularly in the light of methodological advances such as the work of Whitaker and his colleagues (Whitaker and Ojemann 1977; Rapport, Tan, and Whitaker 1980).

If the major questions of the early period were which language returns first and why, two less-treated ones must also be mentioned, since they are being asked again in different forms. Freud asked in 1891, and

Penfield again in 1953, whether there were any differences in the organization of the two languages of the bilingual, but they had answered their rhetorical question with a firm No, both claiming never to have seen cases of truly differential impairment or differential recovery among the cases of polyglot aphasia they had tested. That question, then, remained to be rephrased in terms of the notion of cerebral lateralization for language. The second less-treated question, asked on the basis of the few aphasia cases in which the patient had particular difficulty switching from one language to the other, was: Where is the mechanism that permits healthy bilinguals to switch between their two languages? Lesions both left and right and both posterior and anterior had been postulated in a literature produced since the 1920s (see Paradis 1977). Neurologists dropped out of the debate in the mid-1960s, giving way to psychologists and psycholinguists, whose models for switching behavior are not tied to specific brain localizations.

Psychologists had had an indirect interest in bilingualism since the early part of this century, when the newly developed IQ tests were used to screen immigrants to the United States. By the mid 1950s, the psychologists Ervin and Osgood (1954) advanced the field with their notion of compound or interdependent organization versus coordinate or independent organization to identify the theoretical possibilities of organization of language (more precisely, storage of lexicon) in the bilingual. Lambert and Fillenbaum (1959) applied this notion to a population of bilingual aphasics in Montreal and found that it explained their recovery patterns as well as Ribot's rule, Pitres's rule, or Minkowski's effect ever had.

The Past Decade

In the early 1970s, neurolinguists and neuropsychologists began to enter the arena with a broad range of new questions. Paradis detailed the diversity of recovery patterns in aphasics who exhibited differential recovery, distinguishing six patterns in his 1977 article and a seventh more recently (Paradis, Goldblum, and Abidi 1982). Martin Albert and I pulled together *The Bilingual Brain* (Albert and Obler 1978) out of our work in Jerusalem. I shall now review some of the questions asked in that book in the light of advances in our thinking since its publication.

With respect to the notion of separate versus shared storage of the two languages (the compound-coordinate dichotomy), in *The Bilingual*

Brain we put the word *dichotomy* in quotation marks; we asked not who is and who is not a compound or coordinate bilingual but whether it is not more likely that, to some extent, certain elements of the two languages are organized more in compound fashion while other elements are organized more in independent fashion in the same individual. In the years since the book was published it has become clear to me that certain experimental methods (particularly tests of free association to single-word stimuli) are likely to give evidence for coordinateness whereas others (such as list-recall paradigms) are likely to give evidence for compoundedness. Thus, *compound* is not a label for an individual, but a statement of the organization of—or, better, the psycholinguistic approach for—specific language tasks. Furthermore, compound and coordinate organizational structures coexist within the same individual; I return to this point below in discussing parsimony.

With respect to individual differences in brain organization for language, we treated age of acquisition of the second language. Lenneberg's (1967) notion of the critical period clearly entered the discussion, as did degree of proficiency, for we had come to realize that what one learned could conceivably influence how one's brain was organized for it. Indeed, we went so far in the book as to propose the shockingly neo-Whorfian possibility that differences in the languages learned (more precise, differences in the pairings of languages in the bilingual) might have repercussions on the way language materials were processed in the brain. This has been borne out in the literature since, for visual representations of language more than for auditory processing. Thus, whereas VanLancker and Fromkin (1978) demonstrated that tone is processed predominantly by the left hemisphere in native speakers of Thai, a series of studies on readers of logographic scripts (Hatta 1978; Sasanuma et al. 1977; Nguy, Allard, and Bryden 1980) demonstrated that phonological scripts evidence highest degrees of left lateralization and logographic scripts show more right-hemisphere participation.

Translation, we made a point of saying in *The Bilingual Brain*, is a special skill independent of other language skills; the bilingual, like the monolingual, can live a perfectly healthy life without the least need or desire to translate. We made this point explicit in order to counter the assumptions implicit in many of the psychological studies that translation tasks in some way reflect the essence of bilingualism, and to address those who maintained that the notion of right-hemisphere participation in bilingualism was absurd because everyone knows one learns a second language by linking the newly learned words to their equivalents in the first language. Recent years have borne us out well. Paradis, Goldblum, and Abidi (1982) discovered two cases of aphasia that illustrate our point superbly. Those aphasic individuals had some premorbid experience in conversational translation from each of their languages to the other. With aphasia, each could translate only from the better-spoken language to the one that was otherwise not accessible. That is to say, the translation task facilitated production of a language that was otherwise hard to produce. Schoenle (personal communication) has reported a similar phenomenon in the case of an 84-year-old woman with a dementia of 10 years' duration whose severe anomia in her native German was relieved when she was asked to translate items from English, which she had learned as a young woman. That is, German nouns she could not produce to tactile stimulation or to seeing pictures of the objects were produced in response to hearing the translation equivalent in English.

In the 1978 book we reviewed the literature on the hypothesized switch mechanisms. [Kolers (1966) and other psychologists had by then posited one switch for input and another one for output.] We were disconcerted to find that relatively few aphasics in the literature on polyglot aphasics had evidenced switching difficulties. Anecdotal evidence suggested several cases of dementing patients reverting to their native language even when speaking to grandchildren with whom they had previously spoken only English. (This phenomenon, we have since realized, is not universal; we have seen cases of severe Alzheimer's dementia in balanced bilinguals with virtually no switching difficulties.) In the book we concluded that multiple processes are involved in both conscious and unconscious decisions to switch languages, and that some of these processes involve general cognitive mechanisms (such as those involved in switching from one register to another in the monolingual who may be able to talk to members of such linguistically deviant groups as scientists and foreigners) whereas other processes are more specific to bilingualism. Since writing the book we have developed a model of a monitor system of bilingual processing whereby input material is assigned first to phonemes, regardless of language, and then, as the attempt is made to assign phoneme strings to lexical items, predictions are made, as per the Fodor-Bever-Garrett (1974) analysis-by-synthesis model (Obler and Albert 1978). For the bilingual these predictions include judgments of which language is likely to be being spoken by the speaker, in what specific context, and so on. In addition, the monitor scans incoming data for cues—phonemes or consonant clusters or lexemes peculiar to one of the target languages—and then directs lexical search toward that language. If the search is unsuccessful, the monitor is immediately prepared to direct efforts to search in the other language. This flexible monitor, we maintain, operates for processing input, while greater constraints are laid on production, except perhaps in cultures where mixing two languages is deemed entirely appropriate.

Paradis dissents from the single-switch concept in another way, arguing that the reason no switch has been located anatomically is that the bilingual simply has a larger lexical store than the monolingual, and that within that lexical store associations are naturally formed among words of one language as they are among words of a given semantic field in the monolingual. Although this is undoubtedly true at a certain level, it is hardly a parsimonious explanation; it leaves all levels of processing other than the lexical un-accounted-for. Work clearly remains to be done to test our monitor model and to advance answers to our question of how language organization in the bilingual differs qualitatively from language organization in the monolingual or to Paradis's question whether indeed the bilingual is not just like the monolingual except for happening to control linguistic items that happen to constitute two languages.

From a different perspective, that of language-acquisition theory, in the early 1970s researchers began to ask who are the best second-language learners and why. Sociolinguists' studies on this question regularly suggest that individuals who are happy where they are and who feel good about the language they are learning are most likely to learn it well. Hartnett (1974), by contrast, looked at success under either of two language-teaching methods and was able to correlate success under the deductive method with greater left laterality for language and success under the inductive method with greater right-hemisphere processing; unfortunately, she used the questionable method of observing direction of eye movement as the measure of laterality. However, her study opened the field to neurological answers to the question of who learns languages well or poorly.

As to more direct measures of cerebral lateralization in bilinguals, we concluded in 1978 that bilinguals may well have greater right-hemisphere participation in language processing than monolinguals. We came to that conclusion largely on the basis of a review of our own and others' dichotic and tachistoscopic studies. Even at that time, this body of studies presented a number of contradictions, some maintaining that bilinguals showed greater degrees of left-hemisphere dominance in one language than the other and others supporting the orthogonal notion that bilinguals showed greater right-hemisphere participation for both their languages than did monolinguals for their single language. Such contradictions have continued to manifest themselves up to the present, and they go beyond the difficulties encountered in dichotic or tachistoscopic studies of bilinguals (Obler et al. 1982). The difficulties range from specifics of subject-selection criteria among bilinguals (Who learned what, when and how, and how have they used it?), to problems of creating equivalent stimulus sets for two languages, to problems of statistical manipulation of data, to problems of interpreting findings of no significant difference between the two channels for one or both languages tested.

One set of data pertinent to cerebral lateralization in bilingualism that has been treated in somewhat greater depth since our statements of the 1970s is that of the crossed aphasias. We had noted a higher proportion of presumable crossed aphasias in the 106 cases outlined in the book—10% versus the 2% reported for monolingual populations. In a compilation of 103 cases, Galloway (1982) documented 15% crossed aphasia in the literature on right-handed polyglots, but again raised the caveat that the population was the unusual set of polyglots who got reported—that is to say, there was not only a bias toward cases of differential recovery, but also a bias toward reporting crossed aphasia. When April (1979) discovered two cases of crossed aphasia in right-handed native Chinese speakers in New York, he conducted a survey of two hospitals there, but he was not able to document a higher incidence of crossed aphasia in the population. Thus, studies of crossed aphasia in polyglots give only the weakest support to a notion of greater right-hemisphere participation for language in general in the bilingual population. Moreover, the Wada tests reported by Rapport et al. (1980) speak strongly against any significant long-term contributions of the right hemisphere to language representation in the bilingual, for naming at least; the phenomenon to which their cortical-stimulation data and those of Ojemann and Whitaker (1978) speak is more diffuse representation of lexicon within and around the left-hemisphere language area in the bilingual.

But how then do we account for the case Paradis reported (1981) of "selective crossed aphasia" in a trilingual patient? After removal of a parasitic cyst from the right prerolandic area, the right-handed patient presented aphasic behavior in one of his two early-learned tongues

(Gujarati) and no aphasic disorder in his two other languages (French and Malagasy). In the case of this patient we would likely want to posit a measure of differential localization, since with Paradis we would reject an explanation related to structural differences in the languages in question or a hysterical response. Paradis prefers the explanation of selective inhibition, pointing to the parallels between this case and those more frequent cases in which one language is produced asphasically and another not at all, but to our mind reference to "inhibition" does not answer any questions at all. We may still ask why one language was selectively inhibited, and why, in this case reported by Paradis, from a right-hemisphere lesion.

We must also account for the data from a study in which a verbal manual interference paradigm was employed with 40 bilinguals and matched controls (Sussman, Franklin, and Simon 1982). According to the most recent report by these authors, early bilinguals (those acquiring a second language prior to age 6) revealed left-hemisphere dominance for both languages whereas bilinguals who acquired the second language after age 6 evidenced left-hemisphere dominance for the first language and ambilateral dominance for the second. This study concurs nicely with the first of Vaid's (1983) conclusions in her critical evaluation of 40-odd studies of lateralization in bilinguals based on the more traditional methodologies, tachistoscopic and dichotic testing. Vaid concludes that patterns of cerebral dominance in the bilingual are more likely to resemble those of the monolingual when the second language is learned early. However, she also concludes that for late learners the manner of acquisition is important—not specifically whether a language is learned with or without written input, as we had suggested in the book, or through a translation method, but more generally whether the second language is learned via a formal or an informal method of teaching. She observes that subjects who have learned a second language informally as young adults, especially those who are proficient in the language at the time of testing, are more likely to evidence greater right-hemisphere participation in the second language than in the first, whereas those who learn by formal teaching will show greater left-hemisphere participation in the second language than in the first. This is consistent with our proposal that certain of the processes involved in learning a second language may call on right-hemisphere cognitive skills, while others will rely on left-hemisphere skills.

Wesche and Schneiderman (1980) have reported a correlation of one of the two language-aptitude subtests they administered (that for phonetic script) with higher left-ear scores on a dichotic listening task. The phonetic-script task of the Modern Language Aptitude Test requires subjects to associate and remember single syllables presented auditorally with their representation in an unfamiliar transcription system. The positive correlation of this task with high left-ear scores on the dichotic listening task suggests that these components of second-language-learning aptitude may be linked to right-hemisphere skills.

We have come to believe at this point that language representation *in grosso*, as we know it from aphasiological studies and from tests of sodium amytal injection, is demonstrably "in" the left hemisphere for most bilingual right-handers. At the same time it is clear that bilinguals will use the right hemisphere more than monolinguals will in performing a number of language-related cognitive tasks, such as dichotic and tachistoscopic tasks and speaking while tapping their fingers. This right-hemisphere participation, we believe, is particularly likely in the early stages of language acquisition, as what has come to be termed our stage hypothesis would have it, but may well be possible even in late stages where proficiency is high. The stage hypothesis so neatly demonstrated by Silverberg et al. (1979) showed how, for native speakers of Hebrew, lateralization for tachistoscopically presented Hebrew materials was demonstrably to the left for 13-, 15-, and 17-year-olds, whereas for English stimuli the youngest group who had 2 years' experience learning English showed a left-visual-field effect (LVFE) while the oldest group showed an RVFE but not so strong as that for Hebrew. Follow up studies by Silverberg et al. (1980) suggested that that finding might be related to the difficulty of the task, although the stimuli in the first study were words such as *pen* and *comb* from first-year textbooks. Elsewhere, right-hemisphere processing has been associated with relatively difficult tasks. One question we must now consider, then, is what bilingualism can tell us about right-hemisphere processing of language in humans, especially since the adolescent or adult second-language learner is surely easier to test than the monolingual child whose right hemisphere is also participating in language acquisition and processing.

Directions for Future Research

Three notions represent the cutting edge of thinking about language representation in bilingualism: potency, attrition, and parsimony. Of course these notions have not sprung forth from some brain of Zeus;

rather they are revisings of earlier questions. The concept of language attrition shifts the focus from how language is acquired to how it is progressively lost in ostensibly healthy individuals who have learned a second language but, without a chance to use it, find it is no longer as accessible as it once was. Potency returns to the assumption underlying the major question of the first 80 years of study of aphasia in polyglots: Which language returns first, and why? The interesting question is not the current one of how one language is inhibited in brain damage (after all, we expect language deficits with damage to the left hemisphere), but rather how one language is potent enough to be relatively spared. Parsimony brings a new life to the long nearly dead, certainly well-flogged notion of a potential for compound as compared with coordinate (or subordinate) "storage" of the languages of the polyglot.

Let us turn first to the concept of language attrition, as that concept consists almost entirely of questions on which no data have been brought to bear yet, then turn to potency, a notion for which supporting if not explanatory data may be marshaled, and end with parsimony, for which we may actually discuss a corpus of data.

Richard Lambert and Barbara Freed of the University of Pennsylvania are to be credited with articulating the importance of researching language attrition. In the conference they convened on the topic in May 1980 the questions posed ranged from the linguistic and the socio-linguistic to the political and the economic. The neurolinguistic questions as I framed them then (Obler 1982) include questions of how the attrition of a second language in a healthy person who does not use that language for a period relates to the language loss in aphasia or to the progressive language loss in certain dementias. Do production and perception deteriorate differentially, and if so, why? Do certain word classes or semantic features disappear more rapidly than others, and if so, why? How does accent deteriorate, and why? What features of the individual's neurolinguistic or language history (handedness, gender, age at second-language acquisition, manner of lamguage acquisition, having learned more than one language as an infant) interact with the rate or the pattern of attrition? What factors of the languages in question exacerbate attrition, and what factors hinder it? Do these factors include structural or cognate closeness to the first language, or of a second language to a third? Do they include more or less shared phonology, or morphology, or lexicon, or syntax? What forms of language usage are likely to retard attrition; for example, does practice in reading hinder attrition of speaking or of aural comprehension skills? And what of the critical-age hypothesis: If one learns a language before a certain age or maturational stage, is it protected from subsequent loss? Or is there a critical proficiency, a certain level of skill in a second language, that protects against subsequent loss of that language?

Clearly these questions demand longitudinal studies of large samples of bilinguals and second-language acquirers, as well as intensive studies of individuals. Such a program is gearing up under the Pennsylvania project; at present what we have is a lot of anecdotal evidence that attrition does occur. A colleague of mine, for example, spoke Hebrew fluently until age 5 and then stopped using it altogether. At age 18 she relearned it, and to his day (age 34) she speaks it with a strong American accent. Geschwind reports that Eric Lenneberg learned Portuguese at age 12, when his family moved to Brazil, but continued speaking German with them. From the age of 24, when he arrived in the United States, until the age of 42 he spoke virtually no Portuguese. At age 42, when he returned to Brazil at the time of his father's death, he noted that he had difficulties in comprehending Portuguese and spoke it both with difficulty and with errors. Within a week, however, it had returned to him. Research must explain to us what it means to say that a language "returned' to Lenneberg after a week but that one did not "return" to my colleague.

The closest we come to having data on the critical age is the hypnotic regression studies, which I believe are more telling in their dearth than in their existence. The two cases reported in *The Bilingual Brain* are the only two we could turn up in a more recent computer search of the field, and the one in which a Finnish-born youth could recall only a few Finnish words under hypnotic age regression is decidedly unconvincing. In Fromm's (1970) more persuasive study, a 21-year-old Japanese-American poured forth the Japanese appropriate to his concentration-camp experience at age 7 and then could not comprehend the tape when returned to age 21. John Schumann and his colleagues at UCLA undertook a fairly extensive project to hypnotically regress people who reported having forgotten an early language. They had no trouble finding such people in Southern California, but they canceled the project because these people proved unsusceptible to hypnosis. We may still ask, of course, what neurological mechanisms are responsible for this resistance to hypnosis, but that takes us somewhat far afield.

Let us return, then, to potency. As Goldstein (1948) pointed out, differential recovery in polyglot aphasia is hard to explain under the schema of circumscribed brain centers for each of the various language behaviors, unless, of course, one is willing to posit separate areas for the second language. A few early neuropathologists, among them Scoresby-Jackson (1867), were prepared to assign the extra languages of the intellectuals whose brains they examined to the extra frontal-lobe convolutions the individuals in question had (extra, that is, with reference to those of farm servants whose brains the neurologists had seen). But such explanations were not taken seriously, as the methodological flaws of such research (the problem of matched controls, and the existence of small-brained geniuses) were recognized. Only with the recent cortical-stimulation work of Whitaker and hs colleagues has the notion of somewhat separate locations within the left hemisphere for the languages of the bilingual or polyglot taken on a new respectability, and these investigators have refined the notion, for they present us with data that lexicon for the second language is simply more diffusely represented in the bilingual, around the core language zone of the left hemisphere. Data pertaining to the effects of language potency have come not only from the cases of differential recovery from aphasia but also from the work on recovery from sodium amytal. Rapport, Tan, and Whitaker (1980) reported that in a number of their cases language production (that is, naming) returned differentially, and no simple explanation (e.g., language learned first) predicted which language would return first or last. Obler and Albert (1977) did find that, at least up until age 65, what has been called Pitres's rule—that the most familiar or most used language at the time of the aphasia-producing accident returns first—applies to a significant majority of polyglot aphasics with differential recovery, while the "rule" of Ribot that the first-learned language returns first does not apply with greater than chance frequency. Thus, whatever constitutes potency, usage would seem to be more powerful in creating it than primacy. Indeed, the only model to come along in a while to account for the potency of one rather than another language in the bilingual is Whitaker's automatization model. It will be of interest to determine how age and usage effects interact in this model, and to clarify how automatization in Whitaker's sense differs from the more restricted usage in standard aphasiological terminology; the term as Whitaker employs it should provide for propositional language in the automatized zone. And of course the neurochemical or neurophysiological processes involved in setting up one language as more potent than another remain to be explored.

Finally, then, we may turn to parsimony, an advance I propose on the hopelessly muddled compound-coordinate controversy concerning the storage of language in the bilingual. In order to go beyond that messy concept, we must realize that what is of interest is not the mere storage of lexicon but rather the dynamics of language processing. Psycholinguistic studies of voice onset time (VOT) motivate the model I propose. In work carried out in Jerusalem with Martin Albert and Alfonso Caramazza and with the aid of Edgar Zurif, we replicated a set of VOT studies carried out elsewhere on French-English and Spanish-English bilinguals. What is of interest is not our replication (the perception parts of which were reported in *The Bilingual Brain*) of the earlier findings of Caramazza et al. (1973) and Williams (1977, 1980) but the finer analysis of the data that present a striking picture of parsimony (Obler 1983). The methodology for such a study of adults involves two parts. For VOT production, fluent bilinguals and monolingual controls are taped reading words beginning with the stop consonants (p), (b), (t), (d), (k), and (g) in one language and then in the other language. For the perception component, subjects label taped stop consonants along a VOT continuum synthesized at the Haskins Laboratory. Bilinguals are tested in two environments at two different times, once with Hebrew instructions and answer sheets and once with English instructions and answer sheets.

Voicing of stops begins earlier in Hebrew than in English, and thus for production the mean VOT values of Hebrew monolinguals occur earlier relative to the burst than the means for English. This is true for the bilingual conditions between the Hebrew and English, to a lesser extent, but the difference is significant by an analysis of variance at the .01 level. What is particularly striking, however, is that the differences between the voiced and voiceless members of a pair for the bilingual speakers are even greater than those for the respective monolingual populations. Moreover, we see a consistent skewing for the bilinguals on both conditions toward the later (English) mean for the voiceless consonants, /p/, /t/, and /k/, and toward the earlier (Hebrew) mean for the voiced consonants, /b/, /d/, and /g/. We may consider the perception of voice onset a dual system clearly biased or directed or governed by the consideration of the specific languages in question; the bilinguals have generated a system different from those of the respective monolinguals—a system that analyzes and exaggerates differences between the two languages. Similarly, in the perception data, the bilingual subjects show a broader range on all six instances between the voiceless and the voiced stops than the monolinguals. That is, there is a broader range of perceptual uncertainty for the bilinguals than for either group of monolinguals—specifically, a broader range of acoustic stimuli which they may label either *p* or *b* (depending on context, one would presume).

This maximization of extremes, of course, serves different functions for the production task and for the perception task. For the production task it serves to optimally differentiate a distinctive feature dichotomy; for the perception task it serves to extend flexibility of interpretation. One might say (Obler 1983) that for the perception task the bilinguals present a relatively unified system, whereas the production task they present a relatively dual system. What is now clear is that our fluent bilinguals appear to have developed a system that coincides with that of neither monolingual group; rather, it incorporates the extremes from analogous structures in a way that permits the bilingual to pass as a native speaker of each language with greatest functional ease.

Acknowledgments

This paper has benefited from critical readings by Martin Albert, Joan Borod, Margaret Fearey, and Michael Patrick O'Connor.

References

Albert, M., and L. Obler. 1978. *The Bilingual Brain: Neuropsychological and Neurolinguistic Aspects of Bilingualism.* New York: Academic.

April, R. S. 1979. Concepts actuels sur l'organisation cérébrale du language a partir de quelques cas d'aphasie croisée chez les orientaux bilingues. *Rév. Neurology* (Paris) 135 (no. 4): 375–378.

Caramazza, A., G. Yeni-Komshian, E. Zurif, and E. Carbone. 1973. The acquisition of a new phonological contrast: The case of stop consonants in French-English bilinguals. *Journal of the Acoustical Society of America* 54: 421–428.

Charlton, M. 1964. Aphasia in bilingual and polyglot patients—A neurological and psychological study. *Journal of Speech and Hearing Disorders* 29: 307–311.

Ervin, S., and C. Osgood. 1954. Psycholinguistics: A survey of theory and research problems. In C. Osgood and T. Sebeok, eds., *Psycholinguistics.* Baltimore: Waverly.

Fodor, J., T. Bever, and M. Garrett. 1974. *The Psychology of Language.* New York: McGraw-Hill.

Freud, S. 1891. *On Aphasia. A Critical Study*, E. Stengel, ed. and tr. New York: International University Press, 1953.

Fromm, E. 1970. Age regression with unexpected reappearance of a repressed childhood language. *International Journal of Clinical and Experimental Hypnosis* 18:79–88.

Galloway, L. M. 1982. Towards a neuropsychological model of bilingualism and second language performance: A theoretical article with a critical review of current research and some new hypotheses.

Goldstein, K. 1948. *Language and Language Disturbances.* New York: Grune and Stratton.

Hartnett, D. 1974. The relation of cognitive style and hemispheric preference to deductive and inductive second language learning. Master's thesis, University of California, Los Angeles.

Hatta, T. 1978. Recognition of Japanese Kanji and Hirakana in the left and right visual fields. *Japanese Psychological Research* 20:51–59.

Kolers, P. 1966. Reading and talking bilingually. *American Journal of Psychology* 79:357–376.

Krapf, E. 1955. The choice of language in polyglot psychoanalysis. *Psychoanalytic Quarterly* 24:343.

Krapf, E. 1957. A propos des aphasies chez les polyglottes. *Encéphale* 46:623–629.

Lambert, W., and S. Fillenbaum. 1959. A pilot study of aphasia in bilinguals. *Canadian Journal of Psychology* 13:28–34.

Lenneberg, E. 1967. *Biological Foundations of Language.* New York: Wiley.

Lhermitte, R., H. Hécaen, J. Dubois, A. Culioli, and A. Tabouret-Keller. 1966. Le problème de l'aphasie des polyglottes: Remarques sur quelques observations. *Neuropsychologia* 4:315–329.

Minkowski, M. 1963. On aphasia in polyglots. In L. Halpern, ed., *Problems of Dynamic Neurology.* Jerusalem: Hebrew University.

Nguy, T. V. H., F. A. Allard, and M. P. Bryden. 1980. Laterality Effects for Chinese Characters: Differences between Pictorial and Nonpictorial Characters. Manuscript, University of Waterloo, 1980.

Obler, L. 1982. Neurolinguistic aspects of language attrition. In R. Lambert and B. Freed, eds., *Language Loss.* Rowley, Mass.: Newbury House.

Obler, L. 1983. The parsimonious bilingual. In L. Obler and L. Menn, eds., *Exceptional Language and Linguistics.* New York: Academic.

Obler, L., and M. Albert. 1977. Influence of aging on recovery from aphasia in polyglots. *Brain and Language* 4:460–463.

Obler, L., and M. Albert. 1978. A monitor system for bilingual language processing. In M. Paradis, ed., *Aspects of Bilingualism.* Columbia, S.C.: Hornbeam.

Obler, L. K., R. J. Zatorre, L. Galloway, and J. Vaid. 1982. Cerebral lateralization in bilinguals: Methodological issues. *Brain and Language* 15:40–54.

Ojemann, G. A., and H. A. Whitaker. 1978. The bilingual brain. *Archives of Neurology* 35:409–412.

Paradis, M. 1977. Bilingualism and aphasia. In H. Whitaker and H. Whitaker, eds., *Studies in Neurolinguistics*, vol. 3. New York: Academic.

Paradis, M. 1981. Selective Crossed Aphasia in a Trilingual Patient. Paper presented at Annual Conference on the Neuropsychology of Language, Niagara Falls.

Paradis, M., M. C. Goldblum, and R. Abidi. 1982. Alternate antagonism with paradoxical translation behavior in two bilinguual aphasic patients. *Brain and Language* 15: 55–69.

Penfield, W. 1953. A consideration of the neurophysiological mechanisms of speech and some educational considerations. *Proceedings of the American Academy of Arts and Sciences* 82 (no. 5):201–214.

Pitres, A. 1895. Etude sur l'aphasie. *Révue de Medecine* 15:873–899.

Rapport, R. L., C. T. Tan, and H. A. Whitaker. 1980. Language Function and Dysfunction Among Chinese Polyglots; (1) Wada Testing (4 cases), (2) Cortical Stimulation (3 cases). Paper presented at Academy of Aphasia, Fall River, Mass.

Ribot, T. 1882. *Diseases of Memory: An Essay in the Positive Psychology.* London: Paul.

Sasanuma, W., M. Itah, K. Mori, and Y. Kabayashi. 1977. Tachistoscopic recognition of Kana and Kanji words. *Neuropsychologia* 15:547–553.

Scoresby-Jackson, R. 1867. Case of aphasia with right hemiplegia. *Edinburgh Medical Journal* 12:696–706.

Silverberg, R., H. Gordon, S. Pollack, and S. Bentin. 1980. Shift of visual field preference for Hebrew words in native speakers learning to read. *Brain and Language* 11:99–105.

Silverberg, R., W. Bentin, T. Gaziel, L. Obler, and M. Albert. 1979. Shift of visual field preference for English words in native Hebrew speakers. *Brain and Language* 8 (no. 2):184–191.

Sussman, H. M., P. Franklin, and T. Simon. 1982. Bilingual speech: Bilateral control. *Brain and Language* 15:125–142.

VanLancker, D., and V. Fromkin. 1978. Cerebral dominance for pitch contrasts in tone language speakers and in musically untrained and trained English speakers. *Journal of Phonetics* 6:19–23.

Vaid, J. Bilingualism and brain lateralization. 1983. In S. Segalowitz, ed., *Language Functions and Brain Organization*. New York: Academic.

Wesche, M. B., and E. T. Schneiderman. 1980. Language Lateralization—Language Aptitude and Cognitive Style in Second Language Learning. Paper presented at TESOL International Convention, San Francisco.

Whitaker, H. A. 1978. Bilingualism: A neurolinguistics perspective. In W. Ritchie, ed., *Second Language Acquisition Research*. New York: Academic.

Whitaker, H. A., and G. A. Ojemann. 1977. Graded localisation of naming from electrical stimulation mapping of left cerebral cortex. *Nature* 270 (no. 3):50–51.

Williams, L. 1977. The perceptions of stop consonant voicing by Spanish-English bilinguals. *Perception and Psychophysics* 21:289–297.

Williams, L. 1980. Phonetic variation as a function of second-language learning. In G. Yeni-Komshian, J. Kavanagh, and C. Ferguson, eds., *Child Phonology*, vol. 2, Perception. New York: Academic.

Part III

Neural Issues

Part III explores the anatomical and physiological bases of language. Chapters 11–13 deal with the anatomy of language, the remainder with the physiology. In both areas, but more so in the realm of physiology, the formulations presented are supported by observations now in hand and point to questions that will require more investigation.

Lecours begins the discussion by examining the notion of cerebral dominance for language. He does not dispute this notion; however, he does argue that it may be true of only a portion of the world's population. It is not true of children or of many left-handers, and it may not be true of women. By "not true" is meant, of course, "not the same in." Lecours also stresses the importance of factors other than these endogenous biological determinants of cerebral dominance. In particular, he points out that structural features of languages—the nature of word formation (which he considers to be divided along "agglutinative" and "nonagglutinative" principles) and the nature of suprasegmental prosodic features (tonal versus stress languages)—and certain features of the psychology of language (literacy, especially) may influence the degree of hemispheric specialization. In all, he estimates that no more than a quarter of the world's population show the sort of hemispheric dominance for language.

Furthermore, in the quarter of the world's population that possesses both the endogenous biological and exogenous psycholinguistic factors making for left-hemisphere dominance, exceptions to the rule arise. The three appendixes to Lecours's study present a case of crossed aphasia in a dextral and a discussion and presentation of cases of pure anarthria, which highlight the existence of exceptions to the rule that the left hemisphere is the dominant one in the population Lecours delineates and which illustrate some of the variation in localization of syndromes within specific areas of the dominant hemisphere. Poncet and his colleagues present a case in which the right hemisphere seems to be the language domain in a left-handed woman. This and the crossed dextral aphasia cases present an opportunity to investigate a question that naturally follows the hypothesis that basic language capacities are located in the left hemisphere in an identifiable and well-studied portion of the world's population: In cases where the left hemisphere is "dominant" in this sense, what is the nature of the partitioning of language between the hemispheres? At a yet more fundamental level, we can ask whether the existence of patterns of hemispheric mediation of language other than the "normal" pattern of left-hemisphere dominance is associated with a functional organization of language different from that found in the right-handed, literate, adult population. As John Morton has asked (personal communication), is there any evidence that different neural organizations of language are associated with different functional divisions of language capacities, or is it simply that the same functional architecture of language is distributed differently among different neural loci in different populations? Detailed study of cases such as the crossed dextral aphasics and Poncet's patient can begin to answer such questions. To the extent that answers are currently available (as, for instance, in the chapters here), the impression is that it is only the neural loci and not the functional organization of language that varies. If this is true and holds up as functional abilities of these patients continue to be studied in detail, it must reflect a quite narrow set of possible ways in which neural tissue can support the language function.

The location of linguistic capacities within a hemisphere is also treated by Kertesz in chapter 12. Kertesz describes a number of frequent patterns of recovery of language after aphasic impairment. Clearly, these patterns are orderly. The majority of patients in his series evolved into anomic aphasics; this was particularly true for those who presented as jargon aphasics when their evolution was favorable. Severe Broca's aphasics who are initially very limited in their output can evolve into cases of verbal apraxia, something not seen in the fluent group. If we were to attempt to analyze these symptoms and ascribe to the areas of the brain associated with each symptom a role supporting the function whose disturbance leads to the symptom, we would ascribe different functions to areas of the brain depending on when in the course of the illness we examined the patient. Clearly, either we need to be able to say how lesion extent varies as a function of time (if we want to ascribe these different functions to different areas of the brain) or—and this is more attractive—we would like to be able to characterize the details of lesion development and ascribe to particular features of the lesion (not just its location) the particular interruptions in language functions that are seen at different stages of syndrome evolution. The larger part of Kertesz's chapter is devoted to a description of these details of lesion development. He traces the notions that have been presented about the basis for functional recovery (substitution, reorganization, and so on) and presents a modern recasting of several of the physiological theories about the organic events that could make for evolution of functional capacities. The challenge is to choose among these various possibilities and to relate the description of functional recovery to the physiological and anatomical changes observed in the brain after injury.

We now know that sprouting occurs after neural injury and that jargon aphasia evolves into anomia. Are the two related? If so, how? Is there any way of investigating the question that is imaginable, even if not yet available? Answering such questions will, presumably, take us toward more detailed structure-function correlations.

Galaburda's chapter goes beyond gross anatomy to consider the evolution of language areas on the basis of the types of cells found in species as far back as amphibians. Archicortical and paleocortical structures appear in amphibians and give rise to whole sets of structures on (respectively) the medial and ventral and the dorsal and lateral surfaces of the brain. The former are initially connected to the hypothalamus and involved with internal regulation and motor outflow; the latter are initially connected to olfactory structures and evolutionarily become connected with external receptors and perceptual function. The former are bilaterally connected to their outflow, the latter unilaterally connected. Galaburda traces the development of these primitive moities through reptiles and mammals to man, indicating the increasing cellular complexities seen as the phylogenetic scale is ascended and indicating the basic divisions of cortex that seem to be derived from one or the other system. He then considers the development of asymmetries in the sizes of cortical areas in relation to the origins of cortex, suggesting that cortex derived from the archicortical moiety can reach greater development on the right and that from the paleocortex on the left. This can be related to language. Galaburda distinguishes the lateral, perisylvian language area from a medial language area, which includes the supplementary motor area. The latter is of archipallidal origin, the former of paleopallidal origin. Galaburda suggests that the latter may be more developed on the right than is the case for the perisylvian areas, and that the absence of lasting aphasia following medial lesions that do produce language symptoms is in keeping with this view.

Galaburda also argues that it is possible to explore the connectivity and other cellular features of the human language areas by examining the homologous areas in nonhuman primates. His analysis of the frontal and parietal cortices of the rhesus monkey in terms of cellular structures suggests homologies between human and monkey brain regions. If these homologies do exist, it is possible to study connections in the monkey using tracer methods that cannot be used in humans. Such studies might corroborate more gross observations in humans and suggest the details of connectivity.

The structures identified by Galaburda add significantly to our appreciation of the cellular nature of the areas related to language, a field still in its early stages of study. Yet the greatest challenges surely still lie ahead. How do the changes in cellular structure that are characteristic of human language areas make for language? Galaburda suggests that minor structural changes can have major functional effects; for example, the animal with small webs between its toes sinks in water, whereas its cousin with only slightly larger webs can swim. But what is the content of the analogous structural change in the brain, and what is the mechanism whereby the change could exert its effect? Large webs allow for swimming via their mechanical effects. Presumably, whatever neural changes there are between the primate and the human, they allow for language because of their effects on permissible representations and computations. But how structural properties of the nervous system constrain and permit these sorts of functional capacities is largely unknown.

The final chapters deal with the physiology of language. They approach this area from very different perspectives. In chapter 14, Picton and Stuss provide an extensive critical review of work relating evoked potentials to language. They stress the technical and design difficulties presented by this technique. If an event-related potential (ERP) is measured in response to a linguistic stimulus, we may conclude that some electrophysiological event has taken place in connection with the stimulus. But it is a far from trivial task to rule out noncerebal sources as the generators of such events, and Picton and Stuss are uneasy about accepting many results reported in the literature because of the lack of adequate controls of this sort. If such controls have been run and an ERP has been shown to be cerebral in origin, other questions arise. On the neural side, an ERP, being a volume-conducted electromagnetic wave, cannot reveal the location of its generator source from an analysis of either its configuration or its energy maxima and minima; thus, the determinations of where ERPs originate requires other studies, especially lesion studies of the traditional sort. On the psycholinguistic side, the determination of what aspect of a stimulus or of its processing gives rise to the measured potential requires the presentation of a variable number of comparison and control stimuli. In all, work on ERPs is in its infancy, according to Picton and Stuss, insofar as these potentials are known to be related to language.

There are important reasons to pursue these studies. ERPs are currently the only technique for direct measurement of cerebral functional activity with millisecond-by-millisecond temporal resolution; regional blood-flow studies, positron emission tomography, and other techniques are much less temporally focused. Many purely psychological studies of the "mental chronometry" underlying language processing are being formulated in terms of attainment of stages of language representation on a very short time basis; hypotheses have been phrased in terms of differences as small as 50 milliseconds. Of the measures now in use, only ERPs are likely to be useful for determining cerebral physiological correlates of these psycholinguistic processes. Moreover, taxonomies of ERPs are emerging in nonlinguistic psychological work that may make the interpretation of certain aspects of ERPs (their temporal occurrence as "early" or "late", for instance) possible with respect to language processing. Whatever the difficulties and present limitations of this technique, Picton and Stuss stress its promise for future work.

In chapter 15, Anderson approaches the physiology of the brain in a totally different way. Though not unconnected to empirical observations and empirically justified theories of synaptic function, the model that Anderson discusses in relation to certain aspects of language processing is, in its entirety, far removed from correlation with physiological or anatomical observations of the brain. Its value lies in showing how, given certain views about how information is represented in the brain, a variety of psychological aspects of performance could be produced by the operation of a proposed model of neural function. The model shows a number of properties reminiscent of features of the psychology of language. Anderson's model is one of several that have been proposed and explored, but it is the one that has been most extensively investigated with respect to psychological performances related to language and the only one that has been related to language disturbances. These studies have suggested new mechanisms for the production of certain aphasic symptoms. Other work using Anderson's model has raised questions about basic issues in neuropsychological theory, such as the way evidence from brain-injured patients speaks to the localization or distribution of psychological functions. Perhaps the most important aspect of the approach Anderson takes to modeling neurophysiology is that it shows what a theoretical neurophysiology relating language and the brain could be like. We can begin to see how a set of "neural" elements, each representing a very limited type of information, can be organized and can function so that particular relations hold between input "stimuli" presented to the system and its output "responses," which in some ways correspond to psychological processes. It is possible to think of the inputs and outputs—which are vectors of a certain form in the abstract model—as representations of linguistic or other mental structures, much as the actual inputs into computations accomplished by computers are systematically assigned mathematical or other representational values in the normal course of machine computation.

Of course, in contrast with the operation of a computer, in which case we know how the semantic interpretation of the elementary computing elements has been arrived at (we have assigned it via programs, compilers, etc.), we are ignorant of many aspects of how computational elements are assigned a semantic interpretation in the brain. Our ignorance begins with our lack of certainty as to the nature of the final computational elements; whether they take the form of the vectors Anderson's model uses is a matter for investigation. It extends to the nature of the elementary aspects of the neural "hardware" responsible for computations; Anderson assumes the existence of Hebb synapses, but the evidence for these structures is tenuous, and information pertaining to what other structures and physiological operations might be the basis for computations is just beginning to accumulate in simple nonhuman systems. It is at higher levels of organization—the structure of the language code and its processing, for instance, on the functional side; the brain regions in which language processing takes place, on the neural side—that hypotheses are best developed, and it is here that correlations have been established. The chapters in this part and the preceding part of this volume present discussions of what is and is not known about certain of these structures and their correlations. Anderson, however, is dealing with the more mechanistic question of how these higher-order functional and organic structures emerge from "elemental" operations. In view of the limits of our knowledge of these elemental operations, much of Anderson's work is based on views of these elements that, though they are plausible in terms of current knowledge, are very likely to be changed. It will naturally be necessary to modify some of the basic assumptions of Anderson's model as our knowledge of this level of physiology develops. At the moment, Anderson's model shows us how a complex system can emerge from elementary operations, and how such a system can achieve representations and operations similar to those postulated in certain areas of language functioning.

The perspective on the biology of language that existed in the minds of the conference organizers and volume editors includes integration of studies of function and structure, down to this mechanistic elemental level. Much of what is presented in this volume only points out ways in which such integration could be approached. There are several stimulating and compelling analyses, at both organic and functional levels, and regarding their interaction, presented here and in the work referred to in the studies in this collection. But it is clear that many gaps remain to be filled, as concerns both the separate levels and their interaction. One accomplishment of the past decade has been a greater degree of empirical description and explanation at the functional level regarding aphasic syndromes; this is a first step in one approach to the empirical study of language-brain relationships. A second accomplishment has been the clarification of how linguistic, psychological, and neural levels of description could be integrated in the light of the more detailed accounts of language now available; in particular, there has been a realization of some of the benefits and limitations of the approach to language-brain relations that underlies current clinical approaches to the aphasias. A third step is the more detailed study of the anatomy and physiology of the language areas. The studies in this volume are among those that build on these advances, and they, in turn, point to empirical areas still in need of investigation and to new conceptual approaches to the relation between language and the brain.

Chapter 11

Where Is the Speech Area,
and Who Has Seen It?

André Roch Lecours
Anna Basso
Sylvia Moraschini
Jean-Luc Nespoulous

Cerebral Localization

As far as we know, Gall was the first to suggest the existence of anatomo-functional relationships between precise cortical areas and various modes of human behavior (Gall 1825; Gall and Shurzheim 1810–1819). In spite of his erroneous insistence on the phrenological aspects of his methodology, Gall should therefore be credited with the fundamental discovery that gave birth to neuropsychology.

In his atlas, Gall related speech to a cortical area situated above the orbits, rostral to the anterior perforated space and lateral to the olfactory stalk; this area is numbered XV in his nomenclature, and a drawing is provided showing it on a right hemisphere, which was not meant to be taken as an indication of functional lateralization (Gall and Shurzheim 1810–1819). Bouillaud (1825) then claimed that anatomo-clinical research corroborated Gall's view about the cerebral seat of verbal memory. He was, however, less parsimonious than Gall when it came to actual localization; he contended that the "anterior lobules of the brain" (that is, the frontal lobes in their entirety) were the "organ of articulated speech." We do not know to what extent this notion overlaps with that of the "mental organ for language" (see chapter 1 of this volume).

Pierre Marie (1906a) later wrote that Bouillaud had been misled by Gall's phrenological postulates; he made this assertion somewhat credible by insisting that Gall's localization of area XV was based on his observations of exophtalmy among children of the Black Forest. This was not the only instance when Marie's gift for devastating comments led him both to correct a relatively minor mistake (the attribution of a particular functional role to the orbital surfaces) and to discredit a basically correct notion (functional specialization).

Figure 11.1
Louis-Pierre Gratiolet (1815–1865) and Pierre Paul Broca (1824–1880).

Functional Asymmetry

Dax (1836) may have thought that the cerebral representation of language was asymmetrical, lateralized to an undefined yet considerable part of the left hemisphere, including retrorolandic structures. If so, he should have known that ideas of this sort are to be laid down on paper and made public. Louis-Pierre Gratiolet and Paul Broca also had ideas of their own about asymmetries, anatomical and functional. We will come back to Gratiolet at a later point.

Broca's Area

Broca, who was explicitly and exclusively interested in "the seat of the Faculty of articulated language," saw it a few centimeters away from Gall's area XV, in the third frontal convolution, and he certainly deserves credit for being the first to make clear statements about functional lateralization of language to the left (Broca 1865). We did not find in his writings the passage where he supposedly restricted his area to the foot or *pars opercularis* of the third frontal convolution; to the best of

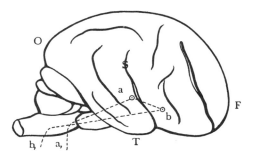

Figure 11.2
Wernicke's speech area in 1874.

our knowledge, the closest he came to this was when he wrote about "the posterior part of the third frontal convolution" (Broca 1865), which can obviously be taken to include *pars triangularis* as well as *pars opercularis*. On the other hand, he certainly predicted (Broca 1861) that a "phrenology of convolutions" was soon going to replace Gall's "phrenology of bumps"; this did occur and is not far from having persisted to this day. Broca also deserves credit for this prediction.

Whatever the exact borders of Broca's Broca's area, if there were any, the anatomical definition of Broca's area has varied, within certain limits, with time and with the accumulation of anatomo-clinical data. Dejerine (1914), for instance, defined "Broca's region" as encompassing—in the left hemisphere—the cape (*pars triangularis*) and foot of the third frontal, the adjacent part of the second, and (possibly) the anterior part of the insula; for some reason, conceivably pertaining to neuropolitics rather than neurosciences, he explicitly excluded the adjacent portion of the precentral gyrus, thus forgetting about his own material and about the canons he himself had defined with regard to the anatomo-clinical method. And Marie (1906b) said, of course, that the third frontal convolution of the left hemisphere has nothing whatsoever to do with language.

Wernicke's Area
One has to agree with Bogen and Bogen (1976) when they suggest that Wernicke's 1874 representation of the speech area in a right hemisphere (figure 11.2) was "more in the nature of a printer's error than anything else," and also when they say that Wernicke's Wernicke's area should not be conceived as restricted to the little circle that Wernicke traced,

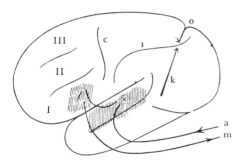

Figure 11.3
Wernicke's speech area in 1881.

on this representation, in the middle of the first temporal convolution. As far as we could judge, however, Wernicke's Wernicke's area was and remained restricted to the first temporal convolution, and by 1881 it had taken its proper place in the left hemisphere (figure 11.3) (Wernicke 1874, 1881).

Like Broca's area, but to a far greater extent and with more persistent consequences, Wernicke's area had made territorial gains with time. in other words, various authors have considered various structures besides the original first temporal convolution to be part or parts of what they have called Wernicke's area (Bogen and Bogen 1976). With regard to structures, this includes or has included the posterior halves or thirds of the second and third temporal convolutions, the insula, the angular gyrus, and the supramarginal gyrus of the left hemisphere. With regard to authors, it is difficult to be sure about who was first to add or delete what; however, it is probably significant that Dejerine (1906) and Marie (1906b) agreed on one point (indeed, on one point only): that the supramarginal and angular gyri of the left hemisphere do belong with the speech area.

The Speech Area

More than a hundred years after the original publications of Broca and Wernicke, and although nearly everybody can talk, relatively few people could answer a question pertaining to tha anatomical components of the speech area. Nowadays, however, students in linguistics, speech therapy, psychology, and (to a lesser degree) anthropology, sociology, and medicine would be among those expected to be capable of answering such a question. Were they asked, whether in French, in Italian, in

Polish, or otherwise, an overwhelming majority of them would no doubt say that the speech area is located in the left hemisphere of the brain and that it is made up of the caudal part of the third frontal convolution, the caudal half or third of the first temporal convolution, the supramarginal gyrus, and the angular gyrus. At this point, a typical student would quote one of Norman Geschwind's papers and complete his answer by adding that the cortical components of the speech area are connected to one another and to other brain structures (Geschwind 1965); if a particular connection were to be mentioned by name, it would invariably be the *fasciculus arcuatus*. This answer now deserves 100% in Rio, Perth, Paris, London (Ontario), Dakar, maybe Tokyo, and conceivably Boston; it summarizes the prevalent doctrine.

If further discussion were requested, our typical student would probably turn out to know that the classical notions of "centers" and "images" have fallen into oblivion [although nearly everybody is interested in "localization," and although one can, without being taxed with simplemindedness, be interested in entities such as "articulems" and "audiospeech traces" (Luria 1964)]; that the prevalent conception of the speech area is founded, essentially, on anatomo-clinical studies of aphasia (which is not the same as if it were founded on, say, electrophysiological or neurosurgical data); and that lesions of the speech area (prevalent conception) are not the only ones that can cause aphasia without being "exceptional." If the last point were to hold the student's attention for a while, he might raise the problem of the effects of lesions of other left-hemisphere structures such as the caudal third of the second frontal convolution, or of lesions of the lingual and fusiform gyri (Lhermitte et al. 1969), or of lesions of the precentral gyrus (Ladame 1902; Lecours and Lhermitte 1976; Levine and Sweet 1982) or of the postcentral gyrus (Luria 1964; Poncet, Ali Cherif, and Brouchon 1978), or else of lesions of the insula and its subcortex (Dejerine-Klumpke 1908; Goldstein 1948), and also the problem of the effects of various subcortical lesions: of the thalamus (Mohr, Watters, and Duncan 1975; Elghozi 1977), of the "lenticular zone" (Marie 1906b; Moutier 1908; Dejerine-Klumpke 1908), or of the caudate nucleus (Damasio and Damasio 1981).

By then, the questioner should be ready to answer a few questions himself. The first one would likely be this: "Why is it—if not because of a mere quarrel on words—that certain cortical areas should be peremptorily excluded from the speech area although lesions of these cortical areas have been shown to cause particular types of aphasia?"

One might decide to argue about this, but our own answer might eventually be that we have no objection to the inclusion of a few more gyri. This answer might lead one to end up quoting Freud's (1891) definition of the speech area as "a continuous cortical region occupying the space between the terminations of the optic and acoustic nerves and the areas of the cranial and certain peripheral motor nerves in the left hemisphere." This can be rephrased in modern terms, and one can add that it is more or less in line with some of Wernicke's original teachings (Wernicke 1874; Freud 1891; Bogen and Bogen 1976; Lecours, Caplan, and Lhermitte 1979). The next question is this: "Why should one keep talking about a two-dimensional speech area when one knows that certain subcortical lesions—deep ones—can cause aphasia?" If one is convinced that one has indeed observed aphasias related to thalamic lesions, caudate lesions, and so forth, the only answer one can propose is that Broca's (1861) "phrenology of convolutions" is no longer entirely satisfying, that someone will sooner or later come up with a clever and aesthetically pleasing three-dimensional representation of the left-hemisphere apparatus subserving speech and language, and that students might eventually be advised to memorize this representation rather than the by-then-"obsolete" two-dimensional scheme.

Anatomical vs. Functional Asymmetry

Unlike his younger colleague Broca, Gratiolet held holistic views on brain function in general and on brain-language relationships in particular (Critchley 1980; Hécaen and Dubois 1969). Nevertheless, he was probably the first to document the existence of systematic asymmetries in the human fetal brain. He observed that at about the end of the seventh (lunar) month of gestation the development of the precentral and postcentral sulci is usually more advanced in the left than in the right hemisphere (Gratiolet 1854) (figure 11.4). It is by reference to these studies that Broca most explicitly postulated, in 1865, that functional lateralization for language is accounted for by an innate biological predisposition.

Broca's postulate in this respect has been the object of renewed attention since the publication of the Geschwind-Levitsky (1968) paper on anatomical asymmetries. As mentioned by Galaburda in chapter 13 below, more has been learned since then on the distribution of such anatomical asymmetries, on the chronology of their apparition, and on their cytoarchitectonic characteristics. Among the classical components

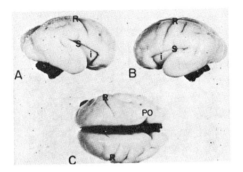

Figure 11.4
Human fetal brain at end of seventh (lunar) month of gestation. R: Rolandic
fissure. S: Sylvian fissure. PO: Parietal-occipital fissure. Note asymmetry between left and right hemispheres (central region and planum temporale).
Case W-109, Yakovlev Collection; published with the kind authorization of
the Armed Forces Institute of Pathology, Washington, D.C.

of the speech area, the supramarginal and angular gyri are structures
whose status in this respect is not all that clear. According to Rubens,
Mahowald, and Hutton (1976), the surface of the inferior parietal lobule
is greater, caudal to the Sylvian fissure, in the right than in the left
hemisphere. On the other hand, Miraglia (1979) found that the textbook
configuration of the inferior parietal lobule was present in only one-
third of the 200 hemispheres that she studied; the rest were variants,
an astonishing number of them. Nonetheless, and although asymmetries
were frequent, no systematic differences were found between right and
left, blacks and whites, males and females, and so forth. With regard
to Wernicke's area proper, the Geschwind-Levitsky (1968) statistical
confirmation of earlier findings (Eberstaller 1884; von Economo and
Horn 1930; Kakeshita 1925; Pfeiffer 1936) has since been reconfirmed
by a number of investigations (Rubens et al. 1976; Tezner 1972; Tezner
et al. 1972; Wada, Clarke, and Hamm 1975; Witelson 1977; Witelson
and Pallie 1973): In the human species the left *planum temporale* usually
covers a greater surface than the right one. Moreover, this particular
macroscopic asymmetry has been shown to be the reflection of a pro-
portional asymmetry in cytoarchitectonic organization (Galaburda et
al. 1978; Galaburda and Sanides 1980; Galaburda, Sanides, and Geschwind 1978). Concerning Broca's area, Wada et al. (1975) found no
statistical difference between left and right when they proceeded to
macroscopic surface measurements of *pars opercularis*; however, a

cytoarchitectonic asymmetry in favor of the left side was observed in this structure by Galaburda (1980) under the light microscope and by Scheibel (1981) under the electron microscope. One might add, considering the subcortical structures said to be related to the speech area, that Eidelberg and Galaburda have compared the volumes of the lateral posterior nuclei of the left and the right thalamus and found it to be significantly greater on the left side in eight of the nine normal human brains that they have studied so far.

Although the information gathered about anatomical asymmetries remains partial, the existence of such asymmetries in relation to the speech area is now generally accepted. Yet, until recently, skeptics could still argue that anatomical and functional asymmetries might very well coexist without being interrelated. This argument has become much less interesting since Ratcliff et al. (1980), who studied in parallel the results of amytal tests (Wada and Rasmussen 1960) and carotid arteriograms (LeMay and Culebras 1972) in 59 patients, demonstrated a "possible relationship" between cerebral dominance for speech and asymmetry between the left and the right *planum temporale.*

Standard Teaching

Current standard teaching about brain-language relationships might thus be summarized as follows:

Language is specifically subserved by certain structures of the left hemisphere, each of which has particular functional attributions. These structures are the posterior parts of the third frontal and first temporal convolutions, and also the supramarginal and angular gyri; they are collectively known as the speech area. This functional lateralization to the left represents the actualization of an innate biological predisposition.

One can teach this and sleep at night. Yet, one then knows that one's teaching does not apply to every member of the human community, which by now, according to UNESCO (1977) projections, should comprise about 4.5 billion persons. There are "exceptions", as one might expect. We will now try to outline a qualitative and quantitative profile of these "exceptions."

Children

It is known that one is usually born with a hefty *planum temporale* on the left side, and that this particular asymmetry has been there since about the end of the seventh (lunar) month of gestation (Tezner 1972; Tezner et al. 1972; Witelson and Pallie 1973). On the other hand, and although not everybody agrees (see Vargha-Khandem and Corbalis 1979), it has also been suggested that a certain degree of functional lateralization to the left already exists at birth in relation to the sounds of language (Entus 1977; Molfese and Molfese 1979; Segalowitz and Chapman 1980). If this is true, one might note that newborns have long been exposed to acoustic stimulations, including linguistic ones, and that the prethalamic acoustic pathway (although not the thalamo-cortical one) shows signs of morphological maturity long before birth (Lecours 1975; Yakovlev and Lecours 1967). Nonetheless, newborn babies do not talk, and wolf-children never really do. One has to be exposed to environmental influences for a while; one has to learn.

Though they might provide and may already have provided fascinating information about the early manifestations of functional lateralization, studies of the behavior of normal human infants have not so far led to new knowledge about the functional genesis of the entity known as the speech area. Most knowledge in this respect has been gathered through studies of childhood aphasia. As pointed out by Woods and Teuber (1978), such studies were first undertaken by Cotard (1868).

Besides showing that there exist major clinical differences between the early and adult forms of aphasia (Alajouanine and Lhermitte 1965), the observation of aphasic children has shown that crossed aphasia following right hemisphere lesions is more frequent in children than in adults. This was interpreted to mean that there is a period at the beginning of one's public life during which lesions of either hemisphere can cause aphasia, and that after this period there is another period during which the right hemisphere can take over linguistic activities rather completely if there are extensive lesions of the left one.

There is disagreement as to the duration of the first of these periods, notwithstanding the fact that it might vary a lot from one child to another. Lenneberg (1967) says it can last as long as 9 years; Hécaen (1976) says as little as 2 years; Woods (1980) goes as far as talking of "the restricted effects of right-hemisphere lesions after age 1." Nor is there agreement on the frequency of crossed aphasia in children: 32% of Basser's (1962) aphasic youngsters were considered to have suffered unilateral right-hemisphere lesions, whereas Woods and Teuber (1978) estimate that less than 10% and probably as little as 5% of the aphasias of childhood result from such lesions. The second period is generally believed to last at least until the age of 8 to 10 or 12, or until puberty.

Whatever one's opinion on the length of these two periods and on the frequency of crossed aphasia during childhood, and if only because the aphasias of children do have a better prognosis than those of adults, one has to agree that research on aphasia has shown that the concept of a speech area does not apply to children and to adults in the same manner. This is no doubt a noteworthy "exception"; according to UNESCO projections, children aged 10 years or less should by now constitute more than a quarter of the total human population.

Non-Right-Handers

Ambidextrals and left-handers constitute another noteworthy "exception." If amytal data obtained by Milner, Branch, and Rasmussen (1964) can be generalized, 64% of them speak with the left hemisphere. Nevertheless, they differ from right-handers in two ways, both suggesting potential differences in cerebral representation of language: Their aphasias are less foreseeable than those of right-handers, clinically and anatomically; and on the whole their aphasias have a better prognosis than those of right-handers (Subirana 1958, 1969; Luria 1970)— especially if they come from families comprising other ambidextrals and left-handers (Brown and Hécaen 1976), as is usually the case. The remaining 36% speak either with the right hemisphere (20%) or with both hemispheres (16%) (Milner, Branch, and Rasmussen 1964), and then they sometimes display intriguing dissociations (Milner, Branch, and Rasmussen 1966).

Now, if the Barcelonians are representative of the species, one out of four human beings should be ambidextral, even if usually right-handed for writing (when literate), and one out of ten should be left-handed (Subirana 1969). This means that the human population should now include between 1.1 and 1.2 billion persons who are more than 10 years of age and whose speech area is not likely to be a standard speech area. And if standards are to be somehow maintained, let us forget about right-handers who belong to families comprising left-handers (Brown and Hécaen 1976).

Illiterates

Another population whose status remains unclear is that of the illiterates. Damasio et al. (1976) have argued that, "since many of the early discoveries of aphasia were made in illiterates," one should not expect to find significant differences between literates and illiterates with regard to the cerebral representation of (oral) language. We do not know exactly what "discoveries" Damasio and his collaborators are talking about, because they do not tell, but we do know that it is often if not usually impossible to tell from the early aphasiogical case reports whether patients were literate or illiterate. By way of exercise, you can try with Leborgne (Broca 1861, 1865). We also know that many more so-called negative cases were reported in Broca's days and till about 1910 than are reported nowadays, so Damasio's argument could be reversed; indeed, one might wonder if exceptional cases were more common when illiteracy was more common.

Let us consider Moutier's dissertation (1908), for instance. When it was published, in the spring of 1908, Moutier was still the faithful and even fervent disciple of Pierre Marie (Lecours and Joanette, in press). Chapters II and III of the dissertation deal with 24 anatomo-clinical case reports. As expected, Moutier begins with three "negative cases," in which no aphasia had occurred although lesions of Broca's area were found at autopsy. The first of these three "negative cases" was that of an illiterate, and possibly the other two as well.[1] Moutier's series comprises only one more case report concerning an illiterate subject; this patient had shown nothing but "transient pure anarthria" following lesions of both Broca's area and Marie's quadrilateral. The remaining 20 anatomo-clinical studies all deal with literate aphasics with left-hemisphere lesions; reading and writing abilities are discussed in all of them.

Ernst Weber (1904) was probably the first to suggest that the acquisition of written language might play a role in the process of functional lateralization for oral language. Critchley (1956), Gorlitzer von Mundy (1957), Eisenson (1964), and Wechsler (1976) have also discussed this issue. A neurologist from Rio de Janeiro, Sergio Novis, told one of us (A.R.L.) that he had seen several illiterate right hemiplegics without aphasia, or with short-lasting aphasia, and a speech therapist from São Paulo, Fernanda P. da Silva, mentioned that it had been easy and rewarding to reeducate within a few months an elderly illiterate right hemiplegic who had had global aphasia.

There are several ways in which one might phrase this problem. Let us try the following: Given that a 5- or 6-year-old child and an adult do not differ much with regard to properly linguistic capacities (not taking encyclopedic knowledge into consideration), that the right hemisphere of a 5- or 6-year-old retains the capacity of taking over language cybernetics in the presence of extensive left-hemisphere lesions, and that it would be loathsome to think that senescence might begin at 5 or 6, what advantage does one's organism seek, after the age of 6, by going further with the process of functional lateralization to the left? Getting more clever at the game of passive transformation and other marginally linguistic practices (Dennis and Kohn 1975; Dennis and Whitaker 1977) would hardly counterbalance the loss of security. The advantage might be a capacity to communicate, influence, lie, and exert power far away in space and time, which until recently could be achieved only by learning to read and write.

One can think of various arguments in this respect. Diachronic studies of childhood aphasia have shown that, if young enough, one can learn or relearn to speak rather well without a left hemisphere; on the other hand, learning to write without a left hemisphere seems to be more problematic (Lawrence 1979; Dennis, Lovett, and Wiegel-Crump 1981). Likewise, it is well known that developmental dyslexia is often associated with nonstandard handedness (Critchley 1964; Hallgren 1950), i.e., lack of a strong functional lateralization to the left. Although they are interesting if one is already interested, arguments of this sort fall short of the mark. One is inclined to come back to data on illiterate adult aphasics.

As far as we know, the problem of aphasia in illiterate adults has been tackled only twice through systematic research. The first study was done by Cameron, Currier, and Haerer (1971) in Mississippi. They considered the cases of 65 patients with right hemiplegia resulting from a left-hemisphere stroke. Sixty-two of these subjects were right-handers and three were left-handers; 37 were "literate," 14 "semi-literate," and 14 "illiterate." Aphasia was considered present even in cases where it had lasted "only a day or two." Twenty-nine of the 37 "literate" patients, 9 of the 14 "semi-literate" patients, and only 5 of the 14 "illiterate" patients were aphasic. Cameron et al. conclude that aphasia following left-hemisphere damage is significantly less frequent in "illiterates" than it is in "literate" subjects and, consequently, that the acquisition of reading and writing is a factor in left brain specialization for language. The second study was done in Portugal by Damasio et al. (1976). They considered "a random series of 247 adults" with unilateral focal brain damage. Vascular disease and tumor are mentioned as possible etiologies. Thirty-eight of the subjects were illiterate (24 with left-hemisphere lesions and 14 with right-hemisphere lesions); the remaining 209 were literate (for some reason, the proportion of lesions of the left and right hemispheres is not given for this group). Twenty-one of the 38 analphabets and 114 of the 209 literate subjects were aphasic. This and other comparisons between their two groups led Damasio et al. to conclude that the aphasias of literate subjects and those of analphabets differ neither as to frequency nor as to "clinical type" and "semiological structure." The latter part of this conclusion can be understood clearly only if one supposes that reading and writing impairments were not considered in the assessment of the 209 literate subjects. The title of the paper accounting for this research is "Brain specialization for language does not depend on literacy."

Now, what we have here is two good papers, founded on serious research, directed at an important problem, and leading to contradictory results and conclusions. In our opinion, the next study should be limited to totally unschooled as well as totally illiterate subjects. Side of lesions should not be a parameter of selection. Above all, the natural evolution of aphasia in illiterates should be taken into consideration, as well as its mere existence. Moreover, it might be wise to exclude tumor cases from such a study, since it is well known that such cases are of little interest and can be misleading in research on functional localization.

Meanwhile, let us substract again. The only statistics that we could find on illiteracy were compiled between 1946 and 1976, and they concern only 158 countries and territories (UNESCO 1977). A conservative estimate founded on these data leads one to conclude that there are now, in these countries and territories, nearly a half-billion right-handed illiterates aged 15 or older.

Polyglots

The first to discuss the problem of cerebral representation of languages in polyglots was perhaps Robert Scoresby-Jackson (1867). His idea was that one's mother tongue is represented in the *pars opercularis* of the third frontal convolution of the left hemisphere, and that other languages are distributed regularly in the *pars triangularis*. The problem of cerebral representation of language in polyglots is reviewed elsewhere in this volume (chapter 10), and we shall therefore limit our own comment to one assertion: Although things are not as simple as Scoresby-Jackson made them two years after Broca's discovery, the notion of a speech area does not apply to bilinguals and polyglots in the same way it applies to unilinguals (Albert and Obler 1978; Galloway 1980; Lebrun 1975; Ojemann and Whitaker 1978; Paradis 1977; Vaid 1983; Vaid and Genese 1980).

We could find no data on the distribution of bilinguals and polyglots throughout the world. On the other hand, we think that they deserve at least a symbolical deputation among our "exceptions." Of course, a fair number of them have already been considered as children, or as ambidextrals, or else as illiterates (since illiteracy is often linked to poverty, which in turn is often linked to a colonial situation and therefore to bilingualism). Be that as it may, we probably remain well within the domain of reality if we further substract 1% of right-handed literates aged 15 or older, nearly 17 million people.

Tone Languages and So Forth

If he was quoted correctly by Gabriel Racle in *Le Monde* of October 19, 1980, Tsunoda (1971, 1978) believes that functional lateralization of oral language to the left is still more redhibitory in the brains of Japanese people living in Japan than it is in those of nearly everybody else. According to Tsunoda, this would, at least in part, be linked to the fact that the Japanese language has its own ways of dealing with vowels. It is indeed a characteristic of this language that it permits the structuration of entirely vocalic words or even sentences; for instance, using only vowels, one can say "ue o ui a o oi ai o ou a iuo" (Racle 1980)—"Disturbed by hunger and disguising his old age, he is in search of love," or something of the sort. Perhaps Tsunoda has identified a pertinent problem.

Other languages also have characteristics that might eventually arouse the interest of aphasiologists. For instance, try to predict lesion localizations and clinical manifestations of agrammatism in speakers of agglutinating languages or in speakers of languages in which grammatical information (at least in certain classes of words) is transmitted through vowels whereas lexical information is borne by consonants. Or else, if you think these are not good examples, consider tone languages. What do we know about the effects of brain lesions in speakers of tone languages? Is it Wernicke's Wernicke's area alone that subserves the discrimination of tonal oppositions? Are there types of aphasia in which tonal discrimination is more specifically involved than other aspects of language production and comprehension? If so, should one expect lesions of some part of the standard speech area to cause an aphasia of this type? Who knows?

It is probably nothing but a coincidence that three out of the nine case reports we could find on aphasia in speakers of tone languages turned out to be cases of crossed aphasia in (bilingual) right-handed adults (Alajouanine et al. 1973; April 1979; April and Han 1980; April and Tse 1977; Hécaen et al. 1971; Lyman, Kwan, and Chao 1938), and this coincidence obviously does not solve the problem—if there is one—of the eventual particularities of the cerebral representation of tone languages. We wrote to a Chinese colleague about this; all he has answered, so far, is that he has seldom seen right-handed aphasics with left hemiplegia. This is comforting, especially if you interpret the word "seldom" in the light of your own experience.

Native speakers of tone languages are indeed numerous. In China alone, according to data gathered by the UNESCO in 1976, there were nearly half a billion, counting only presumed right-handers who were 10 years of age or older. Also, one should not forget that the use of two languages is widespread elsewhere in Asia and in Africa.

Pictographic Writing

As far as we know, Lyman, Kwan, and Chao (1938) were the first (and nearly the last) to report on a case of aphasia in a native speaker of a tone language. It is probable that their paper was also the second to deal with dissociated impairment, as the result of a brain lesion, of pictographic written language. With a left parieto-occipital tumor, Lyman's patient did much better when reading English than when reading Chinese (Mandarin?). If we are not mistaken, the first discussion of such a dissociation in aphasia had been published 4 years earlier, in Japan, by K. Kimura (1934). Most of what is now known about the effects of brain lesions on pictographic writing has been learned through the research led during the early 1970s by Sasanuma and her collaborators (Sasanuma 1975; Sasanuma and Fujimura 1971, 1972; Sasanuma and Momoi 1975). Sasanuma's studies have provided fascinating data on more or less selective impairments of Kana and Kanji processing as the result of various focal brain lesions. The only point we will raise here is that of localization. According to Sasanuma (1975), lesions of the anatomo-functional complex comprising Broca's area, Wernicke's area, "and perhaps the *fasciculus arcuatus*" might be the cause of the phonological disturbance underlying Kana impairment, whereas Kanji impairment, which reflects a disintegration of "the nonphonological aspects of language," might be due to lesions "outside" of "*the* speech area." Concerning this "outside," Sasanuma and Momoi (1975) cautiously tell about left-hemisphere lesions involving the deep white matter "of the second and of the third temporal convolutions" and suggest the possibility of a caudal extension to the "parietal region."

Back to UNESCO (1977) statistics: Without counting children (aged 10 or younger) or illiterates (of whom there are very few in Japan— hardly more than a million in 1976), the Japanese population comprised nearly 58 million presumably right-handed users of Kana and Kanji.[2] This also represents a noteworthy "exception" with reference to the standard Western European–North American speech area.

Where Is the Speech Area, and Who Has Seen It?

All things considered, and although we are obviously not suggesting that it is not a rule within the human species that one should speak mostly with one's left hemisphere, it might be that traditional teaching about the exact localization of *the* speech area applies, strictly, only to a little more than a billion people, all of them adult unilingual literate right-handers speaking a language without tonal oppositions and writing it in a syllabic or alphabetical code. This represents hardly more than a quarter of the human population.[3] Not taking into consideration the long-known (although often underestimated) "exceptions" of children and non-right-handers, traditional views on the speech area probably depend in part on a historical accident, that is, on the geographical situation of those who first "saw" the speech area and wrote about it (Lecours 1980; Lieff Benderly 1981).

Further Exceptions

The last question to be raised is of course the following: How frequent are exceptions to the prevalent doctrine among right-handed literate adult unilingual speaker-writers of a nontonal, nonpictographic language?

Let us begin with those having a dominantly right-hemisphere representation of language, about whom knowledge has been gathered mostly through anatomo-clinical studies of crossed aphasia (Hécaen 1976; Joanette 1980) but also through use of the Wada test (Milner et al. 1964). It seems that there are very few of them, at least within the populations that have been significantly studied so far; indeed, they probably do not exceed 1% of the right-handers (Hécaen et al. 1971; Zangwill 1960). However, we have all noted (no doubt in relation to the generalization of tomodensitometric studies) a recent increase in case reports of crossed aphasia in dextrals; one can think of nearly a dozen new reports published in the two years preceding the Montreal Symposium of May 1981 (Carr, Jakobson, and Boller 1981; Denes and Caviezel 1981; Hyodo et al. 1979; Ikeda et al. 1979; Pillon, Dési, and Lhermitte 1979; Trojanowski, Green, and Levine 1980; Yarnell 1981). To this dozen, one might add a group of five (as yet unreported) right-handed adult aphasics with unilateral right-hemisphere lesions (0.85% of 587 patients with unilateral brain lesions observed in Milano, at the Centro di Neuropsicologia, Clinica delle Malattie Nervose e Mentali) and the most remarkable case that is reported in the second appendix to this chapter. In spite of this apparent upsurge, our impression is that such cases will remain the really exceptional exceptions, as they have always been. One can offer no satisfactory "explanation" of such exceptions; however, in view of the magnificent work of Patricia Goldman Bakič on the anatomical plasticity of the fetal rhesus monkey brain (Goldman 1978; Goldman and Galkin 1978), one might wish to evoke the possibility of prenatal left-brain lesions followed by fully adequate anatomo-functional reorganization.

There are also the exceptions that most of us tend to forget about. By this, we mean those patients with left-hemisphere lesions in localizations that are unexpected in view of their clinical disorders. For example, Lhermitte et al. (1973) reported on a standard right-handed adult who managed to produce a very fluent neologistic jargon in spite of lesions destroying, besides his Wernicke's area, both his Broca's area and its homologue in the right hemisphere. The case of one of the two anarthric patients discussed in the second appendix to this chapter (Mrs. G.) raises the same problem, although less dramatically.

In relation to this problem, the two of us who work in Milan (A. B. and S. M.) reviewed the tomodensitometric examinations of 576 patients with clear clinical evidence of left-brain lesions. Because we were aiming at identifying "exceptions" to the standard doctrine, which we believe has always been and still is mostly grounded on the observation of adults with unilateral left cerebro-vascular lesions, we excluded all subjects with nonvascular disorders. We also excluded subjects with bilateral brain damage (including massive cortical atrophy), subjects with clinical

or radiological evidence of more than one left-hemisphere stroke, subjects whose tomodensitometric data had been obtained less than 15
days after their stroke, and subjects with left-hand dominance as assessed
by means of the Edinburgh inventory (Oldfield 1971). We were thus
left with the tomodensitometric examinations of 267 right-handed adults
with unambiguous clinical evidence of a focal vascular lesion confined
to the left cerebral hemisphere. The notion of "deep anomaly" was
arbitrarily defined as including all cases in which tomodensitometric
anomalies could only be seen, on standard CT pictures, at 5 millimeters
from the cortical surface or deeper (which, of course, is not meant to
suggest that there necessarily was no cortical damage in these cases);
the notion of "cortico-subcortical anomaly" is self-explicit. Lesions
were mapped following the method described by Luzzatti, Scotti, and
Gattoni (1979).

Of the 267 scans, 16 showed no anomaly whatsoever (group A), 55
showed deep anomalies (group B), 62 showed prerolandix and retrorolandic (group C), 52 prerolandic (group D), and 82 retrorolandic
(group E) cortico-subcortical anamolies.

The 16 subjects of group A comprised 10 patients without clinical
evidence of aphasia as well as 1 nonfluent and 5 fluent aphasics; the
main interest of this group is that it witnesses to the limits of our
tomodensitometric methodology (Mori et al. 1977). The 55 subjects
of group B comprised 29 patients without clinical evidence of aphasia
as well as 8 nonfluent and 18 fluent aphasics; this group is of lesser
interest in the present context and will not be discussed further.

In light of the question that we have raised, 33 of the 267 scans
(12.3%), all from patients of groups C, D, and E, are of particular
interest because their tomodensitometric data were grossly different
from what one would have expected given the standard anatomo-clinical
doctrine of aphasia. In summary: Among the 62 patients of group C,
4 displayed the clinical picture of (very) mild Broca's aphasia in spite
of huge pre- and retrorolandic tomodensitometric anomalies. Among
the 52 patients of group D, those with cortico-subcortical anomalies
restricted to the prerolandic region, 7 displayed the clinical picture of
Wernicke's aphasia, 1 that of conduction aphasia, 2 that of transcortical
sensory aphasia, and 4 that of global phasia, all of them with tomodensitometric involvement of Broca's area itself and tomodensitometric
sparing of the retrorolandic components of the speech area, for a total
of 14 "exceptions" (26.9%) (not taking into account a few subjects with
focal prerolandic tomodensitometric anomalies of the speech area but

without clinical manifestations of aphasia). Among the 82 patients of group E, those with cortico-subcortical anomalies of the retrorolandic region, 8 displayed the clinical picture of nonfluent (Broca's) aphasia and 1 that of global aphasia, all of them with tomodensitometric involvement of Wernicke's area itself and/or of the left inferior parietal lobule and sparing of the prerolandic components of the speech area, for a total of nine "exceptions" (11%) (not taking into account a few subjects with focal retrorolandic tomodensitometric anomalies of the speech area but without clinical manifestations of aphasia. Finally, there were 6 patients of group C, D, or E who showed sizable tomodensitometric involvement of some part of the cortical speech area but whose results on our standard aphasia test battery were normal, although all 6 patients stated, when asked, that their disease had indeed been the cause of subtle undefined linguistic difficulties.

Conclusion

It can no longer be doubted that Gall was right when he suggested in 1825 that the human brain is not a functionally homogeneous organ distributing "vital energy" with sovereign indifference. It also can no longer be doubted that the anatomo-clinical method has been and will long remain productive, and that one of the notions it has yielded—that of a "speech area"—is valid. Nonetheless, it is now getting more and more obvious that the biological, psychological, and social determinants of functional specialization interact—in normal or pathological circumstances—in a most complex fashion, and that standard contemporary teaching about the speech area and the mutual relationships of brain and language does not apply uniformly to the whole of mankind. Cheer up! A lot of work remains to be done in our field.

Notes

1. No information concerning written language is given for the second patient, which constitutes an exception among Moutier's personal case reports; the third patient was glaucomatous and blind when Moutier saw him, and the reader is not told about premorbid literacy of illiteracy.

2. Another thing about the Japanese population is that, as pointed out by Makita (1968), the incidence of developmental reading disabilities among Japanese children is about 10 times lower than among occidental children.

3. As thoroughly reviewed by McGlone (1980), a fair number of recent publications have raised the possibility of appreciable sex differences in the cerebral representation of language. For instance, D. Kimura (1981) suggests that the

left prerolandic region might be of greater importance in women than in men with regard to language. One might therefore be tempted to further divide the remaining billion by half. For the time being, we have decided to resist this temptation because, on the one hand, we have often observed jargonaphasic women (with retrorolandic lesions of the left hemisphere) and, on the other hand and mostly, we are ready neither to define two standard speech areas nor to choose whether women or men should be attributed the one standard in case of conflict.

References

Alajouanine, T., M. P. Cathala, J. Métellus, S. Siksou, V. Alleton, F. Cheng, C. de Turckeim, and M. C. Chang. 1973. La problématique de l'aphasie dans les langues à écriture non alphabétique. *Revue Neurologique* 128: 229–243.

Alajouanine, T., and F. Lhermitte. 1965. Acquired aphasia in children. *Brain* 88: 653–662.

Albert, M., and L. Obler. 1978. *The Bilingual Brain.* New York: Academic.

April, R. S. 1979. Concepts actuels sur l'organisation cérébrale du langage à partir de quelques cas d'aphasie croisée chez les orientaux bilingues. *Revue Neurologique* 135: 375–378.

April, R. S., and M. Han. 1980. Crossed aphasia in a right handed bilingual Chinese man: A second case. *Archives of Neurology* 37: 342–348.

April, R. S., and P. C. Tse. 1977. Crossed aphasia in a Chinese bilingual dextral. *Archives of Neurology* 34: 766–770.

Basser, L. S. 1962. Hemiplegia of early onset and the faculty of speech with special reference to the effects of hemispherectomy. *Brain* 85: 427–460.

Bogen, J. E., and G. M. Bogen. 1976. Wernicke's region—Where is it? *Annals of the New York Academy of Sciences* 280: 834–843.

Bouillaud, J. B. 1825. Recherches cliniques propres à démontrer que la perte de la parole correspond à la lésion des lobules antérieurs du cerveau et à confirmer l'opinion de M. Gall, sur le siège de l'organe du langage articulé. *Archives Générales de Médecine* 8: 25.

Broca, P. 1861. Nouvelle observation d'aphémie produite par une lésion de la moitié postérieure des deuxième et troisième circonvolutions frontales. *Bulletin de la Société Anatomique* 6: 337–393.

Broca, P. 1865. Sur le siège de la faculté du langage articulé. *Bulletin de la Société d'Anthropologie* 6: 337–393.

Brown, J. W., and H. Hécaen. 1976. Lateralization and language representation. *Neurology* 26: 183–189.

Cameron, R. F., R. D. Currier, and A. F. Haerer. 1971. Aphasia and literacy. *British Journal of Disorders of Communication* 6: 161–163.

Carr, M. S., T. Jakobson, and F. Boller. 1981. Crossed aphasia: Analysis of four cases. *Brain and Language* 14: 190–202.

Cotard, J. 1868. Etude sur l'atrophie partielle du cerveau. Thesis, Faculté de Médecine, Paris.

Critchley, M. 1956. Premorbid literacy and the pattern of subsequent aphasia. *Proceedings of the Society of Medicine* 49: 335–336.

Critchley, M. 1964. *Developmental Dyslexia.* London: Heinemann.

Critchley, M. 1980. *The Divine Banquet of the Brain.* New York: Raven.

Damasio, A. R., A. Castro-Caldas, A. Grosso, and J. M. Ferro. 1976. Brain specialization for language does not depend on literacy. *Archives of Neurology* 33: 300–301.

Damasio, A. R., and H. Damasio. 1981. The Subcortical Aphasias: Syndrome and Mechanism. Communication to 19th Annual Meeting of Academy of Aphasia, London, Ontario.

Dax, M. 1836 (as printed in 1865). Lésions de la moitié de l'encéphale coïncidant avec l'oubli des signes de la pensée. *Gazette Hebdomadaire de Médecine et de Chirurgie* 32: 259–269.

Dejerine, M. J. 1906. L'aphasie sensorielle. La localisation et sa physiologie pathologique. *Presse Médicale* 55: 437–439.

Dejerine, M. J. 1914. *Séméiologie des affections du système nerveux.* Paris: Masson.

Dejerine-Klumpke, A. 1908. In Deuxième discussion sur l'aphasie. *Revue Neurologique* 16: 974–1023.

Denes, M., and F. Caviezel. 1981. Dichotic listening in crossed aphasia. Paradoxical ipsilateral suppression. *Archives of Neurology* 38: 182–185.

Dennis, M., and B. Kohn. 1975. Comprehension of syntax in infantile hemiplegies after cerebral hemidecortication: Left hemisphere superiority. *Brain and Language* 2: 475–486.

Dennis, M., M. Lovett, and C. A. Wiegel-Crump. 1981. Written language acquisition after left or right hemidecortication in infancy. *Brain and Language* 1: 54–91.

Dennis, M., and H. Whitaker. 1977. Hemispheric equipotentiality and language acquisition. In S. J. Segalowitz and F. A. Gruber, eds., *Language Development and Neurological Theory.* New York: Academic.

Eberstaller, O. 1884. Zur Oberflächenanatomie der Grosshirnhemisphären. *Wiener Medizinische Blätter* 7: 479–482, 542–582, 644–646.

Eidelberg, D., and A. M. Galaburda. 1982. Symmetry and asymmetry in the human posterior thalamus. II: Cytoarchitectonic analysis in normals. *Archives of Neurology* 39: 325–332.

Eisenson, J. 1964. Discussion. In A.V.S. de Reuck and M. O'Connor, eds., *Disorders of Language.* London: Churchill.

Elghosi, D. 1977. Discussion sur le rôle du thalamus dans l'activité du langage. Thesis, Faculté de Médecine, Paris.

Entus, A. K. 1977. Hemispheric asymmetry in processing of dichotically presented speech and nonspeech stimuli by infants. In S. J. Segalowitz and F. A. Gruber, eds., *Language Development and Neurological Theory*. New York: Academic.

Freud, S. 1891. *Zur Auffassung der Aphasien*. Vienna: Deuticke.

Galaburda, A. 1980. La région de Broca: Observations anatomiques faites un siècle après la mort de son découvreur. *Revue Neurologique* 136: 609–616.

Galaburda, A. M., M. Le May, T. L. Kemper, and N. Geschwind. 1978. Right-left asymmetries in the brain. *Reprint Series* 199: 852–856.

Galaburda, A., and F. Sanides. 1980. Cytoarchitectonic organization of the human auditory cortex. *Journal of Comparative Neurology* 190: 597–610.

Galaburda, A., F. Sanides, and N. Geschwind. 1978. Cytoarchitectonic left-right asymmetries in the temporal speech region. *Archives of Neurology* 35: 812–817.

Gall, F. J. 1825. *Sur les fonctions due cerveau et sur celles de chacune de ses parties*. Paris: Baillière.

Gall, F. J., and Shurzheim. 1810–1819. *Anatomie et physiologie du système nerveux en général et du cerveau en particulier*. Paris: Schoell.

Galloway, L. 1980. Towards a neuropsychological model of bilingualism and second language performance. In M. Long, S. Peck, and K. Bailey, eds., *Research in Second Language Acquisition*. Newbury, Mass.: Rowley.

Geschwind, N. 1965. Disconnexion syndromes in animals and man. Part II. *Brain* 88: 585–644.

Geschwind, N., and W. Levitsky, 1968. Human brain: Left-right asymmetries in the temporal speech region. *Science* 161: 186–187.

Goldman, P. S. 1978. Neuronal plasticity in primate telencephalon: Anomalous projections induced by prenatal removal of frontal cortex. *Reprint Series* 202: 768–770.

Goldman, P. S., and T. W. Galkin. 1978. Prenatal removal of frontal association in the fetal rhesus monkey: Anatomical and functional consequences in postnatal life. *Brain Research* 152: 451–485.

Goldstein, K. 1948. *Language and Language Disturbances*. New York: Grune and Stratton.

Gorlitzer von Mundy, V. 1957. Zur Frage der paarig veranlagten Sprachzentren. *Nervenarzt* 28: 212–216.

Gratiolet, L. P. 1854. *Mémoire su les plis cérébraux de l'homme et des primates*. Paris: Bertrand.

Hallgran, B. 1950. Specific dyslexia (congenital word-blindness): A clinical and genetic study. *Acta Psychiatrica et Neurologica Scandinavica* suppl. 65.

Hécaen, H. 1976. Acquired aphasia in children and the ontogenesis of hemispheric functional specialization. *Brain and Language* 3: 114–134.

Hécaen, H., and J. Dubois. 1969. *La naissance de la neuropsychologie du langage*. Paris: Flammarion.

Hécaen, H., G. Mazars, A. M. Ramier, M. C. Goldblum, and L. Mérienne. 1971. Aphasie croisée chez un sujet droitier bilingue. *Revue Neurologique* 126: 319–323.

Hyodo, A., Y. Maki, K. Nakagaura, T. Enomoto, and H. Akimoto. 1979. Computed tomography in crossed aphasia. *Neurological Surgery* 7: 791–796.

Ikeda, Y., M. Higuchi, M. Chin, T. Skimura, and S. Nakazawa. 1979. A case of thalamic germinoma with crossed aphasia in a dextral. *Neurological Surgery* 7: 859–864.

Joanette, Y. 1980. Contribution à l'étude anatomo-clinique des troubles du langage dans les lésions cérébrales de droitier. Ph.D. diss., Université de Montréal.

Kakeshita, T. 1925. Zur Anatomie der operkularen temporalen Region (Vergleichende Untersuchungen der rechten und linken Seite). *Arbeitsblatt des Neurologischen Instituts Wien* 27: 292–298.

Kimura, D. 1981. Sex Differences in Speech Organization within the Left Hemisphere. Research Bulletin 548, Department of Psychology, University of Western Ontario.

Kimura, K. 1934. Aphasia: Characteristic symptoms in Japanese. *Journal of Psychiatric Neurology* 37: 437–459.

Ladame, P. L. 1902. La question de l'aphasie motrice sous-corticale. *Revue Neurologique* 10: 13–18.

Lawrence, B. 1979. Writing patterns of the congenitally neurologically impaired adult. *Brain and Language* 7: 252–263.

Lebrun, Y. 1975. The neurology of bilingualism. *Child Language* 27: 179–186.

Lecours, A. R. 1975. Myelinogenetic correlates of the development of speech and language. In E. Lenneberg and E. Lenneberg, eds., *Foundations of Language Development: A Multidisciplinary Approach*, vol. 1. New York: Academic.

Lecours, A. R. 1980. Corrélations anatomo-cliniques de l'aphasie: la zone du langage. *Revue Neurologique* 136: 591–608.

Lecours, A. R., D. Caplan, and F. Lhermitte. 1979. Zone du langage. In A. R. Lecours and F. Lhermitte, eds., *L'aphasie*. Paris, Flammarion.

Lecours, A. R., and Y. Joanette. In press. François Moutier or "Folds to Filds." In H. Whitaker, B. Joynt, and S. Greenblatt, eds., *Historical Studies in Neuropsychology and Neurolinguistics*. New York: Academic.

Lecours, A. R., and F. Lhermitte. 1976. The "pure form" of the phonetic disintegration syndrome (pure anarthria): Anatomo-clinical report of a historical case. *Brain and Language* 3: 88–113.

Lemay, M., and A. Culebras. 1972. Human brain. Morphologic differences in the hemispheres demonstrable by carotid arteriography. *New England Journal of Medicine* 287: 168–170.

Lenneberg, E. H. 1967. *Biological Foundations of Language.* New York: Wiley.

Levine, D. M., and E. Sweet. 1982. The Neuropathological basis of aphasia and its implications for the cerebral control of speech. In M. Arbib, D. Caplan, and J. Marshall, eds., *Neural Models of Language Process* New York: Academic.

Lhermitte, F., F. Chain, D. Aron, M. Leblanc, and O. Souty. 1969. Les troubles de la vision des couleurs dans les lésions postérieures du cerveau. *Revue Neurologique* 121: 5–29.

Lhermitte, F., A. R. Lecours, B. Ducarne, and R. Escourolle. 1973. Unexpected anatomical findings in a case of fluent jargon aphasia. *Cortex* 9: 433–446.

Lieff Benderly, B. 1981. The multilingual mind. *Psychology Today,* March: 9–12.

Luria, A. R. 1964. Factors and forms of aphasia. In A. V. S. de Reuk and M. O'Connor, eds., *Disorders of Language.* London: Churchill.

Luria, A. R. 1970. *Traumatic Aphasia.* The Hague: Mouton.

Luzzati, C., G. Scotti, and A. Gattoni. 1979. Further suggestions for cerebral CT-localization. *Cortex* 15: 483–490.

Lyman, R. S., S. T. Kwan, and W. H. Chao. 1938. Left occipito-parietal brain tumor with observation on alexia and agraphia in Chinese and English. *Chinese Medical Journal* 54: 491–516.

Makita, K. 1968. The Rarity of Reading Disability in Japanese Children. Communication to Seventh Annual Meeting of Japanese Child Psychiatric Association, Kyoto.

Marie, P. 1906a. Révision de la question de l'aphasie: l'aphasie de 1861 à 1866: essai de critique historique sur la genèse de la doctrine de Broca. In *Travaux et Mémoires,* vol. 1. Paris: Masson.

Marie, P. 1906b. Révision de la question de l'aphasie: la troisième circonvolution frontale gauche ne joue aucun rôle spécial dans la fonction du langage. In *Travaux et Mémoires,* vol. 1. Paris: Masson.

McGlone, J. 1980. Sex differences in human brain asymmetry: A critical survey. *Behavior and Brain Sciences* 3: 215–263.

Milner, B., C. Branch, and T. Rasmussen. 1964. Observations on cerebral dominance. In A. V. S. de Reuch and M. O'Connor, eds., *Disorders of Language.* London: Churchill.

Milner, B., C. Branch, and T. Rasmussen. 1966. Evidence for bilateral speech representation in some non-right handers. *Transactions of the American Neurological Association* 91: 306–309.

Miraglia, S. M. 1979. Contribuçâo ao estudo das variaçâes anatômicas dos sulcos e giros do lôbulo parietal inferior no homen. Thesis, Escola Paulista de Medicina de São Paulo.

Mohr, J. P., W. C. Watters, and G. W. Duncan. 1975. Thalamic hemorrhage and aphasia. *Brain and Language* 2: 3–17.

Molfese, D. L., and V. J. Molfese. 1979. Hemisphere and stimulus differences as reflected in the cortical response of newborn infants to speech stimuli. *Development Psychology* 15: 505–511.

Mori, H., C. H. Lu, L. C. Chiu, P. A. Cancilla, and J. H. Christie. 1977. Reliability of computed tomography: Correlation with neuropathologic findings. *American Journal of Roentgenology* 128: 795–798.

Moutier, F. 1908. *L'aphasie de Broca*. Paris: Steinheil.

Ojemann, G. A., and H. Whitaker. 1978. The bilingual. *Archives of Neurology* 35: 409–412.

Oldfield, R. C. 1971. The assessment and analysis of handedness: The Edinburgh inventory. *Neuropsychologia* 9: 97–113.

Paradis, M. 1977. Bilingualism in aphasia. In H. Whitaker and H. Whitaker, eds., *Studies in Neurolinguistics*, vol. 3. New York: Academic.

Pfeiffer, R. A. 1936. Pathologie der Hörstrahlung und der Corticalen Hörsphäre. In O. Bumke and O. Foerster, eds., *Handbuch der Neurologie*, vol. 6.

Pillon, B., M. Dési, and F. Lhermitte. 1979. Deux cas d'aphasie croisée avec jargonagraphie chez des droitiers. *Revue Neurologique* 135: 15–30.

Poncet, M., A. Ali Cherif, and M. Brouchon. 1978. Néologismes: rôle du cortex pariétal dans le contrôle moteur des organes bucco-phonateurs. Manuscript, Departement de Neuropsychologie et de Rééducation du Langage, Centre Hospitalier Universitaire de la Timone, Marseille.

Racle, G. 1980. Les japonais parlent à gauche. . . . *Le Monde*, October 19, p. 17.

Ratcliff, G., C. Dila, L. Taylor, and B. Milner. 1980. The morphological asymmetry of the hemisphere and cerebral dominance for speech: A possible relationship. *Brain and Language* 11: 87–98.

Rubens, A. B., J. T. Mahowald, and Hutton. 1976. Asymmetry of the lateral (Sylvian) fissures in man. *Neurology* 26: 620–624.

Sasanuma, S. 1975. Kana and Kanji processing in Japanese aphasics. *Brain and Language* 2: 360–383.

Sasanuma, S., and O. Fujimura. 1971. Selective impairment of phonetic and non-phonetic transcriptions of words in Japanese aphasic patients: Kana versus Kanji in visual recognition and writing. *Cortex* 7: 1–18.

Sasanuma, S., and O. Fujimura. 1972. An analysis of writing errors in Japanese aphasic patients: Kanji versus Kana words. *Cortex* 8: 265–282.

Sasanuma, S., and H. Momoi. 1975. The syndrome of Golgi (word-meaning) aphasia. *Neurology* 25: 627–632.

Scheibel, A. B. 1980. Cooperativity in brain function: Assemblies of 30 neurons. *Abstracts*, 6.

Scheibel, A. B. 1981. Structural specifications of cortical speech areas: A Golgi study. Communication to 19th Annual Meeting of the Academy of Aphasia, London, Ontario.

Scoresby-Jackson, R. E. 1867. Case of aphasia with right hemiplegia. *Edinburgh Medical Journal* 12: 696–706.

Segalowitz, S. J., and J. S. Chapman. 1980. Cerebral asymmetry for speech in neonates: A behavioral measure. *Brain and Language* 9: 281–288.

Subirana, A. 1958. The prognosis in aphasia in relation to cerebral dominance and handedness. *Brain* 81: 415–425.

Subirana, A. 1969. Handedness and cerebral dominance. In P. J. Vinken and G. W. Bruyn, eds., *Handbook of Clinical Neurology*, vol. 4. Amsterdam: North-Holland.

Tezner, D. 1972. Etude anatomique de l'asymétrie droite-gauche du planum temporale sur cent cerveaux adultes. Thesis, Faculté de Médecine de Paris.

Tezner, D., A. Tzavaras, J. Gruner, and H. Hécaen. 1972. L'asymétrie droite-gauche du planum temporale: à propos de l'étude anatomique de cent cerveaux. *Revue Neurologique* 126: 444–449.

Trojanowski, J. Q., R. C. Green, and D. N. Levine. 1980. Crossed aphasia in a dextral: A clinicopathological study. *Neurology* 30: 709–713.

Tsunoda, T. 1971. The difference of the cerebral dominance of vowel sounds among different languages. *Journal of Auditory Research* 11: 305–314.

Tsunoda, T. 1978. The left cerebral hemisphere and the Japanese language. *Japan Foundation Newsletter* 6: 1.

UNESCO. 1977. *Statistical Yearbook.*

Vaid, J. 1983. Bilingualism and brain lateralization. In S. Segalowitz, ed., *Language Function and Brain Organization*. New York: Academic.

Vaid, J., and Genese. 1980. Neuropsychological approaches to bilingualism: A critical review. *Canadian Journal of Psychology* 34: 417–445.

Vargha-Khadem, F., and M. Corbalis. 1979. Cerebral asymmetry in infants. *Brain and Language* 8: 1–9.

von Economo, C., and L. Horn. 1930. Neberwindungsrelief, Masse und Rindenarchitecktonik der Supratemporälfläche, ihre individuellen und ihre Seitenunteschiede. *Zeitschrift Neurologische Psychiatrie.*

Wada, J. A., R. Clarke, and A. Hamm. 1975. Cerebral hemispheric asymmetry in humans. *Archives of Neurology* 32: 239–246.

Wada, J., and T. Rasmussen. 1960. Intracarotid injection of sodium amytal for the lateralization of cerebral speech dominance. *Journal of Neurosurgery* 17: 266–282.

Weber, E. 1904. Das Schreiben als Ursache der einseitigen Lage des Sprach-zentrums. *Zentralblatt für Physiologie* 18: 341–347.

Wechsler, A. F. 1976. Crossed aphasia in an illiterate dextral. *Brain and Language* 3: 164–172.

Wernicke, C. 1874. *Der aphasische Symptomenkomplex.* Breslau: Cohn and Weigert.

Wernicke, C. 1881. *Lehrbuch der Gehirnkrankheiten.* Kassel: Fischer.

Witelson, S. F. 1977. Anatomic asymmetry in the temporal lobes: Its documentation, phylogenesis and relationships to functional asymmetry. *Annals of the New York Academy of Sciences* 299: 328–354.

Witelson, S. F., and W. Pallie. 1973. Left hemisphere specialization for language in the newborn. *Brain* 96: 641–646.

Woods, B. T. 1980. The restricted effects of right-hemisphere lesions after age one: Wechsler test data. *Neuropsychologia* 18: 65–70.

Woods, B. T., and H. L. Teuber. 1978. Changing patterns of childhood aphasia. *Annals of Neurology* 3: 273–280.

Yakovlev, P. I., and A. R. Lecours. 1967. The myelogenetic cycles of regional maturation of the brain. In A. Minkowski, ed., *Regional Development of the Brain in Early Life.* Oxford: Blackwell.

Yarnell, P. R. 1981. Crossed dextral aphasia: A clinical radiological correlation. *Brain and Language* 12: 128–139.

Zangwill, O. 1960. *Cerebral Dominance and its Relation to Psychological Functioning.* London: Oliver and Boyd.

Crossed Aphasia in a Standard Dextral

Michele Puel
André Rascol
Alain Bonafe
André Roch Lecours

In chapter 11 the classical conceptions about the left-hemisphere cortical components of the speech area were said to have been grounded in the study of the effects of focal brain lesions in "standard" right-handed, unilingual, literate adults whose mother tongue was neither a tonal or agglutinating one when spoken nor a pictographic one when written. From recent UNESCO statistics (1977), one can figure out that standard subjects of this sort constitute about 25 percent of the human population. We will now report on the clinical and tomodensitometric observation of a patient who clearly did not function as one with a standard speech area although he had all of the above characteristics of standard subjects.

Doctor B, a radiologist, is now 58 years old. His birth was normal and, as far as can be known, took place at the end of a normal gestation. He acquired spoken and written French in due time and without developmental disorders. He considers himself an absolute right-hander, which is supported by a score of 100 on the Edinburgh Inventory (Oldfield 1971). He reports no family history of ambidexterity or left-handedness. Although he was exposed to Egyptian while an infant and to Greek and Latin in the course of his studies in the humanities, Doctor B is functionally unilingual for both spoken and written language. He is normotensive and free of other risk factors. He was also free of neurological disease until the age of 55.

When he woke up on the morning of June 29, 1979, Doctor B was fully conscious but he had become speechless and he had a left hemiplegia, a left somato-sensory deficit, and a left homonymous hemianopia. He was hospitalized a few hours later. Angiography showed a total thrombosis of his right internal carotid artery; his left carotid artery and its branches were found to be radiologically normal.

At the onset of his disease, Doctor B remained speechless for about 24 hours. His aphasia thereafter evolved in two phases, with the first one progressively merging into the second after 2 months or so. The characteristics of the first phase can be summarized as follows.

Spontaneous speech was nonfluent and typically agrammatic, with numerous pauses and hesitations. There was no phonetic disintegration (Alajouanine, Ombredane, and Durand 1939), although buccofacial apraxia was present. The reduced speech output comprised relatively numerous phonemic deviations (Lecours and Lhermitte 1969). Quotation 1, for instance, was the patient's answer, about a week after his stroke, when required to describe his activity of the previous day, and quotation 2 was his attempt at telling about his wife's work as a laboratory physician. (Ellipses indicate relatively long pauses.)

1. Oh, rien. . . . Dormir. . . . Dormir et puis, après, euh. . . . /telefœr/. Et puis. (Oh, nothing. . . . To sleep. . . . To sleep and then, after, /telefœr/. And that's that.)

2. Faire des prises de sang . . . pour savoir le sucre, l'urée. . . . Compter globules rouges, compter globules blancs. (To take blood samples . . . to know the sugar, the urea. . . . To count red cells. To count white cells.)

In quotation 1, /telefœr/ was an obvious phonemic paraphasia, with the target word being *téléphone*, /telefɔn/, i.e., telephone. Likewise, when further asked about the tests usually made in a hospital laboratory, the patient transformed *prothrombine*, /protrɔ̃bin/, i.e., prothrombin, into /loteplɔbir/.

Repetition and reading aloud were abnormal in that they frequently led to the production of phonemic paraphasias.

Naming tests with photographs of common objects revealed moderate word-finding difficulties.

Written expression was fluent and heavily jargonagraphic from the second day of evolution, both in spontaneous production and on dictation. Literal paragraphias, neologisms, and written paragrammatic deviations were frequent. Graphism proper was normal. Quotations 3 and 4 illustrate the patient's spontaneous writing abilities on July 1, 1979:

3. Monelle doit être qu'à devez ou Toulouse par le train de 1er juillet 79. (Monelle must be who at Toulouse by the train of first July 79.)

4. Les infirmières sont transfavorables[1] sous la pysetopente. (The nurses are transfavorable under the pysetopente.)

Written copy was normal in all respects.

But for complex commands, especially those requiring the integration of personal pronouns and morphemic markers of time, comprehension of oral and written language was normal. Multiple-choice pointing and pairing tasks were done normally for single nouns, for sentences comprising two nouns and a preposition such as behind, above, under, etc., and for more complex sentences such as "The horse pulls the boy." (The latter was presented with four drawings in which the same boy and horse were shown in different mutual relationships.)

The clinical changes that led to the second phase of Doctor B's aphasia were the following:

Spontaneous speech became more and more fluent. Agrammatic behavior disappeared gradually, and paragrammatic deviations became progressively more numerous in spontaneous speech. The production of phonemic deviations increased in direct proportion to the increase in fluency; indeed, there came a point when Doctor B's discourse qualified as "phonemic jargon" (Lecours et al. 1981). Quotation 5 illustrates these points; it is excerpted from the patient's narration of "Little Red Riding-Hood" 2 months after his stroke. Arrows indicate moments when the patient changed the program of a sentence as the result of word-finding difficulties. In the English translation, obvious phonemic paraphasias are represented by their corresponding target words set in italics, whereas other neologisms are transcribed as such following the IPA convention.

5. C'est le loup. C'est le loup qui est ↓ qui /daratɔ̃/. Alors, c'est ↓—c'est difficile, hein!—c'est un ↓ disons, un petit ↓ un petit fils qui doit / trãte/ à la tante /malax/ des /mjez/ ↓ des /kɔ̃sifi/ qui sont faits par la /mene/. Et elle s'/amiz/. Elle met une /tamiz/ /rus/ ↓ un /tamɔrs/ avec des ↓ . . . . (It is the wolf. It is the wolf who is ↓ who /daratɔ̃/. Then, it is ↓ —it is difficult, eh!—it is a ↓ let's say a small ↓ a grandson who has to bring to the *sick* aunt some *honey* ↓ some *jams* which are made by the *mother*. And she *plays around* She puts on a *red shirt* ↓ a /tamɔrs/ with some ↓ . . . .)

Written expression remained dyssyntactic and dysorthographic, but the production of written neologisms gradually disappeared as the patient's productions in this respect became proportionally more effective as a communication tool. Dysorthographia itself disappeared progressively at a later stage. Quotation 6 is excerpted from a letter that Doctor B wrote to his speech therapist while vacationing in Nice 11 months after his stroke.

6. Je vous écrire que Nice et je souhaite que la lettre vous trouver en bonne santé. Le voyage s'est bien passé, un peu fatigué plus dans le

train que dans l'auto. Il est un soleil à Juan, mais un gros orage à
Toulon. . . .) (I to write you that Nice and I hope that the letter to
find you in good health. Everything was all right during the trip, a
little tired more in the train than in the car. It is a sun in Juan, but
a big storm in Toulon. . . .)

Graphism proper and written copy remained normal in all respects.

Although some overall improvement took place gradually after a
year of evolution, there were only minimal changes with regard to
repetition, reading aloud, and comprehension of oral as well as of written
language.

Tomodensitometric studies revealed a massive infarction involving
nearly all of the territory of the patient's right middle cerebral artery—
anterior and posterior, superficial and deep. The involved cortical territory comprises the pars opercularis and pars triangularis of the third
frontal convolution as well as the adjacent portion of the second, the
opercular halves of the pre- and post-central gyri, all of the insula, the
first and second temporal convolutions in their entirety, and nearly all
of the inferior parietal lobule. (See figure.) The patient's left cerebral
hemisphere was and remains radiologically normal.

By reference to the standard doctrine, as presented in chapter 11,
Doctor B obviously represents a "further exception," from both the
clinical and the anatomical points of view.

Clinically, evolution from a nonfluent agrammatic to a fluent paragrammatic aphasia is certainly unusual among right-handers (and, as
far as we know, among left-handers as well), whatever side of the brain
lesions. This might teach that agrammatism and paragrammatism are
not by definition mutually exclusive (and God knows that we are not
unconditional fans of Pierre Marie; see Lecours and Caplan, in press;
Lecours and Joanette, in press). Likewise, the coexistence of nonfluent
agrammatic speech and jargonagraphia remains exceptional, although
it has been reported to occur in right-handers with left-brain lesions
(Lecours and Rouillon 1976); in this respect, it certainly is of interest
that this particular dissociation was also observed by Pillon, Dési, and
Lhermitte (1979) in two adult right-handers with crossed aphasia. Another "rarity" observed in this case was the existence of bucco-facial
apraxia without phonetic disintegration, which indicates that (clinically)
the one can go, in certain circumstances, without the other.

On the other hand, crossed aphasia remains definitely exceptional
in absolute right-handers. In a recent review of the literature, Joanette

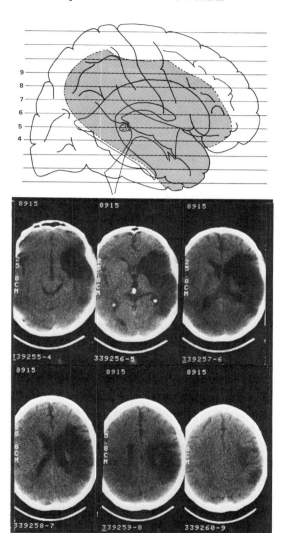

(1981) concluded that only nine reports of clear-cut cases had been published until August 1980. Although their criteria overlapped only in part, Urbain et al. (1978) concurred with that number. Be this as it may, and although the incidence of predominantly right-hemisphere representation of language in adult dextrals has been estimated to be as low as 0.4% (Russel and Espir 1969) and as high as 3% (Hécaen et al. 1971) and even 7% (Milner, Branch, and Rasmussen 1966), cases such as that of Doctor B remain definitely exceptional. This fact is not really challenged by the recent upsurge in tomodensitometric case reports of crossed aphasia in dextrals. (See chapter 11; Yarnell 1981; Assal et al. 1981; Carr, Jakobson, and Boller 1981.)

The fluency that characterized the second phase of Doctor B's language disorder—which had by then become very much akin to conduction aphasia—constitutes an exception among exceptions; the same is true of one of Yarnell's (1981) patients whose case is like Doctor B's in all respects, including lesion localization.

In view of the localization of Doctor B's brain infarct and the absence of phonetic disintegration and of major comprehension disorders throughout the evolution of his aphasia, one might suggest that his cerebral representation for language was bilateral, although the role played by each hemisphere was, at least in part, lateralized.

Whatever increase there may be in reports of crossed aphasia in unilingual literate adult right-handers, it is clear that Doctor B's hemispheric specialization was highly exceptional with regard to language, although apparently more regular with regard to other functions. Now, one can only wonder whether this represents a particular form of genetically determined anatomo-functional organization or whether it witnesses to reorganization related to early, conceivably prenatal, brain damage. In this respect, Goldman's recent demonstrations of the anatomical plasticity of the fetal brain in monkeys (Goldman 1978; Goldman and Galkin 1978) might cautiously be taken as a potential basis for further discussion.

Acknowledgment

The work of André Roch Lecours was supported by grant MT-4210 of the Medical Research Council of Canada.

References

Alajouanine, T., A. Ombredane, and M. Durand. 1939. *Le Syndrome de désintégration Phonétique dans l'aphasie*. Paris: Masson.

Assal, G., E. Perentes, and J. P. Duaz. 1981. Crossed aphasia in right-handed patient. Post mortem finding. *Archives of Neurology* 38: 455– 458.

Carr, M. S., T. Jakobson, and F. Boller. 1981. Crossed aphasia: Analysis of four cases. *Brain and Language* 14: 190–202.

Goldman, P. S. 1978. Neuronal plasticity in primate telencephalon: Anomalous projections induced by prenatal removal of frontal cortex. *Reprint Series* 202: 768–770.

Goldman, P. S., and T. W. Galkin. 1978. Prenatal removal of frontal association in the fetal rhesus monkey: Anatomical and functional consequences in post-natal life. *Brain Research* 152: 451–485.

Hécaen, H., G. Mazars, A. M. Ramier, M. C. Goldblum, and L. Mérienne. 1971. Aphasie croisée chez un sujet droitier bilingue (vietnamien-français). *Revue Neurologique* 124: 319–323.

Joanette, Y. 1981. L'aphasie croisée du droitier. *Revue de la littérature.*

Lecours, A. R., and F. Lhermitte. 1969. Phonemic paraphasia. Linguistic structures and tentative hypotheses. *Cortex* 5, no. 2: 193–228.

Lecours, A. R., and F. Rouillon. 1976. Neurolinguistic analysis of jargonaphasia and jargonagraphia. In H. Whitaker and H. Whitaker, eds., *Studies in Neuro-linguistics*, vol. 2. New York: Academic.

Lecours, A. R., E. Osborn, L. Travis, F. Rouillon, and G. Lavallée-Huynh. 1981. Jargons. In J. W. Brown, ed., *Jargonaphasia.* New York: Academic.

Lecours, A. R., and D. Caplan. In press. Augusta Dejerine-Klumpke or the lesson in anatomy. In H. Whitaker, B. Joynt, and S. Greenblatt, eds., *Historical Studies in Neuropsychology and Neurolinguistics.* New York: Academic.

Lecours, A. R., and Y. Joanette. In press. François Moutier from 'Folds to Folds.' " In H. Whitaker, B. Joynt, and S. Greenblatt, eds., *Historical Studies in Neuropsychology and Neurolinguistics.* New York: Academic.

Milner, B., C. Branch, and T. Rasmussen. 1966. Evidence for bilateral speech representation in some non-right-handers. *Transactions of the American Neurological Association* 91: 306–309.

Oldfield, R. C. 1971. The assessment and analysis of handedness: The Edinburgh inventory. *Neuropsychologia* 9: 97–113.

Pillon, B., M. Dési, and F. Lhermitte. 1979. Deux cas d'aphasie croisée avec jargonagraphie chez des droitiers. *Revue Neurologique* 135: 15–30.

Russel, R. W., and M. L. Espir. 1969. *Traumatic Aphasia.* London: Oxford University Press.

UNESCO. 1977. Statistical Yearbook.

Urbain, E., X. Seron, A. Remits, A. Cobben, M. Van der Linden, and R. Mouchette. 1978. Aphasie croisée chez une droitière. A propos d'une observation. *Revue Neurologique* 134: 751–759.

Yarnell, P. R. 1981. Crossed dextral aphasia: A clinical radiological correlation. *Brain and Language* 12: 128–139.

Localization of Lesions in Pure Anarthria

Michele Puel
Jean-Luc Nespoulous
Dominique Cardebat
André Roch Lecours
André Rascol

Bouillard (1825) was no doubt the first to teach that the localization of encephalic lesions can be deduced on the basis of neurological examination. Broca agreed, and in the second of his few publications on "aphemia" (Broca, 1861) he predicted that knowledge about functional specialization of various parts of the cortical mantle would be acquired through the study of the localization of cortical lesions in precise anatomical terms (he insisted on the importance of giving the names of the involved convolutions in case reports) rather than through the correlative study of behavior and skull bumps. This, of course, did occur, while phrenology gradually fell into oblivion.

Although the doctrine it has yielded concerning the speech area has been the object of unsettled discussions, the anatomo-clinical method has been used fruitfully ever since. Various types of acquired disorders of speech and language have been studied in this manner. One of them selectively impairs the phonetic moment of speech production. It is known as pure anarthria, pure phonetic disintegration, cortical dysarthria, subcortical motor aphasia, or pure motor aphasia. As indicated by these labels, it has been considered to be the result of lesions involving various left-hemisphere structures, such as the pars opercularis of the third frontal convolution (Mohr 1976; Mohr et al. 1978), the U fibers linking this structure to the adjacent part of the precentral gyrus (Dejerine 1906a,b), and the lower third of the precentral gyrus itself (Ladame 1902; Lecours and Lhermitte 1976).

We will now report on two cases of phonetic disintegration, one of them "pure" and the other associated to dysorthographia. Clinical findings will be discussed in relation to localization of lesions as defined through tomodensitometric examinations.

The first case is that of Mrs. G, a unilingual (French) right-handed retired clerk with a tenth-grade education and no family history of ambidexterity or left-handedness. This patient was 71 years old when she suffered a stroke on July 1, 1979, while she was preparing a cup of coffee. She was known to have a mitral stenosis, and her CVA was believed to be of embolic origin. Be this as it may, it resulted in right hemiparesis and speechlessness. The paresis was limited to upper limb and lower face, and it cleared after less than an hour without pyramidal sequelae. Speechlessness lasted a few days and was thereafter progressively replaced by phonetic disintegration (Alajouanine, Ombredane, and Durand 1939; Alajouanine, Pichot, and Durand 1949). Although significant improvement occurred thereafter, the arthric disorder has persisted to this day.

The second case is that of Mrs. M, a unilingual (French) right-handed housewife with a seventh-grade education and no family history of ambidexterity or left-handedness. This patient was also 71 years old when she suffered her stroke on December 6, 1980. She experienced clumsiness of her right hand while feathering a chicken and realized that she had suddenly become speechless. The clumsiness lasted only for a few minutes, but the patient was later observed to have a mild central right facial palsy, which is still to be seen. Total speechlessness lasted 2 days and was progressively replaced by phonetic disintegration (Alajouanine et al. 1939, 1949). Improvement was more rapid and complete than in the case of Mrs. G, but some degree of arthric disorder has persisted to this day.

Both patients were examined after 8 days of evolution. Complete neuropsychological and neurolinguistic testing was done. Phonetic disintegration was found to exist in both patients. Extensive tests, including the token test (de Renzi and Vignolo 1962), showed oral and written comprehension to be normal. Written production was considered to be normal in the case of Mrs. G (figure 1) and mildly dysorthographic in the case of Mrs. M.

In both subjects the arthric disorder was manifest in spontaneous speech, naming, repetition, and reading aloud. The latter two tests were used in order to define the linguistic characteristics of the phonetic disintegration in these patients. One hundred sixty-nine words, chosen according to various phonetic and phonological criteria, were used as stimuli. These were presented one by one, and time for corrections was allowed whenever the subjects so desired. Answers were tape-recorded, and an IPA transcription was done. The methodology used for the analysis of deviations has been described elsewhere (Puel et al. 1981). Involvement of vowels was so minimal in both cases that no pertinent quantitative study could be done in this respect. Furthermore, no significant differences were found between repetition and reading with regard to involvement of consonants; both tests demonstrated that the

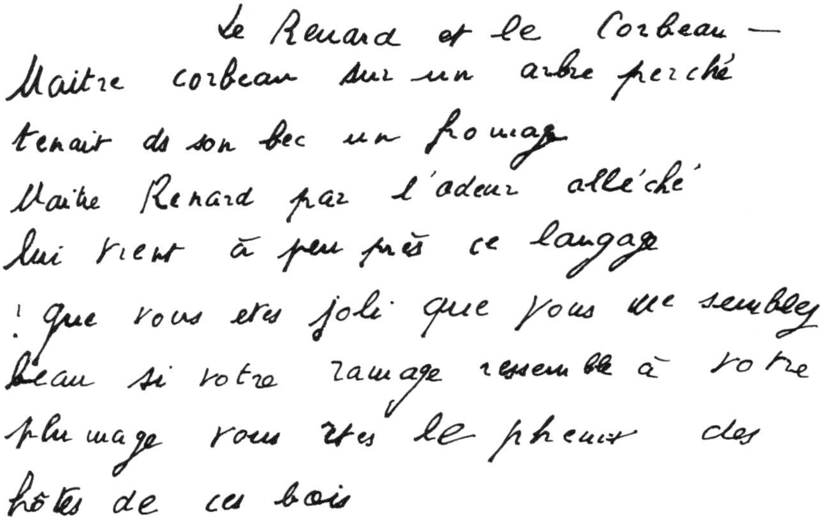

Le Renard et le Corbeau —
Maître corbeau sur un arbre perché
tenait de son bec un fromage
Maître Renard par l'odeur alléché
lui vient à peu près ce langage
! que vous êtes joli que vous me semblez
beau si votre ramage ressemble à votre
plumage vous êtes le phenix des
hôtes de ces bois

Figure 1

two patients initially showed (and still share to this day) a very similar linguistic profile as to the characteristics of their phonetic deviations. The deviations observed in reading are sufficient to illustrate this. The main difference between the two patients was in the overall intensity of anarthria (394 deviations in case G and only 75 in case M). The relative frequency of basic types of errors was the same in the two cases. Only phonemic substitutions and deletions had a high occurrence, with the former (65% in case G and 64% in case M) more frequent than the latter (35% in case G and 32% in case M); phonemic additions did not occur in case G and accounted for only 4% of the errors in case M; metathesis was entirely absent from both corpora. Similarities between the profiles of the two patients become even more striking when one analyzes the 256 substitutions of corpus G and the 48 substitutions of corpus M. Substitution essentially occurred between consonants sharing all but one of their constitutive features (67% of substitutions in case G and 58% in case M). For patient G, substitutions happened 75% of times on the articulation point, which was then pushed 81.8% of times toward the alveolo-dental zone. For patient M, substitutions happened 25% of times on the articulation point, which was pushed 100% of times toward the alveolo-dental zone. When substitution did not happen on the articulation point, substituted phonemes had their articulation point in the alveolo-dental zone 87% of times. This shows that the alveolo-dental zone is of characteristic importance in those two patients with phonetic disintegration. This phenomenon is close to that reported in normal children at a particular stage in speech acquisition (Jakobson 1969; Shibamoto and Olmsted 1978).

Tomodensitometric studies of these two cases were made on an Ohio Nuclear Head Scanner Delta 25; 8-millimeter-thick tomographies were taken before and after injection of contrast. Reading of documents in neuro-anatomical terms was made by reference to control necropsy specimens and was greatly favored, in both cases, by the existence of a degree of cortical atrophy sufficient to permit unequivocal identification of several sulci in the immediate vicinity of brain lesions. But for the cortical atrophy, the right hemisphere was considered to be radiologically intact in both cases. On the other hand, if the brain lesions of our two patients are entirely accounted for by tomodensitometric anomalies, and if our anatomical reading of these anomalies is indeed reliable, it can be suggested that Mrs. G's left-hemisphere softening is limited or almost limited to the caudal third of the second frontal convolution (figure 2) and that Mrs. M's is limited or almost limited to the lower half of the precentral gyrus, apparently with sparing of the opercular margin itself (figure 3).

Both Mrs. G and Mrs. M exhibited anarthria or phonetic disintegration (Alajouanine et al. 1939) as the consequence of a focal brain lesion apparently restricted to a small area of the left frontal cortex and the immediate subcortex. Similarities between the two cases include age, handedness, sex, cultural background, and, above all, very close resemblance in terms of the linguistic profile of the disorder. We do not know to what extent this profile is characteristic of phonetic disintegration; that is, we do not know if there exists one or several semiological forms of the latter. Nonetheless, given the semiological unity of our two cases, and provided that one accepts tomodensitometric data in general (Gado, Hanaway, and Frank 1979; Mazzochi and Vignolo 1979) and our own in particular as a reliable tool for defining anatomo-clinical correlations, these two observations raise interesting questions about the possible localizations of lesions responsible for phonetic disintegration. With respect to case M:

- "Isolated" lesions of the left precentral gyrus have previously been reported to cause anarthria (Bay 1964, 1972; Ladame 1902; Lecours and Lhermitte 1976).

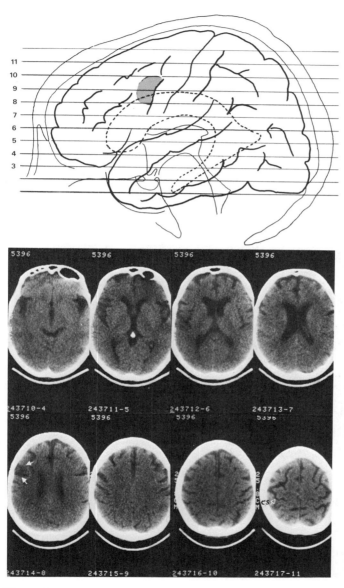

Figure 2

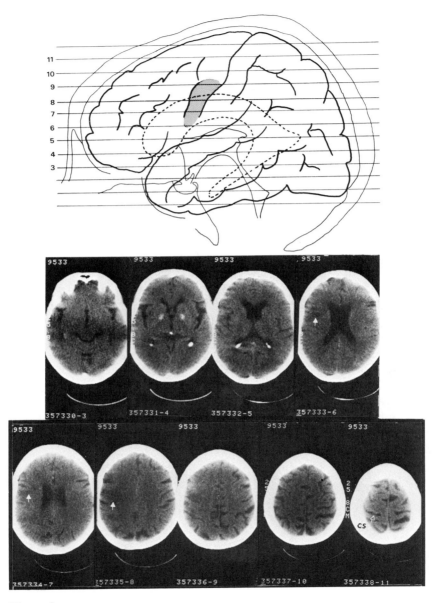

Figure 3

- Given that the opercular third of the precentral gyrus is considered to be the last cortical relay of buccophonatory motricity, one easily accepts that a lesion of this structure should provoke arthric disorders.
- Cases of anarthria have also been reported in which involvement of the precentral gyrus existed although the articulatory perturbation was believed to be the result of damage elsewhere in the immediate cortical vicinity (Bernheim 1901; Dejerine 1906a, 1926).

With respect to case G, on the other hand:

- As far as we know, focal lesions of the caudalmost third of the second frontal convolution of the left hemisphere have never been reported to cause anarthria.
- The only language-related role classically attributed—although seldom without reserve—to this cortical area has to do with written rather than oral speed production (Dubois, Hécaen, and Marcie 1969; Exner 1881).
- Cases of anarthria have also been reported in which involvement of the second frontal convolution existed although the articulatory perturbation was believed to be the result of damage in the immediate cortical vicinity (Dejerine 1926; Souques 1928).

Given the above, one has to conclude either that case G indeed represents an exception or that the opercular third of the precentral gyrus and the caudal third of the second frontal convolution of the left hemisphere belong to a single functional territory. Suggesting at this point that the disorder was of paretic nature in case M and of dyspraxic nature in case G would not settle a thing, since the clinical manifestations were identical in the two patients. On the other hand, one might wish to underline the fact that in case G written expression was spared in spite of a lesion of Exner's region whereas in case M the anarthria was associated with dysorthographia in spite of apparent sparing of this region. We do not really know what to make of these radiological interpretations.

Acknowledgments
The work of Michele Puel, Dominique Cardebat, and André Rascol was supported by grant 80-21 from the Conseil Scientifique de L'Université Paul Sabatier, and that of André Roch Lecours by grant MT-4210 from the Medical Research Council of Canada.

References

Alajouanine, T., A. Ombredane, and M. Durand. 1939. *Le syndrome de désintégration phonétique dans l'aphasie.* Paris: Masson.

Alajouanine, T., P. Pichot, and M. Durand. 1949. Dissociation des altérations phonétiques avec conservation relative de la langue la plus ancienne dans un cas d'anarthrie pure chez un sujet français bilingue. *L'Encéphale* 28: 245–262.

Bay, E. 1964. Principles of classification and their influence on our concepts of aphasia. In *Disorders of Language*, ed. A. V. S. De Reuck and M. O'Connor. London: Churchill.

Bay, E. 1972. Cortical dysarthria. Communication to biennial meeting of Aphasia Research Group of World Federation of Neurology, Lisbon.

Bernheim, F. 1901. De l'aphasie motrice. Thesis, Faculté de Médecine, Paris.

Bouillaud, J. B. 1825. Recherches cliniques propres à démontrer qe la perte de la parole correspond à la lésion des lobules antérieurs du cerveau et à confirmer l'opinion de M. Gall, sur le siège de l'organe du langage articulé. *Archives générales de Médecine* 8: 25–45.

Broca, P. 1861. Nouvelle observation d'aphémie produite par une lésion de la moitié postérieure des deuxième et troisième circonvolutions frontales. *Journal de la Société d'anatomie* 6: 398–497.

Déjerine, J. 1906a. L'aphasie motrice; sa localisation et sa physiologie pathologique. *Presse Médicale* 57: 453–457.

Déjerine, J. 1906b. A propos de la localisation de l'aphasie motrice. *Presse Médicale* 56: 742–743.

Déjerine, J. 1926. *Séméiologie des affections du système nerveux.* Paris: Masson.

de Renzi, E., and L. A. Vignolo. 1962. The token test: A sensitive test to detect receptive disturbances in aphasics. *Brain* 85: 665–678.

Dubois, J., H. Hécaen, and P. Marcie. 1969. L'argraphie pure. *Neuropsychologia* 7: 271–286.

Exner, S. 1881. *Untersuchungen über die Lokalisation der Funktionen in der Gross Hirnrinde des Menschen.* Vienna: Braumuller.

Gado, M., J. Hanaway, and R. Frank. 1979. Functional anatomy of the cerebral cortex by computed tomography. *Journal of Computer Assisted Tomography* 3(1): 1–19.

Jakobson, R. 1969. *Langage enfantin et aphasie.* Paris: Minuit.

Ladame, P. L. 1902. La question de l'aphasie motrice sous-corticale. *Revue Neurologique* 10: 13–18.

Lecours, A. R., and F. Lhermitte. 1976. The "pure form" of the phonetic disintegration syndrome (pure anarthria); anatomo-clinical report of an historical case. *Brain and Language* 3: 88–113.

Mazzochi, F., and L. A. Vignolo. 1979. Localization of lesion in aphasia: Clinical CT-scan correlations in stroke patients. *Cortex* 15: 627–654.

Mohr, J. P. 1976. Broca's area and Broca's aphasia. In *Studies in Neurolinguistics*, ed. H. Whitaker and H. Whitaker, vol. II. New York: Academic.

Mohr, J. P., M. S. Pessin, S. Finkelstein, H. H. Funkenstein, C. W. Duncan, and K. R. Davis. 1978. Broca aphasia: Pathologic and clinical. *Neurology* 28: 311–324.

Puel, M., J-L. Nespoulous, A. Bonafe, and A. Rascol. 1981. Etude neuro-linguistique d'un cas d'anarthrie pure. *Grammatica VII* 1: 239–291.

Shibamoto, J. S., and D. C. Olmsted. 1978. Lexical and syllabic patterns in phonological acquisition. *Journal of Child Language* 5: 417–446.

Souques, J. 1928. Quelques cas d'anarthrie de Pierre Marie. *Revue Neurologique* 2: 319–368.

Reference and Modalization in a Left-Hander with Callosal Disconnection and Right Hemianopia

Michel Poncet
André Ali-Cherif
Yves Joanette
Jean-Luc Nespoulous

As evidenced by handwriting, daily activities, and testing, Liliane S. is left-handed, left-footed, and left-eyed. Her parents and siblings are right-handed; her paternal grandmother is said to have been left-handed. Liliane was considered to be normal in all respects until the age of 14 1/2, when she underwent surgery for an anterior callosal hematoma resulting from the rupture of pericallosal artery malformation. The rostral half of the corpus callosum was sectioned and the hematoma removed. A second and a third intervention were performed 3 and 5 weeks later in order to evacuate and treat a left anterior frontal abcess. Liliane's complex postoperative evolution has been reported in detail elsewhere (Poncet et al. 1978). The present report bears on Liliane's clinical condition 15 months after surgery.

Liliane's main spontaneous complaint when she was 16 years old was her nearly total incapacity to read and write. She was also aware of some degree of word-finding difficulties, which occasionally led to production of more or less adequate circumlocutions, and of a disturbing dissociation in the activities of her left and right hands. She described the latter as follows: "I do what I want with my left hand but I have the impression that my right hand is crippled; my right hand is independent; it is always at rest; it does nothing or, rather, it does what it wants to do; my will is not really in charge of it, as it is for my left hand; if I tell my left hand to take hold of this or that, it does it, whereas my right hand does not" (Poncet et al. 1978). Liliane's mother thought that her daughter acted clumsily on purpose, in order to avoid working at home.

Neurological examination revealed a right homonymous hemianopia without sparing of the macula (figure 1). The cranial nerves were otherwise normal. But for the visual-field defect, there were no primary sensory disorders. Muscle strength and tone were normal and there were no abnormal reflexes. The existence of a diagnostic apraxia was documented, as well as a dissociation between proximal and distal right upper limb muscle control; constructional apraxia was obvious when the patient drew or copied geometrical figures with her right hand but

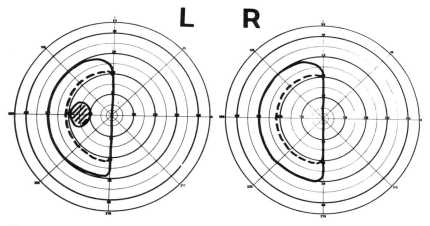

Figure 1
Campimetric visual field examination showing right homonymous hemianopia without sparing of the macula, using 15/1000 (continuous line) and 2/1000 (broken line) stimuli.

not when she drew or copied with her left hand; mnesic performances appeared to be normal (Poncet et al. 1978).

Liliane's spontaneous speech varies according to the topic of conversation. When she is speaking of her everyday activities, for instance, her speech is normal but for a mild word-finding difficulty; however, when she is speaking of a visually apprehended topic such as a film, her speech is limited to some modalizations. The following is an example of the latter condition: "To tell you the truth, the picture, I don't remember it. That's true. I can see things but it is difficult for me to speak." Results of standard repetition tests were normal, but dichotic listening tests revealed anomalies. Writing from dictation was possible only for single words. Dysorthographia was significant and graphism itself was normal when the patient wrote with her left hand, whereas properly graphic disturbances with a much milder dysorthographia were apparent when she wrote with her right hand (Poncet et al. 1978). Copying of single words was normal with the left hand, but grossly distorted with the right, showing both graphic disturbances and dysorthographia (ibid.).

Naming from visual stimuli (objects and images of objects) was adequate, although delayed, for only 9 out of 100 stimuli. However, Liliane regularly commented about the stimuli she could not name. As a rule, such comments were modalizations or personal reactions to stimuli, as in the following examples.

knife: It's a . . . I see what it is. Wait a minute, I know what it is. I can't tell you what it is but I know it. . . .

cigarette: I know what it is. I don't like it. It's not good for me. Wait a minute. I'm afraid of making a mistake. I would rather hesitate. I'm sorry, Doctor, but I can't tell you.

On rare occasions, such typical modalizations intermingled with incomplete referential informations, as in the following.

elephant: I know. I know what it is. I have seen some of them. Not in the street. One can see them on T.V.; they are big.

Naming from tactile stimuli was correct in 42 out of 50 instances when palpation was done with the right hand. Only 2 out of 50 objects were correctly named on palpation with the left hand. Modalization was also the rule in the latter case, as the following examples show.

key: I know what it is. . . . I can see what it is. . . . Wait a minute. . . . I am going to tell you.

coffee cup: I see what it is. I know it. I don't like what you put inside. I can't tell you the name.

Referential informations could also occur exceptionally, as in the following.

watch: It's a. . . . Wait a minute. . . . It's a. . . . I've got one myself . . . not here. . . . It's the time . . . to tell the time.

Naming from auditory stimuli was normal in standard testing. Likewise, Liliane readily named objects that were verbally defined by the examiner. Dichotically presented stimuli revealed gross repetition disturbances under this condition: All sentences and neologisms addressed to the right ear were repeated correctly, but those addressed to the left ear were unnoticed. Words addressed to the right ear were repeated correctly, but only 43% of those presented to the left ear were so. In those cases when words presented to the left ear were not repeated, Liliane quite often showed a most peculiar verbal and gestural behavior. The following illustrates this: For the dichotic pair *pipe* (right ear) and *apple* (left ear), Liliane first repeated the word *pipe* but immediately denied her answer, saying "Euh, sorry, I made a mistake. It's good to

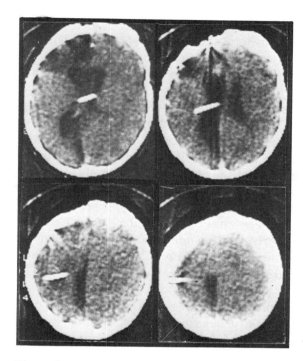

Figure 2
CT scan made 15 months after surgery.

eat, I like it. I don't know the word, sorry sir." At the same time, she acted as if she was biting into an apple held in her left hand. For the dichotic pair *suitcase* (right ear) and *piano* (left ear), Liliane first repeated the word *suitcase* but immediately said "Oh, sorry, excuse me sir, what's the name. I understood. I know. I can't tell the word." At the same time, she acted as if she was playing piano with her left hand.

A CT scan made 15 months after surgery (figure 2) revealed a left anterior frontal low-density zone corresponding to the sequels of the abcess, a hypodense left paramedial zone resulting from infarction in the territory of the left anterior cerebral artery, and a low-density zone in the left medial occipital region probably related to acute ischemia in the territory of the left posterior cerebral artery.

Thus, 15 months after surgery, the clinical neuropsychological signs indicated the presence of a total callosal disconnection syndrome with a right lateral homonymous hemianopia. The hypodensities revealed by the CT scan were compatible with the presence of a vascular lesion involving the corpus callosum's fibers in total or nearly so. These clinical and radiological data allow us to postulate in Liliane the existence of a complete or nearly complete interhemispheric disconnection, even though only an anatomical examination could permit such an affirmation.

In such cases of interhemispheric disconnection, as in split-brain studies (Gazzaniga 1970), it is possible to tackle, among other things, the linguistic abilities of the right nondominant hemisphere. On the whole, these studies insist on the non-negligible language-decoding capacities that appear maximum at the lexico-semantic level (Bradshaw 1980; Gazzaniga and Hylliard 1971; Zaidel 1976); as far as expressive language is concerned, the right nondominant hemisphere reveals itself to be more or less mute in these latter studies. In the present case, the right hemisphere is not properly mute but rather shows a type of language that differs from that of the left hemisphere. This is not surprising, since Liliane is left-handed (Milner, Branch, and Rasmussen 1966). In fact, if one admits that, on visual entry (because of the right hemianopia), left-tactile entry, and left-auditory dichotic entry, Liliane's verbal productions are the result of a right-hemisphere activity, one then can say that the right hemisphere of this left-handed girl is deprived of any referential speech (Nespoulous 1980, 1981) but is capable of modalizing speech (ibid.). Now, the aim of referential speech is to set up particular information or a particular proposition related to persons, objects, and so on, whereas modalizing speech allows the speaking subject to express his own opinions about what he is saying (referential content) or feeling.

With regard to these definitions and to Liliane's verbal performances, is it not then possible to assume that her right hemisphere, although nondominant for language, plays an important part in the actualization of Liliane's self? Two facts can be evoked in support to this hypothesis. The first relates to the diagonistic apraxia Liliane presents: When there is an antagonistic behavior of the two hands, it is Liliane's left hand that acts according to her will and her right hand that acts in opposition. The second fact pertains to the content of her spontaneous speech pertaining to the problems she has in everyday activities. For instance, speaking of her left hand, she says: "I can do what I want with my left hand. When I tell my left hand to take something, it picks up the object, it does what I want." Speaking of her right hand, though, she says: "It's as if my right hand was crippled. My right hand is independent, always at rest. I let it do what it wants, it's beyond my control." It thus appears as if the only hand in total harmony with Liliane's self is her left one.

There is nothing outstanding about the right hemisphere of a young left-handed person taking part in linguistic expression. However, this linguistic expression is apparently dealing exclusively with modalizing speech. Such a dissociation based upon the occurrence of modalizing as opposed to referential speech has already been described in relation to types of aphasia from left-hemisphere lesion (Nespoulous 1980, 1981). What Liliane evidences here is that such a dissociation can oppose one whole hemisphere to the other; in fact, Liliane's intact right hemisphere has such communication motives and/or capacities that the resulting speech is essentially of the modalizing type.

Acknowledgment

The work of Yves Joanette was supported by the Fonds de la Recherche en Santé du Québec.

References

Bradshaw, J. L. 1980. Right-hemisphere language: Familial and non-familial sinistrals, cognitive deficits and writing-hand position in sinistrals, and concrete-abstract, imageable and nonimageable dimensions in word recognition. *Brain and Language* 10: 172.

Gazzaniga, M. S. 1970. *The Bisected Brain*. New York: Appleton-Century-Crofts.

Gazzaniga, M. S., and S. A. Hylliard. 1971. Language and speech capacity of the right hemisphere. *Neuropsychologia* 9: 273.

Milner, B., C. Branch, and T. Rasmussen. 1966. Evidences of bilateral speech representation in some non-right-handers. *Transactions of the American Neurological Association* 91: 306.

Nespoulous, J. L. 1980. De deux comportements verbaux de base: référentiel et modalisateur. De leur dissociation dans le discours aphasique. *Cahiers de Psychologie* 23: 195–210.

Nespoulous, J. L. 1981. Two basic types of semiotic behavior: Their dissociation in aphasia. Publication Series of Toronto Semiotic Circle, II and IV, pp. 173, 213.

Poncet, M., A. Ali Chérif, M. Choux, J. Boudouresques, and F. Lhermitte. 1978. Etude neuropsychologique d'un syndrome de déconnexion calleuse totale avec hémianopsie latérale homonyme droite. *Revue Neurologique* 134: 633.

Zaidel, E. 1976. Auditory vocabulary of the right hemisphere following brain bisection or hemidecortication. *Cortex* 12: 191.

Chapter 12

Determinants of Recovery from Aphasia

Andrew Kertesz

Recovery from cerebral damage is not only relevant to the clinician who has to prognosticate and make treatment decisions, but has become an important subject of research. The study of recovery is intimately connected with the study of brain function itself. Without consideration of recovery, much of the lesion data (especially on humans) remains uninterpretable. Most of the comments that follow will be based on recovery from stroke, because much of my work has been done in this area. Stroke patients recover with a certain regularity which is peculiar to the nature, location, and size of the deficit. Recovery occurs, of course, after trauma and surgical ablations, but the variables in these instances are different and should be dealt with separately.

Concepts of Recovery

First Stage
The first stage of recovery from acute insult to the brain occurs when the initial hemorrhage, edema, or cellular reaction subsides, when the pressure is relieved by surgery or medication, or when the general condition of the patient is stabilized. The significant improvement in the first 1–3 weeks after a stroke must be taken into consideration in establishing a baseline for recovery studies. Few, if any, quantitative studies are available as to the effect of treatment during this acute stage. The initial effect of steroids is apparent in cases of severe trauma with increased intracranial pressure, but it remains controversial and is generally considered ineffective in cases of stroke. Carbon dioxide inhalation to induce vasodilation and pharmacological vasodilators are not considered reliable. Hyperbaric chambers hypothetically increase oxygen intake in the "penumbra" or the halo of ischemic neurons around the actual area of cell death. Barbiturate coma in trauma may promote

recovery. The major stumbling block in the evaluation of these early therapies is the lack of quantification in a rapidly changing situation with multiple variables.

Second Stage

The second stage of recovery from stroke is distinct from the short initial phase. It lasts for months up to a year or more after onset. The restoration of function remains largely unexplained, although a great deal of experimental and clinical evidence has been accumulated dealing with the basic mechanisms of recovery (Stein, Rosen, and Butters 1974; Finger 1978).

Substitution

Repeated observations of recovery of function after experimental ablations in animals led to the idea of a *"sensorium commune."* This idea was championed in the last century by Flourens, who claimed that the cortex has a unitary function capable of a great deal of substitution. Lashley's theory of equipotentiality of function is a modern reiteration of the same principle. One should keep in mind, however, that these theories are based on animal experimentation and that the species differences are often substantial.

Lashley also promoted the principle of the "mass effect." He showed that the relationship between amount of intact cortical tissue and learning capacity is close. He also reviewed the degree of recovery in motor aphasia and the extent of lesions centering in the left frontal convolution in 18 cases and found a high negative correlation of 0.9.

Clinico-pathological evidence since Broca is overwhelmingly against equipotentiality, even in the higher level of cortical function in man. The specificity of the primary sensory, motor, auditory, and visual portions of the cortex were further documented by Fritsch and Hitzig (1870) and Munk (1881). These pioneers of functional localization also recognized that recovery takes place after ablation experiments. When Fritsch and Hitzig removed the motor cortex of dogs—an area they had previously stimulated and found to be specific for certain movements—they thought that recovery occurred through substitution by the homologous contralateral cortex. However, ablation of this contralateral area did not produce a permanent loss of function either. The logical extension of the theory stated that the substituting areas need not be homologous. Munk (1881) called recovery that was mediated by regions of the brain previously "not occupied" *vicarious functioning.* Pavlov argued for a large factor of safety in the nervous system, postulating many potential conditioned-reflex paths that are never used by the normal organism. Lashley (1938), on the other hand, thought that preservation of part of a system that is concerned with the same function is needed for restitution.

Reorganization of Function

A modification of the idea of simple substitution or vicarious functioning is that a new adaptive organization occurs in the remaining systems. This reorganization takes place according to certain principles. Hughlings Jackson stressed the role of the hierarchical organization of the nervous system. When a higher function is impaired, lower levels take over. An example of this is the dependence of various learned functions on more fundamental mechanisms, which need to be intact or restored before recovery takes place in the associated cortical area. Animal experiments demonstrated the dependence of visual cortex on tectal functioning. When a rat is taught to choose the brighter of two lights and the striate cortex is destroyed, the habit is lost; however, the animal relearns just as quickly as a normal. Now, if pretectile nuclei are destroyed, no learning occurs, so we are justified in ascribing the recovery to substitution by the pretectile region. Reverse ablations, however, do not affect function. In other words, destroying the pretectile region does not effect visual learning. Similar reverse ablations were done by Bucy (1934) on the premotor cortex in monkeys, without altering motor function.

Instead of an absolute loss after a lesion, some functions show a certain amount of inconsistency and fluctuation. This *Funktionswandel,* as it is known in the German literature, is especially evident for visual stimuli after occipital lesions, but Henry Head observed it for tactile stimuli and we have seen it with aphasics. Therefore, the number of items in a test and the consistency of performance are crucial when actual measurements are made. The fluctuation of function suggests that associative mechanisms are present and operating, but under altered conditions such as a lowered level of excitability.

Functional Compensation

Functional compensation is understood in terms of a behavioral rather than a neural model. Lashley summarized the animal and human experiments of the 1920s and 1930s by saying that simple sensorimotor disorders and even amnesia for words can be restored by persistent training and strong motivation. However, defects in the capacity of organization, lowered level of abstraction, slowing of learning, and loss of interest did not seem to improve with time or training. Substituted maneuvers or tricks have been described and documented by Sperry (1947). Instead of rerouting connections, the brain-damaged organism develops new solutions to problems using residual structures. Luria et al. (1969) formulated the theory of retraining, which claims that the dynamic reorganization after injury is promoted by specific therapy. Recovery by the use of alternate pathways may occur in man; for example, the ipsilateral corticospinal system may be used by the speaking hemisphere to carry out commands in patients with callosal lesions (Geschwind 1974).

Diaschisis
The best-known physiological explanation of second-stage recovery is diaschisis (von Monakow 1914). This is an essentially physiological concept related to the well-established principle of spinal shock and the return of function following it. Von Monakow thought that when a sizable portion of brain tissue ceases to function this not only produces a direct behavioral deficit but also affects the remaining parts of the brain that were functionally connected to the lesioned area. The sudden cessation of this input creates a loss of function in structurally intact areas. In time, these structures assume their function or independence, thereby providing a recovery from initial diaschisis. The physiological mechanisms and the pharmacodynamics of recovery from diaschisis have been elaborated far beyond this initial formulation.

Diaschisis is more marked after sudden lesions, such as infarcts, or trauma. Tumors often grow slowly, minimizing the effect of diaschisis. A similar phenomenon was shown in animal experiments by Ades and Raab (1946), who called it the "serial lesion" effect. Two-stage removal was much less damaging than when the same area was destroyed all at once. Dax (1865), in describing the 40-odd observations of left-hemisphere lesions with aphasia (in a paper published posthumously by his son, after Broca's discovery), also suggested that left-hemisphere disease may not alter speech if the lesion developed slowly. Von Monakow (1914) also noted that aphasia may be absent in Broca's area in cases of slowly growing tumors. Neural shock or diaschisis is avoided in slowly growing or multistage lesions.

Regeneration
Axonal and collateral sprouting, or regeneration, in the mammalian central nervous system is not comparable to that in the peripheral nervous system. Axonal growth has been found, however, in ascending catecholaminergic fibers. Growing axons tend to invade vacant terminal space (Schneider 1973). The recently discovered phenomenon of collateral sprouting could be the underlying mechanism of recovery in certain instances (Liu and Chambers 1958). Both axonal and collateral sprouting have been demonstrated in the CNS by Moore (1974). Collateral sprouting, whether from intact axons or from collaterals of the damaged axons, seems to be more important.

The significance of these mechanisms remains in doubt because of the complexity of the mammalian CNS and the glial and connective-tissue scars that form after damage and prevent the regenerating axons from reinnervating the area. We need to know more about the actual mechanisms whereby axonal regeneration occurs in certain directions before we can hope to influence the process pharmacologically.

Denervation Hypersensitivity
Denervation hypersensitivity, discovered by Cannon and Rosenbluth (1949) and elaborated recently by Glick and others, explains recovery in at least two, not entirely complementary ways. Glick postulates that loss of input into an area can be compensated for by an increased sensitivity of the postsynaptic membrane to a decreased amount of transmitter substance. The process of recovery will be effected by the relatively few remaining or regenerating neurons to allow a greater physiological action.

The opposite effect of denervation hypersensitivity has also been argued. The initial hypersensitivity would induce the inhibition of function (diaschisis), and collateral sprouting would put an end to the denervation and the accompanying inhibition.

Cerebral Blood Flow and Neurotransmitters
A less accepted but possible explanation is that bilateral reduction of cerebral blood flow following cerebral infarct could be one of the phenomena underlying diaschisis (Meyer et al. 1970). This is known to persist for up to 2 months. Release of catecholamines during this period has also been demonstrated, but whether it is a cause or an associated phenomenon is largely unknown (Meyer et al. 1974).

Pharmacological Aspects

The pharmacological aspects of recovery are most complex. Cholinergic agents (Ward and Kennard 1942), anticholinesterases (Luria et al. 1969), and amphetamines (Braun et al. 1966) accelerate recovery; barbiturates (Watson and Kennard 1945) slow recovery. Catecholamines may act as inhibitors (Meyer et al. 1974) and bicuculline as a disinhibitor of recovery (Duffy et al. 1976). Neuropeptides and beta blockers are the latest to join the list of agents that are considered to influence recovery.

Neural Plasticity

Plasticity or transfer of function occurs much more easily in the immature nervous system. Kennard and McCulloch (1943) found that unilateral lesions in the precentral areas in infant animals have minimal effects compared to those in adults. Goldman (1974) called attention to the changing consequences of early injury at different stages of ontogenetic development. Children up to age 10 or 12 recover from aphasia even after hemispherectomy (Basser 1962); this indicates the plasticity of the right hemisphere in assuming the function of speech. Recent evidence indicates, however, that one hemisphere (usually the left) is dominant for speech at birth. After early left hemispherectomy, subtle deficiencies of verbal function persist and there is a reduction in overall intellectual capacity due to "crowding" in the right hemisphere. The functional plasticity of the young may depend on the microneurons of type II Golgi cells (Hirsch and Jacobson 1974). These cells remain adaptive while the long axon cells responsible for the major transmission of information in and out of the CNS are under early and exacting genetic control and specification. The state of flexibility for these microneurons may be terminated in the teenage years in man by hormonal changes; this may explain the age limit of relocalization of language. Acquisition of a foreign language without an accent also appears to be limited by puberty. Pertaining to this is the experimental evidence that song acquisition by birds depends on hormone levels (Nottebohm 1970; chapter 4 in this volume).

Specific Factors in Aphasia

The Etiology of Language Impairment

It must be recognized that strokes, tumors, trauma, and infection affect the brain differently. Recovery from various etiologies is also substantially different, and studies that do not distinguish among the various etiologies of brain damage cannot be considered optimal. Studies of a large number of war-injured patients after World War II formed the basis of the first recovery studies. Butfield and Zangwill (1946) were pioneers in this field. Wepman and Luria found considerable recovery in post-traumatic patients. Marks, Taylor, and Rusk (1957) and Godfrey and Douglass (1959) found less recovery in stroke patients. Kertesz and McCabe (1977) contrasted post-traumatic recovery with stroke recovery and observed complete recovery in more than half of the post-traumatic aphasics. The age factor is interacting to some extent, because most of the post-traumatic patients tend to be younger. However, the ages were scattered considerably in trauma, and this is certainly not the entire explanation. The extent and severity of the deficit was also variable in our closed head injury auto accident population. Only a few of our subjects had contusion or subdural hemorrhage. In a recent study of head injury and recovery from aphasia following closed head injury, Levine et al. (1980) found positive correlation between the extent of recovery and the duration of post-traumatic amnesia.

Subarachnoid hemorrhage affects the cerebral cortex in a variety of ways. The most devastating effect is the vasospasm, which produces ischemic changes and at times a large infarction. The prognosis following this type of lesion is similar to that of a large stroke. However, when there is no actual infarction, the initial effect of the vasospasm and the obtundation subsides quickly and the recovery from the mild aphasic impairment can be rapid and complete. Since the advent of isotope and CT scanning, we can determine whether a patient has sustained an infarction, and the extent of the infarction, with ease. The CT scan is also useful in detecting intraparenchymal bleeding, which is often a complication of subarachnoid hemorrhage. There are late effects of obstruction of the subarachnoid space in this disease that produce hydrocephalus and obtundation but not a focal deficit leading to aphasia. At times, patients have postoperative aphasic symptoms, but these tend to be mild unless a major infarct has been induced by the operative procedure. Some of our worst jargon aphasics and global aphasics were seen after the rupture of middle-cerebral aneurysms and after surgery (Kertesz and McCabe 1977).

Initial Severity
Initial severity has been observed to be highly predictive of outcome (Godfrey and Douglass 1959; Schuell, Jenkins, and Jimenez-Pabon 1964; Sands, Sarno, and Shankweiler 1969; Sarno, Silverman, and

Sands 1970; Gloning et al. 1976; Kertesz and McCabe 1977). Unfortunately, it is not always taken into consideration in studies of recovery; for instance, one of the major studies of language therapy in aphasia did not control for initial severity among treated and nontreated patients (Basso, Capitani, and Vignolo 1979). The most severely affected patients, although they have a long way to go, show little gain whether treated or untreated (Sarno et al. 1970). The extent of the improvement may also be small in mildly affected patients, because, although they recover quite well, they do not have much room for recovery because they were relatively good to begin with. This ceiling effect was pointed out in Kertesz and McCabe 1977. There is no entirely satisfactory way to cope with the effect of initial severity in studies. Covarying initial severity when examining factors in recovery is a less than perfect solution, since it tends to diminish the effect of this most important variable. It is more important to match patients for severity at the beginning of a study when comparing the effect of various forms of treatment or parameters.

Type of Aphasia
Differences between types of aphasias occur in recovery, and these differences interact with initial severity to a considerable extent. Head (1926) recognized that some types of aphasia improve more rapidly than others. Weisenburg and McBride (1935), Butfield and Zangwill (1946), Messerli, Tissot, and Rodriguez (1976), and Kertesz and McCabe (1977) considered Broca's or "expressive" aphasics to improve most, whereas Vignolo (1964) considered expressive disorders to have poor prognosis (probably because he included global aphasia among the expressive disorders). Basso et al. (1979) did not find any difference between fluent and nonfluent aphasics, but it is likely that if they had not lumped global and Broca's aphasics together such a difference would have been found. The variability of conclusions reflects problems of classification; for example, "expressive disorder" includes many different kinds of aphasics. There is general agreement, however, that global aphasics have poor prognosis.

Language Components
There is evidence that not only various types of aphasics but also various components of language recover differently. Of course, types of aphasia and various language components are interdependent, and sometimes this must be taken into account in the analysis. Kreindler

and Fradis (1968) found that naming, oral imitation, and comprehension of nouns showed the most improvement. Gains in comprehension were greater than in expressive language in the study of Broca's aphasics by Kenin and Swisher (1972), but no difference was found by Sarno and Levita (1971).

Lomas and Kertesz (1978) studied various language components in four groups of 31 untreated aphasics separated by criteria based on factors independent of those examined. The highest overall recovery scores were obtained by the low fluency high comprehension group. Comprehension as examined by Yes-No questions and sequential command and repetition were the most improved components; word fluency was the least improved. In fact, word fluency remained impaired while all other language factors improved, indicating that it measures something in addition to language; this is corroborated by the observation that it is often impaired in nonaphasic brain-damaged subjects. The groups with low initial comprehension showed recovery in Yes-No comprehension and repetition tasks; the others with high comprehension recovered in all tasks except word fluency.

Time Course
The time course of recovery follows a certain regularity, and there is considerable agreement on this in the literature. A large percentage of stroke patients recover a great deal in the first 2 weeks. The greatest amount of improvement occurs in the first 2 or 3 months after onset (Vignolo 1964; Culton 1969; Sarno and Levita 1971; Basso, Faglioni, and Vignolo 1975). Our recovery study with the Western Aphasia Battery (Kertesz and McCabe 1977) confirmed this. Beyond 6 months the rate of recovery drops significantly (Butfield and Zangwill 1946; Sands et al. 1969; Vignolo 1964; Kertesz and McCabe 1977). Recovery after a year does not seem to occur spontaneously in the majority of cases (Culton 1969; Kertesz and McCabe 1977), but some reports note improvement in patients under therapy for many years after the stroke (Schuell et al. 1964; Marks et al. 1957; Smith et al. 1972; Broida 1977).

Age
Although the clinical impression that younger patients recover better has produced a widespread belief that age is an important factor, we did not find such a trend statistically significant (Kertesz and McCabe 1977) even when we excluded the post-traumatic group, which had a mean age well below that of the infarct group. There was a negative

Table 12.1
Mean recovery scores 3 months after onset.

		N	Mean	S.D.	t	P
Content	Males	89	1.9	2.2		
	Females	69	1.8	2.4	0.07	>0.1
Fluency	Males	89	1.3	1.6		
	Females	69	1.3	1.9	0.20	>0.1
AQ	Males	89	17.4	15.6		
	Females	69	16.4	16.8	0.38	>0.1
Comprehension	Males	89	1.7	1.8		
	Females	69	1.5	1.6	1.02	>0.1
Repetition	Males	89	2.0	2.5		
	Females	69	1.7	1.7	1.00	>0.1
Naming	Males	89	1.9	1.8		
	Females	69	2.3	2.0	1.30	>0.1

Values computed are the differences at 0 and 3 months for content, fluency, AQ, comprehension, repetition, and naming. Groups comprise Broca's, Wernicke's, anomic, global, and conduction aphasics.

correlation between age and initial recovery rate, but it just missed being statistically significant. This was also the case in the studies of Sarno and Levita (1971), Culton (1971), and Smith et al. (1972). We observed remarkably good recovery in elderly patients, whereas some of the young will remain severely disabled.

Sex

In our initial study (Kertesz and McCabe 1977), we did not find a significant sex difference in recovery, even though recent reports suggest that females may have less language impairment after brain damage. More recent studies of age and sex differences in the incidence of aphasia suggest that the higher incidence of aphasia in males is accounted for by the higher incidence of strokes in males (Kertesz and Sheppard 1981). When actual severity of impairment is compared, there is no significant difference. We also utilized some of our more recent data on recovery to further analyze sex differences in the rate of recovery. These results show no significant sex difference for the total aphasic population (table 12.1). The more bilateral or diffuse representation of language in women would lead to the expectation of better recovery rates from aphasia in women, but this is not substantiated by our data.

Handedness

Most of our data is related to right-handers, and very few studies were able to collect enough left-handers to draw significant conclusions. The smaller than expected number of left-handers in our population of aphasics (Kertesz and Sheppard 1981) may be related to the more bilateral cerebral organization of left-handers. Some not entirely quantitative data from Subirana (1969) and Gloning et al. (1969) indicate that left-handers recover better from aphasia than right-handers. Gloning et al. (1969) also suggested that left-handers are likely to become aphasics regardless of which hemisphere is damaged. This is somewhat contrary to our experience; one would expect a higher incidence of aphasia in left-handers in that case. According to Geschwind (1974), right-handers with a history of left-handedness among parents, siblings, or children recover somewhat better than right-handers without such family history. However, reliable quantitative data to support this contention is lacking.

Evolution of Aphasic Syndromes

Alajouanine (1956) distinguished four stages of recovery in severe expressive aphasia: differentiation by intonation, decreased automatic utterances, less rigid stereotypic utterances, and only volitional slow, agrammatical speech. Observation of recovering Broca's aphasics indicates yet another stage, characterized by slowing, increased pauses, word-finding difficulty, and articulation problems (verbal apraxia). Kertesz and Benson (1970) pointed out the predictable pattern of linguistic recovery in jargon aphasia. Copious neologistic or phonemic jargon is replaced by verbal paraphasias or semantic jargon, and eventually anomia or (more rarely) "pure" word deafness develops. The extraordinary phenomenon of overproduction is replaced by anomic gaps.

The evolution of aphasic syndromes and the patterns of transformation from one clinically distinct group into another, as defined by the subscores on subsequent examinations, were also studied (Kertesz 1981). It appeared that anomic aphasia is a common end stage of evolution, in addition to being a common aphasic syndrome *de novo*. Four of 13 Wernicke's aphasias, 4 of 8 transcortical and isolation aphasias, 4 of 17 Broca's aphasias, 2 of 8 conduction aphasias, and 1 of 22 global aphasias evolved into anomic aphasia. Complete recovery from aphasia depends on the difficult definition of normal language function. We used an arbitrary cutoff (aphasia quotient) of 93.8. This was the actual mean of the standardization group with brain-damaged but clinically judged nonaphasic patients. Final AQs indicated that 12 anomic aphasics, 5 conduction aphasics, 2 transcortical sensory aphasics, and 1 transcortical motor aphasic reached this criterion of recovery. Although this represents only 21% of the total of 93, it is 62.5% of the conduction aphasics, 50% of the transcortical aphasics, and 48% of the anomic patients. Reversal of the direction on evolution from the patterns described may occur with an extension of a stroke or tumor or Alzheimer's disease; for instance, if an anomic aphasic becomes nonfluent or the jargon becomes more neologistic, one of the above should be considered.

Recovery rates appeared to relate clearly to the time of examination after onset. The degree of improvement was by far the greatest between the first test within 1 1/2 months of onset and 3 months later, but some recovery was noted at all subsequent intervals. The differences in recovery rates among various types of aphasia are evident. The highest rates of recovery were shown by Broca's and conduction aphasics, the lowest by untreated global and anomic aphasics. (Recovery rate should not be confused with level of recovery, which is much higher among anomic aphasics.)

Recovery of Reading, Writing, and Calculation

We examined reading, writing, and calculation at successive intervals in 69 aphasics (Kertesz 1979). Not all of them could be followed for a whole year; therefore, the numbers and starting points varied in each interval. When recovery rates for reading, writing, and calculation were examined in different groups of aphasics, they were found to parallel language recovery, with some important variations.

We followed 20 aphasics with the reading and writing tasks, as well as the other subtests of the Western Aphasia Battery for each of the 0–3, 3–6, and 6–12 month intervals. This group consisted of 3 global, 6 Broca's, 3 Wernicke's, 2 conduction, 1 transcortical sensory, 1 transcortical motor, and 4 anomic aphasics. Levels of significance in a repeated-measures t test, calculated for all subtests, indicated highly significant recovery in the first 3 months, followed by moderate recovery in language, reading, and writing but no significant recovery in calculation. In the third interval, the recovery rates were similarly low. The writing subtest scores were consistently lower in all intervals, even though the recovery rates are parallel with the rest of the parameters, reflecting the usually severe involvement of writing in aphasia.

Recovery of Praxis

Recovery from apraxia had not been studied systematically until we followed 50 aphasics with an apraxia battery (Kertesz 1979). Buccofacial, intransitive, transitive, and complex performances were examined at 0, 3, 6, and 12 months after stroke. Recovery of praxis closely paralleled language function in all aphasic groups. In global aphasics, praxis recovered somewhat better in the initial interval, along with better comprehension, but the differences were not statistically significant. In the Broca's aphasics, on the other hand, praxis recovered less.

Recovery of Visuospatial Function

Relatively little has been written on recovery of nonverbal function. Campbell and Oxbury (1976) used block design and Raven's matrices in right-hemisphere-damaged patients at 3–4 weeks and 6 months after stroke. Those with neglect on the initial test of drawing remained impaired 6 months later, even though the neglect has resolved. Persistent inattention has been described up to 12 years after onset. In Bond's (1975) study of post-trauma patients, nonverbal intellectual function took longer to recover than verbal intelligence. Mandleberg and Brooks (1975) found that recovery of performance IQ continued over 3 years after severe trauma.

To measure the recovery of nonverbal performance in aphasics (Kertesz 1979), we used Raven's colored progressive matrices, block design, and drawing. Results showed that recovery rates were lower than in language subtests for the Raven's matrices for the first 3 months and continued to rise at the same rate between 3 and 6 months, in contrast with language, which reached a plateau after 3 months. In all, we followed 20 cases at all three intervals. In each individual aphasic group we had larger numbers, although the dropout rate was higher. The correlation of nonverbal visuospatial function recovery rates with the recovery of various language parameters revealed that the correlations were significant in the first 3 months. Comprehension, naming, and reading correlated most significantly.

We also had 19 right-hemisphere-damaged patients who were not aphasic. Their recovery rates on Raven's colored progressive matrices, drawing, and block design were similar to the recovery curves of language in aphasics. The curves were flatter, however; this may indicate that we had a better group to start with. These results are only preliminary, and further investigation with the goal of a more comprehensive survey of right-hemisphere recovery patterns is underway.

Relationship of Lesion Size and Location to Recovery
The extent of language recovery in CT-scanned patients was calculated between the initial examination and a followup examination a year later (Kertesz, Harlock, and Coates 1979). Positive correlation was found between recovery of comprehension and lesion size, and a trend of negative correlation was present for the other parameters. This can be best understood if one considers that the comprehension deficit caused by a large lesion often recovers considerably more than other parameters, and that the greater the deficit initially the more room there is for recovery. Small lesions, on the other hand, often result in aphasia with little if any comprehension deficit, and the extent of the recovery is small, with higher scores to begin with. The negative correlations between recovery rates of fluency, repetition, naming, and the total aphasia quotient and the size of the lesion indicated that the larger the lesion the less recovery occurs in total language deficit in parameters other than comprehension. Yarnell, Monroe, and Sobel (1976) correlated the size of chronic CT lesions with the outcome of aphasia in 16 cases and obtained similarly negative trends of correlation. Mohr et al. (1978) pointed out that large parasylvian lesions produced persistent chronic Broca's aphasia but smaller lesions affecting the third frontal convolution were associated with milder, rapidly ameliorating motor aphasia.

The Effect of Therapy
Treatment of aphasics is a well-established practice, even though only a few studies have attempted to control the results and spontaneous recovery. Butfield and Zangwill (1946) and Vignolo (1964) studied treated and untreated patients and did not find a significant difference between these groups. Similarly, Sarno et al. (1970) and Kertesz and McCabe (1977), when comparing treated and untreated global aphasics, did not find any difference. There are two controlled studies in the literature that suggest that therapy is effective. Hagen (1973) used only 10 treated aphasics, matched their severities and types to those of 10 untreated aphasics, and found significantly better recovery in language formulation, speech production, reading comprehension, spelling, and arithmetic in the treated group, mostly in the first 6 months of treatment. Auditory comprehension, reading, and visual motor abilities improved equally in the treatment and control groups. Basso et al. (1979) had a large number of control subjects and treated patients, but they were not randomized. The control subjects could not come for therapy for logistical or family reasons. The design lumped all aphasics in two

categories, possibly allowing more globals or severe jargon aphasics to be included in the untreated sample. This is bound to occur in therapy situations—therapists are less likely to persist with global and severe jargon patients, who do not show promise of progress, so they are put in the untreated group. Another important difficulty with a retrospective study like that of Basso et al. is that, instead of controlling the time of inclusion after onset, it examined the effect of this variable in a factor-analytic statistic. This allowed the treated group, in which early patients with the greatest spontaneous recovery were included, to be favored. To design a study of treatment of aphasia, it is essential to match patients for initial severity type, etiology, and time after onset.

There was a recent cooperative study of aphasia therapy (Wertz et al. 1981) in which the patients undergoing stimulation therapy individually were compared with patients undergoing social-interaction group therapy. No statistically significant differences were found, although the individual-treatment group was considered to do better. Although Wertz et al. did not have controls, they felt that improvement between 26 and 48 weeks in both groups was related to therapy because they expected no spontaneous recovery at that time.

In a multicenter trial, David et al. (1982) compared speech therapists and untrained volunteers for 155 stroke patients. They used the Functional Communication Profile as a basic measure. Patients were entered at any time later than 3 weeks after onset and were randomized to the two treatment groups. Treatment was only for 30 hours over 15–30 weeks. (A small group of 9 late referrals also showed improvement.) This was taken as a period beyond spontaneous recovery, and therefore any recovery could be attributed to treatment. David et al. concluded that there was no significant difference between the treatment groups. They felt that the improvement was the result of the support and stimulation provided by the speech therapist and the volunteer.

The preliminary results of our controlled study of recovery were presented at the 1981 Academy of Aphasia Meeting (Shewan and Kertesz 1981). In this study the recovery in each interval was covaried for initial severity. The three randomized groups of language-oriented therapy, stimulation therapy, and untrained therapists did not differ significantly. Further details of this study are in the process of preparation for publication.

Conclusion

Recovery from cerebral deficit is a complex phenomenon. We have just begun to understand some of the basic mechanisms. The physiological, pharmacological, and morphological phenomena studied have to be matched to certain recently measured clinical patterns of second-stage recovery. The clinical studies indicate a certain degree of uniformity regardless of the parameters studied. The time course of recovery from stroke is quite characteristic, for instance. The steepest recovery occurs in the first 2–3 months, then the rate falls off to reach a plateau at 6–12 months. Individual variations are considerable, but the recovery curves are similar in language and in right-hemisphere deficits. Factors clearly affecting recovery are initial severity, nature of damage, lesion size, and possibly lesion location, all interacting in a complex way. The effects of age, sex, and handedness are less certain. The effect of therapy remains controversial and difficult to measure. As knowledge of the other factors increases, more accurate studies considering those factors can be designed to address some of these issues. The clinical patterns are more compatible with models of reorganization, redundancy, substitution, and diaschisis than with those of regeneration in the central nervous system.

References

Ades, H. W., and D. H. Raab. 1946. Recovery of motor function after two-stage extirpation of area 4 in monkeys. *Journal of Neurophysiology* 9: 55–60.

Alajouanine, T. 1956. Verbal realization in aphasia. *Brain* 79: 1–28.

Basser, L. S. 1962. Hemiplegia of early onset and the faculty of speech with special reference to the effects of hemispherectomy. *Brain* 85: 427–460.

Basso, A., E. Capitani, and L. A. Vignolo. 1979. Influence of rehabilitation on language skills in aphasic patients: A controlled study. *Archives of Neurology* 36: 190–196.

Basso, A., P. Faglioni, and L. A. Vignolo. 1975. Etude coutrôlée de la reéducation du langage dans l'aphasie: Comparaison entre aphasiques traités et non-traitée. *Revue Neurologique* 131: 607–614.

Bond, M. R. 1975. Assessment of psychosocial outcome after severe head injury. In *Outcome of Severe Damage to the Central Nervous System*. Amsterdam: Elsevier.

Braun, J. J., P. M. Meyer, and D. R. Meyer. 1966. Sparing of a brightness habit in rats following visual decortication. *Journal of Comparative and Physiological Psychology* 61: 79–82.

Broida, H. 1977. Language therapy effects in long term aphasia. *Archives of Physical Medicine and Rehabilitation* 58: 248–253.

Bucy, P. C. 1934. The relation of the premotor cortex to motor activity. *Journal of Nervous and Mental Disease* 79: 621–630.

Butfield, E., and O. L. Zangwill. 1946. Re-education in aphasia: A review of 70 cases, *Journal of Neurology, Neurosurgery and Psychiatry* 9: 75–79.

Campbell, D. C ., and J. M. Oxbury. 1976. Recovery from unilateral visuospatial neglect. *Cortex* 12: 303–312.

Cannon, W. F., and A. Rosenbluth. 1949. *The Supersensitivity of Denervated Structures*. New York: Macmillan.

Culton, G. L. 1969. Spontaneous recovery from aphasia. *Journal of Speech and Hearing Research* 12: 825–832.

Culton, G. L. 1971. Reaction to age as a factor in chronic aphasia in stroke patients. *Journal of Speech and Hearing Disorders* 36: 563–564.

David, R., P. Enderby, and D. Bainton. 1982. Treatment of acquired aphasia: Speech therapists and volunteers compared. *Journal of Neurology, Neurosurgery and Psychiatry* 45: 957–961.

Dax, M. 1865. Lésions de la moitié gauche de l'encéphale coincidant avec l'oubli des signes de la pensée (lu a Montpellier en 1836). Gaz Hebd 2ème série.

Duffy, F. H., S. R. Snodgrass, J. L. Burchfiel, et al. 1976. Pharmacological reversal of deprivation amblyopia in the cat. Paper presented at Twenty-eighth Annual Meeting of American Academy of Neurology, Toronto.

Finger, S. 1978. Environmental attenuation of brain lesion symptoms. In *Recovery From Brain Damage*, ed. S. Finger. New York: Plenum.

Fritsch, G. T., and E. Hitzig. 1870. Uber die elektrische Erregbarkeit des Grosshirns. *Archiv für Anatomie und Physiologie* 300–332.

Geschwind, N. 1974. Late changes in the nervous system: An overview. In *Plasticity and Recovery of Function in the Central Nervous System* New York: Academic.

Gloning, L., K. Gloning, G. Haub, et al. 1969. Comparison of verbal behavior in right-handed and non-right-handed patients with anatomically verified lesions of one hemisphere. *Cortex* 5: 43–52.

Gloning, K., R. Trappl, W. D. Heiss, and R. Quatember. 1976. Prognosis and speech therapy in aphasia. In *Neurolinguistics, 4, Recovery in Aphasics*, ed. Y. Lebrun and R. Hoops. Amsterdam: Swets & Zeitlinger.

Godfrey, C. M., and E. Douglass. 1959. The recovery process in aphasia. *Canadian Medical Association Journal* 80: 618–624.

Goldman, P. S. 1974. An alternative to developmental plasticity: Heterology of CNS structures in infants and adults. In *Plasticity and Recovery of Function in the Central Nervous System*. New York: Academic.

Hagen, C. 1973. Communication abilities in hemiplegia: Effect of speech therapy. *Archives of Physical Medicine and Rehabilitation* 54: 454–463.

Head, H. 1926. *Aphasia and Kindred Disorders of Speech.* Cambridge University Press.

Hirsch, H. V. B., and M. Jacobson. 1974. The perfect brain. In *Fundamentals of Psychobiology*, ed. Gazzaniga and Blakemore. New York: Academic.

Kenin, M., and L. Swisher. 1972. A study of pattern of recovery in aphasia. *Cortex* 8: 56–58.

Kennard, M. A., and W. S. McCulloch. 1943. Motor response to stimulation of cerebral cortex in absence of areas 4 and 6 (*Macaca mulatta*). *Journal of Neurophysiology* 6: 181–190.

Kertesz, A. 1979a. *Aphasia and Associated Disorders.* New York: Grune and Stratton.

Kertesz, A. 1979b. Visual agnosia: The dual deficit of perception and recognition. *Cortex* 15: 403–419.

Kertesz, A. 1981. Evolution of aphasic syndromes. *Topics in Language Disorders* 15–27.

Kertesz, A., and D. F. Benson. 1970. Neologistic jargon: A clinicopathological study. *Cortex* 6: 362–386.

Kertesz, A., W. Harlock, and R. Coates. 1979. Computer tomographic localization, lesion size and prognosis in aphasia. *Brain and Language* 8: 34–50.

Kertesz, A., and P. McCabe. 1977. Recovery patterns and prognosis in aphasia. *Brain* 100: 1–18.

Kertesz, A., and A. Sheppard. 1981. The epidemiology of aphasic and cognitive impairment in stroke—Age, sex, aphasia type and laterality differences. *Brain* 104: 117–128.

Kreindler, A., and A. Fradis. 1968. *Performance in Aphasia. A Neurodynamical, Diagnostic and Psychological Study.* Paris: Gauthier-Villars.

Lashley, K. S. 1938. Factors limiting recovery after central nervous lesions. *Journal of Nervous and Mental Disease* 88: 733–755.

Levine, D. N., R. Calvanio, and E. Wolf. 1980. Disorders of visual behavior following bilateral posterior cerebral lesions. *Psychological Research* 41: 217–234.

Liu, C. N., and W. W. Chambers. 1958. Intraspinal sprouting of dorsal root axons. *Archives of Neurology* 79: 46–61.

Lomas, J., and A. Kertesz. 1978. Patterns of spontaneous recovery in aphasic groups: A study of adult stroke patients. *Brain and Language* 5: 388–401.

Luria, A. R., V. L. Naydin, L. S. Tsvetkova, et al. 1969. Restoration of higher cortical function following local brain damage. In *Handbook of Clinical Neurology*, vol. 3, ed. Vinken and Bruyn. Amsterdam: North-Holland.

Mandleberg, L. A., and D. M. Brooks. 1975. Cognitive recovery after severe head injury. *Journal of Neurology, Neurosurgery and Psychiatry* 38: 1121–1126.

Marks, M. M., M. L. Taylor, and L. A. Rusk. 1957. Rehabilitation of the aphasic patient: A survey of three years' experience in a rehabilitation setting. *Neurology* 7: 837–843.

Messerli, P., A. Tissot, and J. Rodriguez. 1976. Recovery from aphasia: Some factors of prognosis. In *Neurolinguistics, Vol. 4, Recovery in Aphasics*, ed. LeBrun and Hoops. Amsterdam: Swets and Zeitlinger.

Meyer, J. S., Y. Shinohara, T. Kanda, et al. 1970. Diaschisis resulting from acute unilateral cerebral infarction. *Archives of Neurology* 23: 241–247.

Meyer, J. S., K. M. A. Welch, S. Okamoto, et al. 1974. Disordered neurotransmitter function. *Brain* 97: 655–664.

Mohr, J. P., M. S. Pessin, S. Finkelstein, et al. 1978. Broca aphasia: Pathologic and clinical aspects. *Neurology* 28: 311–324.

Moore, R. Y. 1974. *Central Regeneration and Recovery of Function*. New York: Academic.

Munk, H. 1881. *Ueber die Funktionen der Grosshirnrinde. Gesammelte Mitteilungen aus den Jahren 1877–1880*. Berlin: Hirschwald.

Nottebohm, F. 1970. Ontogeny of bird song. *Science* 167: 950–956.

Sands, E., M. T. Sarno, and D. Shankweiler. 1969. Long-term assessment of language function in aphasia due to stroke. *Archives of Physical Medicine and Rehabilitation* 50: 202–207.

Sarno, M. T., and E. Levita. 1971. Natural course of recovery in severe aphasia. *Archives of Physical Medicine and Rehabilitation* 52: 175–179.

Sarno, M. T., M. Silverman, and E. Sands. 1970. Speech therapy and language recovery in severe aphasia. *Journal of Speech and Hearing Research* 13: 607–623.

Schneider, G. E. 1973. Early lesions of superior colliculus: Factors affecting the formation of abnormal retinal projections. *Brain Behavior and Evolution* 8: 73–109.

Schuell, H., J. J. Jenkins, and E. Jimenez-Pabon. 1964. *Aphasia in Adults: Diagnosis, Prognosis and Treatment* New York: Harper & Row.

Shewan, C. M., and A. Kertesz. 1981. Language Therapy and Recovery from Aphasia. Paper presented at Academy of Aphasia, London, Ontario.

Smith, A., R. Chamoux, J. Leri, et al. 1972. Diagnosis, Intelligence and Rehabilitation of Chronic Aphasics. Final Report, Department of Physical Medicine & Rehabilitation, University of Michigan.

Sperry, R. W. 1947. Effect of crossing nerves to antagonistic limb muscles in the monkey. *Archives of Neurology and Psychiatry* 58: 452–473.

Stein, D. G., J. J. Rosen, and N. Butters, eds. 1974. *Plasticity and Recovery of Function in the Central Nervous System*. New York: Academic.

Subirana, A. 1969. Handedness and cerebral dominance. In *Handbook of Clinical Neurology*, vol. 4, ed. Vinken and Bruyn. New York: Elsevier.

Vignolo, L. A. 1964. Evolution of aphasia and language rehabilitation: A retrospective exploratory study. *Cortex* 1: 344–367.

Von Monakow, C. 1914. *Die Lokalisation im Grosshirn und der Abbau der Funktionen durch corticale Herde*. Wiesbaden: Bergmann.

Ward, A. A., Jr., and M. A. Kennard. 1942. Effect of cholinergic drugs on recovery of function following lesions of the central nervous system. *Yale Journal of Biological Medicine* 15: 189–229.

Watson, C. W., and M. A. Kennard. 1945. The effect of anticonvulsant drugs on recovery of function following cerebral cortical lesions. *Journal of Neurophysiology* 8: 221–231.

Weisenburg, T., and K. McBride. 1935. *Aphasia*. New York: Commonwealth Fund.

Wertz, R. T., M. J. Collins, D. Weiss, J. F. Kurtzke, et al. 1981. Veterans Administration cooperative study on aphasia: A comparison of individual and group treatment. *Journal of Speech and Hearing Research* 24: 580–594.

Yarnell, P., P. Monroe, and L. Sobel. 1976. Aphasia outcome in stroke: A clinical neuroradiological correlation. *Stroke* 7: 516–522.

Chapter 13

The Anatomy of Language: Lessons from Comparative Anatomy

Albert M. Galaburda

Comparative studies in neuroanatomy for the purpose of clarifying the anatomical substrates underlying language function in humans have not been performed to a major extent recently. This is due mostly to the notion that the absence of language in animals makes animal experimentation and animal models irrelevant. Nevertheless, for almost two centuries it has been known that the brains of most land mammals, especially those of nonhuman primates, show striking similarities to the human brain. These similarities are most clearly visible in the cerebral architecture of many cortical areas. Furthermore, insofar as fiber connections can be shown in lesioned human brains in which degenerating neurons can be impregnated with silver, the pattern of connections thus demonstrated often matches well with connections outlined in animals by experimental lesions and the injection of anterograde and retrograde tracers. Even a novice can accept architectonic homologies of gigantopyramidal area 4 in the human and the monkey brain by virtue of the presence in both species of unusually large pyramidal neurons in cortical layer V in the precentral region. With only a little more training, one can easily distinguish structural equivalencies among species for visual area 17 with its tripartite layer IV and Meynert pyramids, or for auditory area 41, with its rainshower arrangement of delicate granular neurons. There are many other examples. Likewise, the tracing of connections of a centrifugal area such as area 4 shows in both human and nonhuman brains that the axons innervate the motor neurons of the ventral horn of the spinal cord. In a parallel fashion, centripetal cortical sensory areas such as 17 and 41 receive projections from equivalent thalamic nuclei in all mammalian species studied. The clearer specification of architectonic areas and connection pathways, especially in studies made on nonhuman primate brains, have contributed significantly to our understanding of the possible anatomical systems underlying such functions as ocular dominance (Hubel and Wiesel 1974), memory (Mishkin 1982), and pyramidal motor function (Kuypers 1962) in man.

Language is a typically human behavior, yet precursors of language and anatomical precursors may be found in nonhuman primates. Furthermore, it is conceivable that, although the precursors of language may have little to do with human language, they may be carried out by structures that do not differ greatly from the human language areas. By analogy, an animal with small webs between its toes may sink in water, whereas its cousin with slightly bigger webs can swim. A totally new function develops with a trivial anatomical change.

Furthermore, homologous areas in humans and nonhuman primates may share a vast number of functions except for language. Take for example the case of reading. It is well known that there are certain structures in the human brain, mostly in the parietal lobe, that when lesioned result in disorders of reading. Yet until recently nearly all of the human population was illiterate. Thus, it is likely that reading requires certain neurophysiological capabilities mediated through the parietal lobe—capabilities that can be used (and are probably used) to handle other tasks too. Therefore, it is conceivable that many of those capabilities exist in the parietal lobe of the monkey, short of the degree of development required for the emergence of language in that animal. Therefore, it is not surprising to find that, when single neuronal units are sampled in the inferior parietal lobule, many cells show response to multiple sensory modalities (Hyvärinen 1981). This may be a requirement for the intermodal processing needed in reading. Furthermore, stimulation in areas homologous to the human area for vocalization (the anterior cingulate region and the anterior pericingulate area) produces similar vocalization in nonhuman primates (Jürgens and Ploog 1970). These observations, coupled with the great difficulty of obtaining detailed specification of connections and physiology in the human brain, suggest that increased efforts be made in the direction of animal experimentation for the purpose of clarifying brain-behavior relationships with respect to language function. One of the most striking brain-behavior relationships in human language has to do with brain asymmetry and lateralization of language to the left hemisphere in the majority of people. Analysis of nonhuman brains may provide information about the possible origin of this anatomical asymmetry.

Insight into the Possible Nature of Duality and Cerebral Asymmetries

Architectonics refers to the study of the arrangement of cells or other units in the nervous system as seen under lower-power microscope magnification. In the cortex, the prominent architectural feature is the arrangement of cells in layers and columns. On the basis of specific layered appearance, cell sizes, packing densities, and special cellular features, the cortex can be divided into areas of different architectural structure. If cellular perikarya are stained, the analysis is termed cytoarchitectonics; if differences in patterns of myelination are sought, it is termed myeloarchitectonics; if blood vessels are stained, angioarchitectonics; and so on.

A considerable amount of work has been done on comparative architectonics in primates. Sanides (1972) described in great detail some of the changes that occur in architectonic patterns and shifts in architectonic topography from primitive insectivores through prosimians and other primates, including man. According to Sanides, primates have the highest development of the isocortex (neocortex, six-layered cortex), as manifested by greater differentiation and size. The isocortex is also highly variable, and it is the part of the cortex that is most likely to show interspecies differences.

Sanides has claimed that the isocortex has its phylogenetic and ontogenetic origin in primitive cellular moieties making up the cortical plate. These primitive moieties comprise archicortical (hippocampal) and paleocortical (olfactory and insular) layered structures, as seen in amphibians. Both archicortex and paleocortex give rise to the nearly six-layered cortex (proisocortex), present in reptiles. Proisocortex in turn gives rise to isocortex (true six-layered cortex), seen indisputably only in mammals. As already noted, the amphibian brain does not contain true cortex, although the two primitive moieties represented by archipallium and paleopallium are already present. Although the detailed physiology and connections of these two early structures are not known, the available evidence in amphibians and reptiles suggests that these structures possess a separate set of connections and different functions. The archipallium appears to receive its major input from the hypothalamus, but also receives some projections from the thalamus, most of which carry somesthetic and visual information (Northcutt and Kicliter 1980). The lateral pallium, on the other hand, appears to receive most of the olfactory input and does not have direct connections with the hypothalamus. The output of the paleopallium seems to be via the archipallium. (An additional interesting observation is that the connections to the medial cortex are bilateral, whereas those of the lateral cortex are by and large unilateral. This may be relevant to the issue of brain asymmetry.)

The different arrangement of the medial and lateral cortices provides for the possibility that the archipallium, because of its hypothalamic connections, responds to changes in the internal environment as well as to changes in some aspects of the external environment (mostly visual), while the paleopallium responds only to changes in the external environment. Thus, for instance, changes in the body related to changes in the body fluids and to visceral activities would stimulate the archipallium to cause exploration, avoidance, and changes in posture and attention. These movements would not be obviously explained by an outsider's assessment of the animal's external environment. On the other hand, the paleopallium would more clearly be activated by events (olfactory) occurring in the animal's external environment. A dichotomy of activities would thus be set up, with one type comprising reflexes and elaborations of reflexes to changes in the internal (most often chemical) milieu and the other type comprising reflexes and elaborations of reflexes to changes in the external (initially, as in the case of the frog, olfactory) milieu.

In the reptile it is possible to view a further elaboration of the dichotomy mentioned above. Here the medial cortex again mediates the output from the lateral cortex. In lizards, the main portion of visual projections to the cortex is directed to a region lying between the medial and lateral cortices, known as the dorsal cortex (Butler 1980). This is the region of the reptilian brain most similar to the mammalian isocortex, and it probably represents the proisocortical association areas for the medial and lateral pallia.

With the passage from the reptile to the mammalian line, the first isocortex makes its appearance, and it is here that it is possible to distinguish clear anatomical trends beginning both from the medial archicortex and from the lateral paleocortex.

The architectonic characteristic of the medial system is best characterized as pyramidal. Thus, cortical areas located on the medial surface or medially on the convexity tend to be rich in moderately sized and large pyramidal neurons. Striking examples of this feature are provided by the appearance of the dorsal portion of motor area 4 (gigantopyramidalis) and pericingulate supplementary motor areas.

With increasing topographic removal from the medial influence, the archicortically derived cortices become progressively less pyramidal and more granular, thus exhibiting the influence of the lateral paleocortical system. As in the amphibians, the paleocortex in mammals begins with structures receiving the main portion of the olfactory input: the pyriform cortex, the tuberculum olfactorium, and the rostral amydala. The influence of these primitive cortical moieties is seen in the basal and opercular frontal lobe, in the anterior insula, and (in primates) in the temporal pole—all areas with characteristically granular appearance despite their primitive status. Whereas the medial pyramidal influence peaks with the development of the dorsal motor area 4, the granular development reaches its maximum within the perisylvian somesthetic auditory and occipitopolar visual representations.

Sanides has pointed out that the medial influence in the brain of the monkey is seen most strongly beginning in the medial hemisphere, around the dorsum and as far down as the principal sulcus. The lateral olfactory insular influence reaches up from the undersurface of the frontal lobe to the level of the principal sulcus from below. The insula and its opercular derivatives all show the influence of the olfactory moieties. In the parietal lobe, the medial pyramidal and lateral granular influences border on the banks of the intraparietal sulcus (Eidelberg and Galaburda 1983). In the temporal lobe, the olfactory insular influence reaches down to the second temporal gyrus.

Both somesthetic and motor representations show a medial influence dorsally and a lateral influence ventrally. The auditory representation is, to our knowledge, bound solely to the olfactory-insular influence. The visual representation receives its medial influence from the retrosplenial cortices and receives its lateral influence from the temporal pole via the second temporal gyrus.

Architectonic analysis of the human brain discloses a similar arrangement. The medial influence affects the architecture of the superior and middle frontal gyri, the superior parietal lobule, the retrosplenial visual cortex, and the inferomedial temporal lobe. The lateral influence is manifested in the orbitofrontal cortex, the inferior frontal gyrus, the inferior parietal lobule, the occipital pole, the lateral visual areas, and the first and second temporal gyri.

Cerebral Asymmetries

Asymmetries have been demonstrated in the brains of human and nonhuman primates (Galaburda et al. 1978) and, more recently, in the brains of other mammals. In the human brain, most of the documented asymmetries have been restricted to regions surrounding the sylvian fissures. Thus, in the frontal lobe, asymmetries have been demonstrated to occur in the inferior frontal gyrus, whereby the left side appears to be more convoluted and complex (Falzi, Perrone, and Vignolo 1982), and the main architectonic area of the pars opercularis of the inferior frontal gyrus, which is a markedly granular frontal field, tends to be larger on the left side. Likewise, the left parietal operculum appears to be better developed than the right, as is suggested by the more horizontal and longer extent of the left sylvian fissure. Furthermore, granular architectonic area 39 (Brodmann), lying on the angular gyrus, tends to be larger on the left (Eidelberg and Galaburda 1983).

Below the sylvian fissure in the temporal operculum, a structure known as the planum temporale lying posteriorly from Heschl's gyrus is larger on the left side in the majority of brains (Geschwind and Levitsky 1968). Accompanying this gross anatomical asymmetry in the planum temporale, auditory architectonic area Tpt is also more often larger on the left side (Galaburda, Sanides, and Geschwind 1978).

Asymmetries in the rates of development of perisylvian structures have also been demonstrated. The frontal and parietal opercula and the sulci in the vicinity of the sylvian fissures, including Heschl's gyrus and the planum temporale, appear first on the right side. Finally, an asymmetry has been shown to be present in the human posterior thalamus. Eidelberg and Galaburda (1982) have shown significant left-sided preponderance in the lateralis posterior nucleus, an area that is richly connected with the posterior perisylvian cortex (especially the inferior parietal lobule).

The greater size of the perisylvian cortices on the left may not be a hard and fast rule. Von Economo and Horn (1930) found that the primary auditory cortex might be, on the average, larger on the right. This finding, however, has not been tested in enough brains to document a population bias. On the other hand, in a recently completed study on the parietal lobe (Eidelberg and Galaburda 1983) a pyramidalized cortex on the dorsal lip of the inferior parietal lobule, but clearly belonging to the medial trend, was found to be consistently larger on the right side. Finally, analysis of Fontes's fetal-brain photographs (1944) discloses that, contrary to the situation in perisylvian sulci, the dorsomedial sulci seem to appear first on the left side, thus suggesting a slower development of right medial cortical areas. One might be able to conclude that slowly developing areas, such as perisylvian cortex on the left and parasaggital cortex on the right, grow larger.

On the basis of the above-mentioned observations a notion can be created that suggests that asymmetries in the cortex may be related, at least in part, to the phylogenetic origin of the particular cerebral cortex in question. Thus, it may appear that the medial dorsal cortices, which are more closely related to the primitive archipallium, can reach better development on the right side, whereas the lateral perisylvian cortices, which relate more to the paleopallium, are better developed on the left side.

I have stated that the architectonic influence may be greater on the right side by the medial cortices and on the left side by the lateral cortices. A parallel dichotomy of influence may exist at the functional level. The lateral cortices, which derive from the primitive olfactory system, may be better suited for the processing of interactions with the external environment, whereas the medial cortices, which derive from the primitive archipallium, may be better able to process interactions with the organism's internal milieu. The bias in the representation of the two types of cortices in the two hemispheres is consistent with the notion that the left hemisphere has dominance over language, while the right hemisphere rules over motivational and attentional aspects of behavior.

Because the lateralization of the two cortical prototypes is not complete (both hemispheres contain both types), it cannot be expected that lateralization of function will be complete. Also, because of the tendency toward bilaterality of cortical connections of the medial system, as compared with the unilaterality of the lateral system, it might be expected that the lateral system is functionally more lateralized than the medial. Clinical observations in favor of this notion abound. Unilateral lesions producing hemiplegia recover in the trunk and lower limbs (dorsomedial) much sooner and more completely than in the hands (ventrolateral). Recovery from medial lesions producing aphasia is usually fast, whereas perisylvian lesions tend to produce longer-lasting deficits. The blood supply to the medial regions is also less strongly lateralized, as evidenced by the anatomy of the anterior cerebral and anterior communicating arteries and by the related observation that infarctions are far less common in these regions than in the perisylvian cortices.

In summary: The lateralization of the cerebral cortex may stem from an evolutionary difference in the influence of the medial and lateral cortical archetypes on the two hemispheres, and the more striking lateralization of ventrolateral perisylvian regions may relate to the ancient bilaterality of connections for the medial regions and unilateral connections for the lateral regions. There is as yet no explanation for the tendency to lateralize in the first place.

Language Areas

Two sets of language areas can be distinguished in the left cerebral cortex. One set is located in the medial portion of the hemisphere surrounding the cingulate gyrus and is part of what are referred to as the supplementary motor and sensory areas. The other set is in a perisylvian location and is distributed from the frontal operculum through the parietal operculum and around the posterior end of the sylvian fissure to the temporal operculum and the superior temporal gyrus. The exact anatomical extension of the medial speech areas, the supplementary speech areas, is not known. Braak (1979) has shown in lipofuscin preparations that the supplementary motor region contains peculiarly staining pyramids. He has also described a magnopyramidal zone within this region that may correspond to the supplementary speech area. Both in animals and in man, stimulation of this region produces vocalization (Penfield and Roberts 1959). Destruction of the supplementary speech areas in man produces a disorder that is characterized primarily by paucity of speech and failure of speech initiation but fails to show the more typical paraphasic disturbances of perisylvian lesions (Rubens 1976). A large medial lesion encompassing most of the supplementary motor and sensory regions produces a syndrome in which the patient has no spontaneous speech and no comprehension of spoken language, although he tends to echo the examiner's remarks. Unilateral left-side lesions produce only transient phenomena, and it requires bilateral lesions to produce permanent mutism (Freemon 1971). This is in marked contrast with the situation seen with unilateral perisylvian lesions, which (especially in right-handers) tend to produce longer-lasting aphasic disturbances. It is, therefore, apparent that lateralization for language is not as striking in the medial zone as in the perisylvian location, and that the influence of the right hemisphere is also more significant in the medial zone than in the perisylvian location. It appears, furthermore, that the disturbances of language associated with medial lesions do not reflect an abnormality in auditory decoding and motor encoding, but rather one in motivation and attention. The pyramidal appearance of these supplementary motor and sensory cortices, as well as their location, suggests their intimate relationship to the primitive archipallium. It may follow that the medial language representation would be expected to be more developed on the right side. Although no such right-sided lateralization is evident from lesion studies and stimulation, the fact that the medial system is more bilateral suggests that the right hemisphere has a greater role in medial language localization than the right hemisphere does in the perisylvian region.

The perisylvian language cortices have a variety of architectonic and connectional characteristics. First, they all show the granular influence of the orbito-insular primordium. In the frontal lobe the most granular cortices are found in the inferior frontal gyrus, a site of language localization. Likewise, the most granular cortices in the parietal lobe besides the somesthetic postcentral area 3 are found in the inferior parietal lobule—especially in area 39 in the angular gyrus, which is also relevant for language function. In the temporal lobe the best-developed granularity is seen in the superior temporal gyrus, especially posteriorly, where again language localization is demonstrated.

The organization of the perisylvian cortex in nonhuman primates is similar to that in a human brain. The frontal cortices below the principal sulcus in the rhesus monkey are the most granular of the frontal cortices. Likewise, the granularity in the inferior parietal lobule in this animal is greater than that in the superior parietal lobule. The granular influence in the temporal lobe is likewise best seen in the superior temporal gyrus of this animal, primarily again in the posterior one-third to one-half of this gyrus. Although the architectonic details between the species vary, the cortices of the perisylvian region in the human and the nonhuman primate retain similar interrelationships. For instance, granularity and the strength of pyramids in layer IIIc increase in the rhesus monkey as one moves from the cortex in front of the inferior limb of the arcuate sulcus to that lying in back of it in the same manner as when one moves from the pars triangularis of the human frontal lobe backward toward the pars opercularis. In the rhesus monkey the best-developed IIIc pyramids in the inferior premotor region are found in a small area of cortex in the caudal bank of the inferior-limb arcuate sulcus (Galaburda and Pandya 1982). In the human brain the best development is seen in the pars opercularis. Thus, on architectonic grounds, the monkey and human inferior prefrontal cortices can be thought to be homologous in structure. In the superior temporal region, parallel trends of differentiation can be distinguished in the human and nonhuman brains, although cortical thickness, cell packing density, and cell size differ. In the nonhuman primate brain, the largest IIIc pyramids are seen in an area on the superior temporal gyrus known as PaAlt (Pandya and Sanides 1973; Galaburda and Pandya 1983), while the greatest parietal influence is seen in an area caudal to PaAlt known as Tpt. In the human brain the greatest development of IIIc pyramids is seen in an area on the superior temporal gyrus known as PaAi, and the largest parietal influence is seen in an area caudal to PaAi also known as Tpt. Areas PaAlt and Tpt in the monkey and PaAi and Tpt in the human exhibit a similar topograhic relationship with the primary auditory cortex. Similarly, in the inferior parietal lobule several areas can be distinguished in both species. In both species the greatest degree of layered homotypical differentiation of the cortex is found in the posterior half of the inferior parietal lobule, in an area termed PG by Von Economo (1929) (and 39 by Brodmann) in both the human and the nonhuman brain.

Thus, it is possible on architectonic grounds to distinguish homologous structures in human and nonhuman-primate brains even though their functions may be behaviorally distinct. The architectonic homology, however, permits the notion that these structures may be connected in a similar way among the primate species. It is for this reason that architectonically controlled connectional studies in nonhuman primates may provide useful information as to the possible connectional organization of the language areas in the human brain. It has been found in the rhesus monkey that injection of horseradish peroxidase in the small field that is architectonically homologous to the field on the human pars opercularis and that lies caudal to the inferior limb of the arcuate sulcus shows labeled neurons in the posterior part of the superior temporal gyrus (area Tpt) in the monkey. Similarly, injections of isotope-labeled proline-leucine into area Tpt of a rhesus monkey results in labeled terminations in the same inferior premotor field (Galaburda and Pandya 1982). Thus, a connection between area Tpt and the granular inferior premotor field can be established in the rhesus monkey, and it is similar to the proposed connection between the posterior and anterior perisylvian speech areas in the human brain (Geschwind 1970)—namely, the arcuate fasciculus. The arcuate fasciculus has been demonstrated in blunt dissections as well as in silver-impregnation studies in the human brain on numerous occasions, but injection studies in the rhesus monkey allow much finer localization of the origins and terminations of the fibers connecting these two architectonic areas.

Connectional studies in the monkey have outlined additional fiber connections between the posterior superior temporal region and the inferior parietal lobule, as well as with the monkey's posterior pericingulate region, which is homologous to the supplementary sensory area (Pandya and Seltzer 1982). Likewise, injections and lesions in the inferior postarcuate region of the monkey have outlined connections to areas 6 and 4 as well as to the anterior pericingulate region, which is homologous to the supplementary motor area (Pandya, Hallett, and Mukherjee 1969; personal observations).

Insofar as the connectional anatomy of language areas in the human brain can be specified from the analysis of lesioned brains and from blunt dissection, it shows striking similarities to that of the monkey. These anatomical observations, coupled with evidence from behavioral studies, stimulation, evoked responses, and single-unit recordings from the brains of animals, suggest that the basic physiological processes carried out by these anatomically similar structures may share many features. For instance, stimulation of the anterior pericingulate region in the monkey produces vocalization (Jürgens and Ploog 1970); single-unit recordings from the inferior premotor and prefrontal region in the rhesus monkey disclose sensory-dependent cells, reflecting parietal and temporal relations (Rizzolatti et al. 1981); single-unit recordings in the inferior parietal lobule show multimodally responsive neurons (perhaps permitting visual naming when it occurs in man) (Hyvärinen 1981); and recordings from cells in area Tpt in the monkey show cells that are predominantly auditory but also offer responses from somatosensory and visual stimuli (Leinonen, Hyvärinen, and Sovijarvi 1980) (perhaps allowing auditory comprehension when it occurs in man).

In conclusion: The detailed analysis of extant animal brains, especially those of nonhuman primates, with special attention to the phylogenetic issues, is apt to increase our knowledge of the organization of the brain for language function in the human. It is only through these efforts that potentially useful anatomical models can be devised. The discovery of animal models offers a strong possibility for uncovering the preadapted physiological processes that, in the human, contribute to the activities of language.

Acknowledgments

This work was supported in part by NINCDS grant NS14018, by a Biomedical Support Grant to the Beth Israel Hospital, Boston, and by grants from the Underwood Co. to the Orton Society and from the Essel Foundation.

References

Braak, H. 1979. The pigment architecture of the human frontal lobe. *Anatomy and Embryology* 157:35–68.

Butler, A. B. 1980. Cytoarchitectonic and connectional organization of the lacertilian telencephalon with comments on vertebrate forebrain evolution. In *Comparative Neurology of the Telencephalon*, ed. S. O. E. Ebesson. New York: Plenum.

Eidelberg, D. and A. M. Galaburda. 1982. Symmetry and asymmetry in the human posterior thalamus. I. Cytoarchitectonic analysis in normal persons. *Archives of Neurology*. 39:325–332.

Eidelberg, D., and A. M. Galaburda. 1983. Inferior parietal lobule: Divergent architectonic asymmetries in the human brain. *Archives of Neurology* (in press).

Falzi, G., P. Perrone, and L. Vignolo. 1982. Right-left asymmetry in anterior speech region. *Archives of Neurology* 39:239–240.

Fontes, V. 1944. Morfologia do Cortex Cerebral. Monografia do Boletim do Instituto do Antonio Aureleo Da Costa Ferreira (Lisbon).

Freemon, F. R. 1971. Akinetic mutism and bilateral anterior cerebral artery occlusion. *Journal of Neurology, Neurosurgery, and Psychiatry* 34:693–698.

Galaburda, A. M., M. LeMay, T. C. Kemper, and N. Geschwind. 1978. Right-left asymmetries in the brain. *Science* 199:852–856.

Galaburda, A. M., and D. N. Pandya. 1982. Role of architectonics and connections in the study of primate brain evolution. In *Primate Brain Evolution*, ed. E. Armstrong and D. Falk. New York: Plenum.

Galaburda, A. M., and D. N. Pandya. 1983. The intrinsic architectonic and connectional organization of the superior temporal region of the rhesus monkey. *Journal of Comparative Neurology* (in press).

Galaburda, A. M., F. Sanides, and N. Geschwind. 1978. Human brain: Cytoarchitectonic left-right asymmetries in the temporal speech region. *Archives of Neurology* 35:812–817.

Geschwind, N. 1970. The organization of language and the brain. *Science* 170:940–944.

Geschwind, N., and W. Levitsky. 1968. Human brain: Left-right asymmetries in temporal speech region. *Science* 161:186–187.

Hubel, D. H., and T. N. Wiesel. 1974. Sequence regularity and geometry of orientation columns in the monkey striate cortex. *Journal of Comparative Neurology* 158:267–294.

Hyvärinen, J. 1981. Regional distribution of functions in parietal association area 7 of the monkey. *Brain Research* 206:287–303.

Jürgens, J., and D. Ploog. 1970. Cerebral representation of vocalization in the squirrel monkey. *Experimental Brain Research* 10:532–554.

Kuypers, H. G. J. M. 1962. Corticospinal connections: Postnatal development in the rhesus monkey. *Science* 138:678–680.

Leinonen, L., J. Hyvärinen, and A. Sovijarvi. 1980. Functional properties of neurons in the temporo-parietal association cortex of the awake monkey. *Experimental Brain Research* 39:203–215.

Mishkin, M. 1982. A memory system in the monkey. *Philosophical Transactions of the Royal Society of London B* 298:85–95.

Northcutt, R. G., and E. Kicliter. 1980. Organization of the amphibian telencephalon. in *Comparative Neurology of the Telencephalon*, ed. S. O. E. Ebesson. New York: Plenum.

Pandya, D. N., M. Hallett, and S. K. Mukherjee. 1969. Intra- and interhemispheric connections of the neocortical auditory system in the rhesus monkey. *Brain Research* 14:49–65.

Pandya, D. N., and F. Sanides. 1973. Architectonic parcellation of the temporal operculum in rhesus monkey and its projection pattern. *Zeitschrift für Anatomie und Entwicklungsgeschichte* 139:127–161.

Pandya, D. N., and B. Seltzer. 1982. Intrinsic connections and architectonics of posterior parietal cortex in the rhesus monkey. *Journal of Comparative Neurology* 204:196–210.

Penfield, H., and L. Roberts. 1959. *Speech and Brain-Mechanisms.* Princeton, N.J.: Princeton University Press.

Rizzolatti, G., C. Scandolara, M. Matelli, and M. Gentilucci. 1981. Afferent properties of periarcuate neurons in macaque monkeys. II. Visual responses. *Behavior Brain Research* 2:147–163.

Rubens, A. B. 1976. Transcortical motor aphasia. In *Studies in Neurolinguistics*, volume 1, ed. H. Whitaker and H. A. Whitaker. New York: Academic.

Sanides, F. 1972. Representation of the cerebral cortex and its areal lamination patterns. In *The Structure and Function of Nervous Tissue*, volume 5, ed. G. H. Bourne. New York: Academic.

von Economo, C. 1929. *The Cytoarchitectonics of the Human Cerebral Cortex.* London: Oxford University Press.

von Economo, C., and L. Horn. 1930. Ueber Windungsrelief, Masse und Rindenarchitektonik der Supratemporalflaeche, ihre individuellen und ihre Seitenunterschiede. *Zeitschrift für die gesamte Neurologie und Psychiatrie* 130:678–757.

Chapter 14

Event-Related Potentials in the Study of Speech and Language: A Critical Review	Terence W. Picton Donald T. Stuss

Language is closely linked to the structure and function of the human brain. Behavioral studies of language in normal subjects and in patients with brain damage have led to theories about the localization of language processes in the brain. Physiological recordings might be extremely helpful in evaluating and extending these theories. Physiological studies have an advantage over behavioral studies in that they can examine the processes of language while they are occurring and not just when a behavioral response is produced. Physiology can therefore provide a line of evidence toward understanding language that is convergent with evidence from linguistics and neuroanatomy.

The most effective physiological tool for studying human language is the *event-related potential* (ERP). Event-related potentials are changes in the electrical activity of the nervous system that are temporally assocated with physical stimuli or psychological processes. They may be recorded in response to sensory stimuli, in which case they are called *evoked potentials.* They may also occur in the absence of external stimuli, either in association with a psychological process or in preparation for motor activity; in these cases they are called *emitted potentials* (Weinberg et al. 1974). ERPs can also be classified as exogenous or endogenous (Sutton et al. 1965). Exogenous potentials are largely determined by the physical parameters of a stimulus, whereas endogenous potentials are mainly affected by its psychological context. Evoked potentials may be exogenous or endogenous; usually the earlier components of an evoked potential are stimulus-determined and the later components more variable to psychological manipulation. Emitted potentials are always endogenous.

Event-related potentials are usually too small to be accurately evaluated in human scalp electroencephalograms (EEGs). They are lost in the wealth of electrical activities generated by the brain as it performs its many other functions. The most commonly used technique for distinguishing the ERP or "signal" from the background EEG or "noise" is signal averaging. Electrical recordings taken at the same time in relation to repeated events are averaged together. The background activity, being random with respect to the particular event being examined, will tend to cancel itself during the averaging process, whereas the ERP, being the same for every event, will remain constant during the averaging process. With averaging the ERP will become more and more recognizable in the decreasing EEG background. One main assumption underlying the use of averaging is that the ERP remains constant from trial to trial. Changes in the ERP, particularly in its latency, will distort the averaged waveform. This may be a particular problem in the study of language, since it is difficult to repeat a linguistic process in exactly the same way over many trials. Techniques exist to help control for variations in latency (Ruchkin and Sutton 1979), but these involve further assumptions about the ERP and require a fairly large ERP waveform. They have not yet been used in ERP studies of language.

Scalp-recorded ERPs register the electrical fields generated in the brain and propagated to the recording electrodes by volume conduction. Four corollaries to this basic principle are essential to understanding ERP studies of language.

First, it is important to realize that the fields recorded at the scalp can be generated by many noncerebral sources as well as by cerebral sources, in which case they are considered artifacts. The eyes, the scalp muscles, the skin, and the tongue are all sources of artifactual contamination in the recording of cerebral ERPs from the scalp (Picton 1980). These are particularly troublesome in the evaluation of language processes, because the tongue and the facial muscles are used in speech and the eyes are moved in reading.

Second, scalp ERPs can record only from those cerebral processes that generate electrical fields at a distance. Such fields are generated by neural events that are temporally synchronous and spatially oriented. Many cerebral processes occur, however, in neurons that are not similarly oriented or synchronously activated. The resultant "closed fields" cannot contribute to the scalp recording, although they may be essential to the process being assessed. The ERP cannot provide an exhaustive view of brain activity (Donald 1979). When the ERP changes, one can be sure that something is happening in the brain; however, cerebral processes can occur without changing the ERP. It is not difficult to hypothesize that many language processes can occur in neuronal systems that generate closed fields and that are therefore inaccessible to scalp recording.

Third, it is difficult to localize the cerebral source of a scalp-recorded field. An extensive evaluation of the field's distribution over the scalp and some idea of the locations and orientations of the possible generators are essential (Picton et al. 1978), but even then the source of a scalp-recorded potential may remain unknown. A generator is not necessarily located where the amplitude of the response is largest. The visual-evoked potential to a half-field pattern-reversal stimulus is recorded with greater amplitude over the occipital scalp ipsilateral rather than contralateral to the side of stimulation. This paradoxical localization occurs because the field generated in the medial surface of one occipital lobe is oriented laterally and picked up with larger amplitude over the opposite side (Blumhardt, Barrett, and Halliday 1977). Any attribution of an ERP asymmetry to a similar asymmetry of hemispheric processing must therefore be made very cautiously.

Fourth, any ERP recorded at the scalp can be generated by several sources with temporally and spatially overlapping fields. Each peak recorded in the scalp ERP does not necessarily reflect a separate cerebral process. The component structure of the ERP has to be determined in relation to how its waveform changes with experimental manipulation. A component is thus a "source of controlled, observable variability" (Donchin, Ritter, and McCallum 1978). One means of evaluating the component structures of the ERP (and, by inference, those of the underlying cerebral processes) is the *principal-component analysis* (PCA) (Glaser and Ruchkin 1976; Donchin and Heffley 1978; Picton and Stuss 1980; Rösler and Manzey 1981). This yields a set of components contributing independently to the experimental variance of the recorded ERP waveform. These components are often simplified using a "varimax rotation." A component is described by means of a set of "loadings" or "coefficients" that give the relative contribution of each time point in the ERP to the component. Each of the ERP waveforms entered into the PCA is then multiplied by these coefficients to give a "component score." Component scores can then be evaluated in much the same way as simple measurements of peak amplitude. An analysis of variance on the component scores will demonstrate how significantly a particular component is affected by the experimental manipulations. The technique is powerful and has been used extensively in ERP studies of language. Unfortunately, it is easy to overestimate the significance of the results, since the same variance that is used to determine the components is also used to evaluate their significance. Because the PCA distributes all of the recorded variance among the derived components, some of the components may be related only to the residual EEG noise present in any EEG tracing. Therefore, the results of PCAs should be accepted as definite only if they can be replicated using different sets of data; otherwise, they merely serve to describe the experimental variance. It is also important to realize that the relationship between the components derived from the PCA and the underlying cerebral generators is complex. A PCA component is not necessarily generated in a single brain region, nor is it necessarily generated within the brain. It may represent the combined activity of several independent brain regions, or it may be artifactual.

Several methodological requirements must therefore be fulfilled in order to allow appropriate interpretation of ERPs in the context of language. First, the linguistic process being examined must be carefully determined by the experimental manipulations and must be repeated to allow for averaging of the ERP. Second, the possibility of noncerebral artifacts must be rigidly controlled by instructions to the subject, by monitoring, and by discarding contaminated trials from the average. Whatever protocols for artifact rejection are used, the monitor should still be averaged in order to rule out subtle effects that are not obvious in single-trial waveforms. Third, the logic of a nonexhaustive measurement must be followed. The absence of ERP findings cannot be used to disprove the existence of a cerebral process, but the presence of ERP changes must be accounted for in any theory of the underlying physiology. Fourth, the component structure of the ERP and the localization of the generators for the components will require scalp distribution analysis. Inferences back to generator sources must be made cautiously and in association with the results of other experiments on lesions or intracerebral recordings. Finally, the component structure of the language-related potentials and hence the partial structure of the physiological processes underlying language must be carefully determined and replicated.

There are many caveats to the use of ERPs in the investigation of speech and language. Nevertheless, ERPs remain the only technique for millisecond-by-millisecond analysis of language physiology. Furthermore, ERPs can at times be recorded without requiring any behavioral response. This unobtrusiveness may be helpful when behavioral responses would alter the timing or the strategy of linguistic perception. ERPs can provide evidence for the existence of complex independent processes leading up to relatively simple perceptual responses, and they can indicate the relative timing of these processes (Donchin 1981; Lawson and Gaillard 1981). In association with an increasing knowledge about generator sources, they may help localize linguistic processes. With further understanding of the relationship between particular ERP components and linguistic processes, they may help in the measuring of such processing.

Speech-Related Potentials

Paradigms

Many experiments have investigated the scalp-recorded potentials that precede speech. The goal of these studies has been to localize the cerebral origin of speech and to use this localization in a clinical setting. One important clinical application would be to determine cerebral dominance. Two basic paradigms have been used. In one, the subject repeats an utterance in a self-paced manner. The analysis is triggered by and carried out both before and after the utterance. The activity prior to the trigger can be averaged by using either a tape recorder that is played back in reverse direction or some on-line delay system such as a tape loop or a "rotating buffer" within the computer. This process of backward or "opisthochronic" (Shibasaki et al. 1980) averaging can show the "readiness potential" or *Bereitschaftspotential* (BP) preceding speech as well as the actual "motor potential" (MP) controlling the speech muscles (Deecke, Grözinger, and Kornhuber 1976). The second paradigm involves "forewarned cued language acts" (House and Naitoh 1979). A warning stimulus (S1) precedes a second imperative stimulus or cue (S2) that requires a speech response. Between the stimuli there develops a slow baseline shift that reflects the subject's preparation to respond to the second stimulus. This baseline shift is usually negative in polarity and occurs only if the subject pays attention to the association between the warning and imperative stimuli; it is therefore called the *contingent negative variation* (CNV) (Walter et al. 1964). The CNV is probably composed of several subcomponents, among which are an orienting response to the first stimulus and a preparatory process for the upcoming response (Gaillard 1978). In both paradigms it is usual to compare right and left scalp locations and to compare speech production with nonlinguistic oral movements.

Early Studies

Ertl and Schafer (1967) published the first report of cortical activity preceding speech. They recorded from the right central region a positive-negative wave in the 150 milliseconds preceding speech. McAdam and Whitaker (1971) used a much wider frequency bandpass and found slow negative potentials of maximum amplitude over Broca's area prior to the spontaneous production of polysyllabic words. These speech potentials were larger over the left hemisphere than over the right, whereas the potentials preceding analogous nonspeech acts such as coughing or spitting were symmetrical. This lateralization of the slow negative wave preceding speech was confirmed in a CNV paradigm by Low and his colleagues (1973, 1977). They reported a left-sided lateralization of a prespeech negativity in most right-handed normal subjects and an opposite lateralization for left-handed subjects. They further found that the lateralization of the negative wave accurately predicted cerebral dominance as determined using intracarotid amytal (Wada 1949) in 12 out of 15 patients.

Stuttering

One of the initial applications of these ERP asymmetries was in the study of hemispheric dominance (or lack thereof) in stutterers. Unfortunately, the results have not been consistent. Zimmerman and Knott (1974) found a prespeech CNV of greater amplitude over the left inferior frontal regions in 4 out of 5 normal speakers. They could determine no clear lateralization for this wave, however, in 9 right-handed stutterers, despite the fact that the subjects could perform the required speech task without hesitation. Pinsky and McAdam (1980), however, found no asymmetry of the prespeech CNV in normal subjects and a greater left-sided CNV in stutterers. They also found a left-sided predominance for the readiness potential preceding speech, but there was no difference between the normal subjects and the stutterers in this asymmetry. One of the problems with these studies may be that they do not record the ERPs when the subject is actually stuttering.

Artifacts

After these initial reports it soon became apparent that there were multiple noncerebral potentials that could also be recorded prior to speech production. Horizontal eye movements are often associated with linguistic processes (Kinsbourne 1972). Since these eye movements generate large fields, they could easily contaminate the responses recorded from the inferior frontal regions. Most of the early studies had not averaged horizontal-eye-movement monitors (Anderson 1977). Studies controlling for horizontal eye movements failed to show any lateralization of the frontal slow potential shifts prior to speech in normal subjects (Michalewski, Weinberg, and Patterson 1977; Curry, Peters, and Weinberg 1978). These studies used a CNV paradigm with a verbal warning stimulus. It is therefore possible that a slow negative shift may have begun before S1. No lateralization was found, however, when a nonverbal S1 was used in either normal subjects prior to speech or deaf subjects prior to bimanual language acts in American Sign Language (House and Naitoh 1979). Pinsky and McAdam (1980) found no significant asymmetry of the CNV but did find a significant left-sided predominance for the readiness potential preceding speech. There are some intriguing relations between horizontal eye movements, corneo-retinal potentials, speech, and cortical negative waves that may require further evaluation (Anderson 1977). Because of the positive corneo-retinal potential, a lateral eye movement to the left is associated with a left-positive and right-negative wave in the inferior frontal regions. There may also be a left-frontal negativity generated by activity in the frontal eye-movement area associated with the control of the eye movement. The possibility that there are interactions between the slow negative wave generated in the frontal speech and frontal eye-movement areas must also be considered.

Furthermore, the muscles involved in speech, such as the tongue, facial, masseter, and temporalis muscles, also generate large electromyographic (EMG) potentials. Ertl and Schafer (1969) recognized this soon after their initial report and stated that it was difficult to distinguish cortical from muscular speech-related potentials. Morrell and Huntington (1971) measured no hemispheric asymmetry of the speech readiness potential prior to the initial EMG burst preceding phonation. It has recently been determined that the muscle potentials preceding speech can be quite asymmetric and can occur up to a second before the acoustic utterance (Grabow and Elliot 1974; Szirtes and Vaughan 1977). Other possible sources of artifact that are very difficult to control are the potentials associated with respiration, which of course have to be synchronized with speech reproduction (Grözinger et al. 1977, 1980). Indeed, it appears to be impossible to eliminate these muscular or respiratory potentials from the averaged waveform of the speech-related potential.

One approach to controlling for the concomitant muscle potentials is to try to subtract their contribution from the EEG recording. Levy (1977, 1980) monitored the horizontal electro-oculogram and the lip EMG activity together with several channels of EEG. By means of an analysis of covariance, the relative contributions of each could be subtracted from the scalp-recorded data. The results of such an analysis showed a left inferior frontal negativity preceding the articulation of sequentially complex utterances. These utterances were not necessarily linguistic: a multiple huff, a multiple puff, and a multisyllabic word. The problem with such an analysis is that there may be muscles active that are not picked up in the monitoring channels and therefore not compensated for in the analysis of covariance. This has been shown by Brooker and Donald (1980a,b), who recorded over many different muscles and obtained high correlations between the inferior frontal EEG recording and many of the muscle sites. Furthermore, different muscles may contribute different amounts of potential at different times prior to the speech act—for example, temporalis and masseter muscles become active earlier than tongue and lip muscles. The various covariance analyses performed by Brooker and Donald showed no significant hemispheric asymmetry in the potentials preceding speech.

Summary
The following conclusions appear warranted. First, although they may be difficult to record from the scalp, there must be specific speech-preparatory potentials generated in the human brain prior to and during speech. Fried et al. (1981) have recorded such potentials from the exposed cortex during neurosurgical operations. Whether these potentials are lateralized to the left hemisphere and localized to the inferior frontal regions will need to be evaluated. Certainly many other brain regions will also be active in the control of speech production (Grözinger, Kornhuber, and Kriebel 1979). Second, it is unknown whether the potentials occurring prior to or during speech will be different from those required for other complex facial, vocal, or respiratory motor acts. It is possible that speech is plugged into the final common pathways of the motor and premotor cortex in a similar way to other complex motor activities, although it may be "programmed" in a region of the brain specific to speech and language. The examination of this programming will require more complex experimental paradigms than the repetitive production of a single utterance. Third, any scalp recording of such speech-related potentials must monitor and control all non-

cerebral sources of electrical activity. It will be impossible to eliminate from the recording the muscle potentials that occur before and during speech, since "muscular activity of the vocal apparatus is an inevitable part of the act of speaking" (Brooker and Donald 1980b). It is possible that multichannel monitoring of speech muscle activity can be used to determine what part of the EEG recording is cerebral and what part is muscular in origin. How this will be done, however, is not readily apparent.

Potentials Evoked by Speech Sounds

Auditory Evoked Potentials

Many different auditory evoked potentials can be recorded from the human scalp (Davis 1976; Picton and Fitzgerald 1982). A series of small vertex-positive waves generated in the brain stem can be recorded in the first 10 msec after a transient stimulus. Between 10 and 50 msec, there occur several "middle latency" components, the most prominent of which is a vertex-positive wave with a peak latency of 30 msec. These waves are often difficult to evaluate because of the multiple scalp-muscle potentials that can be reflexively evoked in this latency range. A set of components called P1, N1, P2, and N2 can be recorded between 50 and 500 msec. These have maximal amplitudes at the vertex and typical peak latencies of 50, 100, 180, and 250 msecs, respectively. In the temporal regions several components occur at the same time as the vertex N1 response (Wolpaw and Penry 1975; Picton et al. 1978; Wood and Wolpaw 1982). Since the temporal components may also be picked up from the ear and mastoid regions, they are best recorded using a noncephalic reference (Wolpaw and Wood 1982). McCallum and Curry (1980) have identified three separate N1 components: a frontotemporal N1a with a peak latency of 75 msec that is greater in amplitude over the dominant left hemisphere independent of the stimulated ear, a vertex N1b with a peak latency of 106 msec, and a temporal N1c with a peak latency of 129 msec that is greater in amplitude over the hemisphere contralateral to the stimulated ear. When the auditory stimulus is improbable, informative, and task-relevant, an additional complex of waves occurs in the response. This endogenous complex consists of a frontocentral N2 component, a large parietocentral P3 component occurring with a peak latency of 300 msec or more, and a slow wave that is negative in the anterior scalp regions and positive posteriorly (Donchin et al. 1978).

Early Studies
There have been many studies of the evoked potentials to speech sounds. The usual paradigm involves recording from homologous scalp locations over the left and right hemispheres. The goal has been to lateralize and possibly also localize the physiological processes that are specifically evoked during the perception of speech stimuli. The many methodological problems involved in this kind of research can be introduced through an evaluation of three early ERP studies of speech perception published in *Science.*

Using a rather unusual electrode derivation (2 cm anterior to the external acoustic meatus referred to 2 cm lateral to the midsagittal line on the coronal plane), Cohn (1971) found that loud clicks elicited a large early positive wave at 14 msec over the right but not over the left hemisphere. Single-syllable words elicited later and somewhat more symmetrical response, although most of the subjects showed left-hemisphere predominance. This early positive wave was probably a microreflex recorded from the masseter muscle (Bickford 1972; Picton et al. 1974). Microreflexes occur with high-intensity rapid-onset auditory stimuli and vary in amplitude with the degree of tension in the muscle activated. The speech sounds were therefore less likely to elicit such microreflexes than the loud clicks. The asymmetry of the elicited reflex may have been due to some attention-related asymmetry in the muscle tone (Baribeau-Braun 1977).

Morrell and Salamy (1971) reported that the response to nonsense syllables had a larger N1 (90 msec) component over the left hemisphere than over the right hemisphere in temporoparietal regions but not in frontal or Rolandic regions. This asymmetry did not occur for the succeeding P2 component, which was actually smaller over the left hemisphere. It is difficult to determine how much of this difference was related to asymmetrical processing and how much to anatomical differences between the hemispheres (cf. Geschwind and Levitsky 1968), because no recordings were made of the responses to nonspeech stimuli using the same electrode derivations. Subtle differences in the location or orientation of the auditory cortices between the hemispheres could have caused the ERP effects in the absence of any asymmetry in the processing of the sensory information. These anatomical differences may nevertheless be important and may be related to hemispheric dominance. Davis and Wada (1977) have found that the coherence between the occipital and temporal evoked potentials to clicks is greater over the dominant hemisphere. This finding suggests hemispheric differences in the connections between occipital and temporal regions in the processing of simple nonspeech stimuli.

Matsumiya et al. (1972) investigated the effects of stimulus meaningfulness on the hemispheric asymmetry of the evoked potential. They recorded from Wernicke's area and from a homologous region of the right hemisphere in reference to the ipsilateral parietal regions. They found that a postive "W wave" at 100 msec was most left-sided when meaningful speech was processed. The asymmetry was evaluated using the ratio of the response amplitude on the left to the sum of the amplitudes over both hemispheres; a ratio greater than 0.5 indicated left-side predominance. A similar but less striking hemispheric asymmetry was seen for actively attended nonspeech sounds. Because it occurred long before the end of the words, Matsumiya et al. concluded that this component could not have been related to the meaning of the stimulus but rather must have reflected the mental set in the brain for analyzing meaning. Recent behavioral evidence has, however, suggested that on-line semantic access can occur long before the end of a word (Marslen-Wilson and Tyler 1980). The electrode derivations used in this study were quite unusual, and it is difficult to relate these findings to others or even to determine at which of the two electrodes the changes occurred. There is probably a larger N1 component at the parietal location than at Wernicke's area, and the W-wave probably represents a scalp gradient for an N1 component that changes with the meaningfulness of the stimuli. It would be far clearer if evoked potentials from different brain regions were compared using a common reference electrode.

Principles

These initial studies point to certain principles that must be followed in this kind of experiment. Some comparisons should be made between linguistic and nonlinguistic perceptual processing. Without such comparison, one cannot determine how much of an asymmetry is specific to language processing and how much is just anatomical. The linguistic and nonlinguistic stimuli should not differ consistently in simple physical attributes, such as intensity or frequency contents, since any differences noted in the evoked potentials might then be caused by these physical differences and not by the linguistic or nonlinguistic nature of the stimuli. Recordings should be made from homologous areas over the two hemispheres and referred to a single reference electrode, preferably one that is relatively inactive with respect to the measurements being examined. If both electrodes used in a recording derivation are active and neither one is common between the hemispheres, one will never know the relative contribution of each electrode to any recorded asymmetry.

One early study following these principles was published by Wood, Goff, and Day (1971). They presented syllables, each consisting of a stop consonant and a vowel, at a rate of one per 5 seconds. In one condition the subjects were requested to discriminate between the two equiprobable stop consonants /ba/ and /da/. In a second condition the subjects were asked to discriminate between two different fundamental frequencies (104 Hz and 140 Hz) for the same syllable, /ba/. The /ba/ stimulus with a fundamental frequency of 104 Hz was exactly the same in both conditions. Evoked potentials were recorded from central and temporal locations with a common linked-ear reference. The left-hemisphere responses to the 104-Hz /ba/ stimuli differed between the two conditions, whereas the right-hemisphere responses showed no change. The main difference was that the left-hemisphere N1 peak was larger during the discrimination of stop consonants than during the discrimination of the fundamental frequency. There were also significant differences noted later in the waveform, but these could well have been due to artifacts from the discriminative button press and were therefore discounted. These findings were replicated using other stop consonants and other nonspeech control conditions (Wood 1975).

The differences noted by Wood and his colleagues were very small. They evaluated for significance at each of the averaged time points in the recorded waveforms. This procedure has been criticized as being too susceptible to chance effects (Friedman et al. 1975a). If one performs 100 tests using a criterion of $P < 0.05$, one should find that about 5 trials show "significant" effects in the absence of any actual differences. Attempts to replicate the findings of Wood and his colleagues have not been successful. Grabow et al. (1980b) replicated the Wood experiment but found no significant differences between the tasks in either the left or the right hemisphere prior to latencies where motor responses might interfere with the results. However, they did note a definite hemispheric asymmetry in the temporal regions such that the left temporal response was smaller than the right, independent of the discrimination task in which the subject was involved. Such an asymmetry was suggested in the 1975 data of Wood, though not in the 1971 data of Wood et al. Grabow postulated that this might have been related to the presence of more ongoing activity in the left hemisphere during both discrimination tasks. Any stimulus arriving at the cortex of the left hemisphere would find fewer cells to activate. No asymmetry, however, was noted when subjects listened to nonsense words in an experiment (Grabow et al. 1980a) replicating the 1971 report of Morrell and Salamy, although this study used electrodes that were not exactly located at the mid-temporal regions.

Hemispheric Asymmetries

There have been many other studies of the evoked potentials to speech sounds. Some have not made any comparison between linguistic and nonlinguistic perceptions but have simply evaluated the hemispheric asymmetry of the speech-evoked potential. Haaland (1974) recorded the evoked potentials from right and left frontal and temporal regions in response to consonant-vowel-consonant monosyllables. He failed to replicate the N1 asymmetry reported by Morrell and Salamy, perhaps because his electrodes were not located at the same temporoparietal regions. Furthermore, he found a significant right-side predominance for the P2 component that was stable through monaural, dichotic, and diotic presentations. Genesee et al. (1978) found no amplitude asymmetries between left and right central regions in the responses of bilingual subjects to monaural or binaural words. They did, however, note a significantly earlier latency for the N1 component over the left hemisphere, particularly in those subjects who had become bilingual at an early age.

Other studies comparing the responses to speech and nonspeech sounds have lacked the exact acoustic control of the Wood experiment. The effects reported may therefore have been caused by the acoustic differences between the stimuli or by the specifically linguistic processing of the speech sounds. Because of this problem, comparisons have been made using interhemispheric ratios so as to remove any general effects of stimulus differences. Greenberg and Graham (1970) studied the evoked potentials to consonant-vowel syllables and piano notes during response-association learning. The responses were larger over the left hemisphere (an asymmetry that was probably related to the right-earlobe reference) except when the tonal associations became well learned, at which time the tones evoked larger responses from the right hemisphere. Neville (1974) found that the P1–N1 amplitude of the response to dichotic verbal stimuli was larger over the left centrotemporal region than over the right, whereas there were no significant asymmetries for dichotic clicks. More striking, however, was Neville's finding that the verbal stimuli evoked responses with earlier latencies in the left hemisphere and the clicks evoked responses with earlier latencies in the right hemisphere. However, Galambos et al. (1975) were unable to find significant hemispheric asymmetries in the responses to either speech sounds or tones.

Molfese (1977) and Molfese, Freeman, and Palerno (1975) compared the responses at right and left temporal regions to syllables, words, musical chords, and noise bursts. In adults, a small left-sided predominance was found for the responses to speech stimuli and a larger right-sided predominance for the responses to "mechanical" stimuli. Larger asymmetries were found in infants and children, perhaps because of maturation of the corpus callosum and other commissural pathways. The responses in the infants were quite different from those recorded in adults, however, and it is questionable whether they can be meaningfully compared. Taub, Tanguay, and Clarkson (1976) replicated this asymmetry of the adult response to musical chords, although their differences only reached a significance level of $P < 0.05$ and were apparent only when the left ear was stimulated. However, Tanguay et al. (1977) could find no significant hemispheric asymmetries in the potentials evoked by stop consonants. Hinks, Hillyard, and Benson (1978) reported that the amplitude of the N1 component evoked by verbal stimuli was significantly larger over the left central region than over the right, but this left-sided predominance was even greater for nonverbal stimuli.

Although the extent of artifact contamination is probably much less for the evoked potential than for the potentials preceding speech, it is still important to rule out the possibility of artifacts. Anderson (1977) recorded horizontal eye movements while subjects listened to binaural syllables. Though the fieldspread of the eye-movement potentials to temporal and parietal regions is not as great as to the frontal regions, they can still contaminate posterior recordings. None of the studies discussed in this section have adequately monitored lateral eye movements. Most have not recorded eye movements at all, some have monitored only vertical eye movements, and occasional studies (e.g. Hink et al. 1978) have used a diagonally placed pair of electrodes to evaluate both vertical and horizontal eye movements. Any study of hemispheric asymmetries should record vertical and horizontal eye movements using separate channels, one recording from above to below the eye and the other recording from one external canthus to the other.

In summary: There is little that we can conclude about the evoked potentials to speech sounds. They appear quite similar to those evoked by other auditory stimuli. The differences that have been reported are small and inconsistent from study to study. The absence of any obvious speech-specific response pattern may be explained in three ways. First, linguistic processes may be largely unrepresented in the scalp-recorded ERP. Speech processing may be done by neuronal networks that do not generate electrical fields that are easily picked up at the scalp. It is possible that the scalp-recorded potentials may reflect general information-handling operations rather than the specific content or modality of the information being processed (Schwartz 1976). Even if there is some degree of localization that is recordable in the speech-evoked ERP, this localization may be small and difficult to discern in the larger nonlateralized components of the response. Second, there may be no lateralization of speech perception in the brain for the type of stimuli that are now used in ERP research. Speech is usually perceived as a continuous stream of auditory information that is quite different from the series of discrete stimuli used in ERP studies. Third, the potentials specifically related to speech perception may be more easily evoked by stimuli that are more meaningful than consonant-vowel nonsense syllables. These three possibilities will be considered more closely in the following sections.

Principal Components of the Speech Response

It is possible that the cerebral processes underlying language generate very small potentials at the scalp. These language-specific potentials may not be recognizable by the usual peak-measurement techniques but may be accessible to the more powerful technique of principal-component analysis (PCA). Molfese and his colleagues have studied the principal-component structure of the evoked potentials to speech and speechlike stimuli in an attempt to determine where in the brain the different cues for speech are processed (Molfese, Molfese, and Parsons 1982) and how the specialization of this processing develops from infancy (Molfese and Molfese 1982). A major part of this research has investigated how the human brain processes syllables with different "voice-onset times" (VOTs). Human perception is such that, when voicing begins 40 msec after a bilabial consonant, a perception of /p/ occurs, whereas a VOT of 20 msec or less brings about the perception of /b/. The first study (Molfese 1978a) recorded evoked potentials to free-field /pa/ and /ba/ stimuli with VOTs of 0, 20, 40, and 60 msec. Evoked potentials were recorded from the midtemporal regions of each hemisphere in reference to linked ears and were averaged over 16 repetitions for each VOT. The average waveforms were then submitted to a principal-component analysis and a sex × VOT × hemisphere repeated-measures analysis of variance was performed on the component scores. The first and fourth factors showed a VOT × hemisphere interaction such that they were significantly different across the 20–40-msec category boundary over the right hemisphere but not over the left. These and other factors showed significant VOT effects over the left hemisphere, but the differences were not specifically located at the boundary for the categorial perception. Molfese and Hess (1978) used the same paradigm in preschool children. In this experiment the first factor showed significant VOT effects both within and across the phoneme boundaries. The second factor, however, was specifically affected by the category boundary only when recorded from the right hemisphere. Molfese and Molfese (1979a) recorded the ERPs of infants 2–5 months old. The second factor on this analysis showed a significant hemisphere × VOT interaction such that in the right hemisphere there were significant differences across the phoneme boundary. No such effects could be found, however, in newborn infants (Molfese and Molfese 1979b). In order to evaluate the nature of the VOT discriminations, Molfese (1980a) performed further studies using tonal stimuli with different tone-onset times (TOTs) (Pisoni 1977). In this study recordings were made from central, parietal, and posterior temporal regions as well as from the midtemporal regions. The eighth factor in the PCA showed a hemisphere × TOT interaction such that the right hemisphere responded to the TOT stimuli as if they consisted of two categories separated between 20-msec and 40-msec TOTs. Another factor recorded maximally from parietal and central regions distinguished among all TOT stimuli. Other factors showed differences among electrode locations. These results led to three conclusions: that the ability to discriminate VOT stimuli across phoneme boundaries "is not innate and may require some maturation or experiential influence in order to develop or become functional" (Molfese and Molfese 1979a), that "sometime between 4 years and adulthood either the bilateral VOT process becomes lateralized to the RH or it is replaced by a different mechanism which is restricted to the RH" (Molfese 1980a), and that the discrimination of VOTs is performed by a "temporal processing mechanism within the RH" that is "equally sensitive to such cues in nonspeech acoustic stimuli" (Molfese 1980a).

The interpretation of these results is extremely difficult. The major problem is in the evaluation of the components derived from the PCA. As mentioned, PCA is a data-driven analysis. It will find components that differ across the experimental manipulation, but will make no clear decision as to whether such components are related to the noise or to the signal even when a subsequent analysis of variance is performed on the component scores. This is a particular problem when noise levels are high, as must be the case when ERPs are averaged over small numbers of trials. With averaging performed over only 16 trials (Molfese 1978a,b) a background EEG of about 20 μV would be reduced to 5 μV—the same amplitude as the recorded ERP. One means of evaluating the validity of a component derived through PCA is to replicate that component in different sets of data. The components showing right-hemisphere category-boundary effects in the work of Molfese and his colleagues, however, differ markedly among the different studies. The major peaks in the factors are the following.

Molfese 1978a: factor 1 at 430 msec, factor 4 at 135 msec
Molfese and Hess 1978: factor 2 at 444 msec
Molfese and Molfese 1979a: factor 2 at 920 msec
Molfese 1980a: factor 8 at 355 msec.

The morphologies are as different as the peak latencies. The age of the subjects and the nature of the stimuli may have caused some differences in the factors obtained. Nevertheless, there is as yet no replication of the components and therefore no evidence that they are not caused by the random variability of noisy data.

A second difficulty with the results concerns the interpretation of the findings in terms of cerebral processes. There were no monitors for any of the noncerebral potentials that could be recorded from the scalp. It is therefore impossible to determine whether any of the components, even if they were validly correlated to the experimental manipulation, might not have been generated by such extracerebral sources as the eyes and the scalp muscles. Another problem with the interpretation concerns the nonexhaustive nature of scalp electrical recordings. That the electrical recordings from the left hemisphere do not differentiate across phoneme boundaries does not mean that processes performing such a discrimination do not occur in the left hemisphere. They might occur in neuronal networks that do not generate electrical fields that can be recorded at the scalp. Furthermore, processes occurring in the left hemisphere may be maximally recorded over the right hemisphere.

Molfese and his colleagues have also performed a series of experiments investigating the physiological processes underlying the perception of stop consonants. They are all susceptible to the same criticisms as the studies of VOT. Molfese (1978b) compared the ERPs to phonetic (/b/ and /g/) and nonphonetic stop-consonant transitions. The results suggested left-hemisphere mechanisms for discriminating both the phonetic and the nonphonetic transitions. The presence of such effects in preterm infants (Molfese and Molfese 1980) suggested an innate ability of the left hemisphere to discriminate these transitions. Molfese (1980b) recorded ERPs to /b/ and /g/ sounds that were associated with different vowels and found two components of the scalp-recorded ERP that differentiated the consonants independent of the associated vowels. These components, however, were not at all similar to those reported by Molfese (1978b) as dissociating between /b/ and /g/.

Auditory Probes of Language Function

One of the difficulties in studying the cerebral processes of language by means of the ERPs is that the very nature of the stimuli evoking the ERPs may militate against their being processed in a specifically linguistic manner. The perceptual processes for speech and language are usually invoked by progressing and continually changing sounds. They may not be easily elicited by discrete and repetitive stimuli. Several studies have therefore used the auditory evoked potentials to probe ongoing language functions in the brain rather than specifically to evoke those functions. Shucard et al. (1977, 1981) have reported that the evoked potentials to brief tones may be used as probes of lateralized cerebral function. Pairs of irrelevant tones were intermixed with either verbal or musical material to which the subject was listening attentively. Electrical recordings were taken from the midtemporal regions in reference to the vertex (1977) or to linked ears (1981). In the ear-reference recordings the auditory evoked potentials were significantly larger over the right hemisphere than over the left during the verbal task. There were no significant asymmetries in the musical task. The asymmetry during the verbal task was more prominent for the second of the tones in the pair and for a temporal-negative wave with an average peak latency of 300 msec. These results suggest that the auditory impulses reaching the cortex during the ongoing verbal task find the left hemisphere "busy" and elicit from it a smaller response. There may have been subtle acoustic interactions between the probe clicks and the ongoing auditory stimuli. It would be difficult, however, to explain the asymmetry on the basis of some acoustic masking effect. Furthermore, Papanicolaou, Eisenberg, and Levy (1982) have demonstrated a similar left-hemispheric attenuation of the evoked potential to a click probe during the covert rehearsal of a verbal passage but not during the covert rehearsal of a musical passage.

Seitz (1976) used a somewhat more complex probe task. Clicks were presented in one ear and complex sentences in the other ear. The clicks occurred before, at, or after a "major constituent break" in a complex sentence ("My old torn shirt / is in the wash"). He found that the latency of the N1 component recorded between vertex and mastoid was significantly shorter for the clicks that followed the major constituent break. This effect occurred only when the subject paid close attention to the linguistic structure of the sentence in order to write it down afterward, and not when the subject had only to mark the click on a prepared script of the sentence. These findings accord well with behavioral studies demonstrating the perceptual migration of clicks toward the major constituent break of a concomitant sentence (Seitz and Weber 1974).

Meaningful Stimuli

It is possible that speech sounds that are more natural and meaningful than syllables or nonsense words may elicit more language-specific ERPs. Friedman et al. (1975a) found a greater left-hemisphere amplitude for the late positive component in the response to spoken words that the subject was listening to. Hillyard and Woods (1979) reported larger left-hemisphere N1 amplitudes for spoken words of poetry than for pure tones. Brown and his colleagues have performed a large series of experiments to evaluate the cerebral processes underlying the perception of meaning in speech stimuli. The initial experiment (Brown, Marsh, and Smith 1973) found a significant difference between the evoked potentials to the same word presented in different contexts ("Sit by the fire" vs. "Ready, aim, fire"). The differences were measured by calculating correlation coefficients between the recorded waveforms, low coefficients indicating dissimilarity. The dissimilarities between word meanings were greatest over the left anterior scalp region. It is difficult, however, to determine whether these differences were related to the acoustic or the semantic context of the word. Therefore, in their second experiment, Brown, Marsh, and Smith (1976) used a single phrase ("It was lead") and the subjects were instructed to perceive the stimulus word as a verb ("The horse was led") or a noun ("The metal was lead"). The evoked potentials were analyzed using correlation techniques and stepwise discriminant analysis. Both techniques indicated that the evoked potentials to the two meanings of the stimulus word were significantly dissimilar in the left anterior scalp region. A later report (Brown et al. 1982) indicated that this dissimilarity was less striking in stutterers and that male and female subjects differed in the scalp locations where noun-verb dissimilarities occurred. Unfortunately, the techniques assessing for dissimilarities used the entire waveform, and it is difficult to determine where in the response the differences occur. A principal-component analysis of the 1976 data (Brown, Marsh, and Smith 1979) found three components, with peak latencies at 150, 230, and 370 msec, that were significantly affected by stimulus meaning. None of these components, however, specifically localized to the left anterior scalp region. Correlational techniques are very susceptible to latency differences; a coherence analysis (Davis and Wada 1977) might be a more appropriate technique for assessing waveform similarities.

Megela and Teyler (1978) looked at the semantic effects from a different viewpoint. They found that the ERPs to words of similar meaning but different sound (*rock* and *stone*) were more dissimilar than the ERPs to homonyms with different meanings ("shuffle the deck" and "walk on the deck"). The semantic determinants of the ERP waveform are therefore much less powerful than the acoustic determinants. Teyler and his colleagues (Teyler et al. 1973; Roemer and Teyler 1977) have also examined the relations between meaning and the auditory evoked potential using a neutral stimulus. Subjects were asked to think of one of two different meanings of a word (a rock or to rock) when a click stimulus was presented. In different conditions they either spoke the word or just thought of it. The click-evoked potential was greater in amplitude over the left central area than over the right, particularly for the noun meanings. These data are difficult to interpret, however, because of the large individual differences and because there was no control for speech-muscle artifacts.

Brown and his colleagues (Brown and Lehmann 1979; Brown, Lehmann, and Marsh 1980) further investigated the scalp topography of noun-verb homophones ("a pretty rose" and the "the boatman rows") using twelve scalp electrodes. They evaluated the topographical maxima (positive peaks) and minima (negative peaks) of the evoked potentials at each time point in the waveform and in those portions of the waveform defined by a principal-component analysis. The fields were measured relative to the average value at all electrode locations at that time. The spatiotemporal peaks of the responses differed between the noun and verb contexts. In the majority of cases the verb maxima were more posterior than the noun maxima in the first 175 msec of the recording and the verb minima were more anterior to the noun minima over the first 330 msec of the recording. These results were not due to eye movements and were replicated in a 37-electrode array. Furthermore, very similar results were obtained using Swiss-German homophones in a separate group of Swiss subjects, and also by having these subjects hear a modulated 500-Hz tone as noun or verb. These results are impressive, but they are somewhat difficult to fit with other data because they treat the scalp topography of the ERP independent of its morphology over time.

Conclusions

The ERP indices of speech perception appear to be quite small. There are no consistent findings to date, although there are trends suggesting a lateralization to the left hemisphere. Perhaps there has been too much concern with the localization of language processes and too little concern with the timing and structure of the physiological processes leading to speech perception. The indices of speech perception may be subtle and easily missed if the ERP analysis is restricted to a few waveform peaks or a few scalp locations. Noncephalic reference electrodes may be more useful in assessing the temporal regions of the brain than ear or mastoid references. Techniques such as principal-component analysis and multichannel scalp topography may be helpful in separating language-specific processes from the more general perceptual processes. These techniques must be used only on data that have been adequately collected in terms of artifact rejection and noise attenuation. Most important, the results of these analyses must be replicated to ensure their validity.

Potentials Evoked by Visually Presented Linguistic Stimuli

Visual Evoked Potentials

The visual evoked potential is somewhat more complex than the auditory. The waveform recorded from the occipital regions in response to a flash of moderate to high intensity usually contains prominent negative and positive components with approximate peak latencies of 90 and 120 msec (Kooi and Marshall 1979; Hillyard, Picton, and Regan 1978). These waves often contain several peaks. They are usually followed by a later negative wave peaking at about 200 msec. This waveform varies greatly both among subjects and among stimuli, perhaps because of the wide extent and complexity of the visual cortices and their proximity to the recording electrodes. Changes in the intensity of the stimulus can cause large changes in the latency of the response. Many visual stimuli combine changes in luminance with changes in contrast. One technique of decreasing the interaction between these parameters is to use "constant-luminance" stimulation, holding the overall luminance of the screen constant as the pattern on the screen changes. The visual evoked potential is particularly sensitive to the specific contrast pattern of the stimulus and varies greatly with its location in the visual field. It is therefore essential to control both the contour structure and the retinal location of visual stimuli if valid comparisons are to be made between the responses to linguistic and nonlinguistic stimuli. The input to the visual system can easily be affected by closure of the eyelids, changes in accommodation, or alterations of gaze. These peripheral mechanisms of visual attention must be monitored closely during any experiment evaluating visual evoked potentials.

The scalp distribution of the visual evoked potentials is quite complex. The visual evoked potentials recorded from the frontocentral regions of the scalp differ from those recorded occipitally. They may derive partly from the inverted field spread of the response in the occipital regions and partly from other more anterior cerebral generators. The vertex negative wave peaking at about 150 msec may be homologous to the N1 component of the auditory response. It is more susceptible to increasing rates of stimulus presentation and to attention than the occipital components (Hillyard et al. 1978). An N2–P3 complex occurs in the response to a task-relevant informative visual stimulus. The P3 wave is similar to that recorded in the auditory system, whereas the N2 component has a more posterior scalp distribution (Simson, Vaughan, and Ritter 1977).

Verbal and Nonverbal Stimuli
There have been several studies comparing evoked potentials with linguistic and nonlinguistic visual stimuli. Buchsbaum and Fedio (1969, 1970) presented three-letter words made up of dotted letters or nonsense stimuli composed by shifting the dots that made up the letters of the words. These stimuli were presented to the right or the left visual field, and evoked potentials were recorded from the right and left occipital regions. Correlation measurements were obtained between replications and between evoked potentials for different hemispheres, fields, or stimulus types. The differences noted in the evoked potentials between words and nonsense stimuli were greater over the left hemisphere and

greater when stimuli were presented in the right visual field. Although these results show highly significant differences in the processing of the stimuli, it is impossible to determine what these differences represent in terms of brain physiology. Just because the recorded waveforms at one electrode location differ between words and nonsense stimuli, it does not mean that there are different processes going on in the underlying cortex. Potentials contributing to the recorded differences may be generated in distant regions of the brain, or even by noncerebral generators. Somewhat similar results were obtained when flash-evoked potentials were studied in relation to attention. An occipital N140 component was increased in amplitude by attention; this change was most prominent when stimuli were presented to the right visual field and when recordings were taken from the left temporoparietal scalp (Buchsbaum and Drago 1977). There are intricate relations between attention and language. It is often difficult to invoke the one without the other. Control studies wherein attention is directed to nonlinguistic aspects of the stimuli would be helpful in delineating the neurophysiological correlates of language from those of attention.

Rugg and Beaumont (1978, 1979) compared the evoked potentials to letters and shapes. The subjects were requested to detect those letters having the "ee" sound in their name (bcdegptv) in a block of letters and those shapes that were symmetrical about the midline in a block of symmetrical and asymmetrical shapes. There were no significant hemispheric differences related to these different tasks. However, an early positive wave at about 150 msec occurred at a shorter latency and a negative wave at about 180 msec was of greater amplitude over the right hemisphere for all of the stimuli. These findings may have been caused by asymmetrical early processing of the stimuli. Since the amplitude asymmetry was similar across the two tasks, however, it could also have reflected a simple anatomical asymmetry without functional significance. The skulls of these subjects may have, for example, been thinner on the right—a finding that has been used to explain the right-sided predominance of the alpha rhythm (Leissner, Lindholm, and Peterson 1970). The P3 component was larger and earlier to the spatial stimuli, probably because these stimuli were distinguished more quickly.

Ledlow, Swanson, and Kinsbourne (1978) recorded the evoked potentials to letter pairs—name matches (Aa) or physical matches (AA)—presented to either the right or the left visual field. Asymmetrical responses were obtained, but it is difficult to determine how much of the asymmetry was anatomical rather than physiological in origin.

Neville (1980) recorded evoked potentials to two different four-letter words presented simultaneously to the two visual fields. The N1 component was significantly larger over the left parietal than over the right parietal scalp. There was no significant asymmetry at other electrode locations, and the asymmetry was not present if the stimuli were sufficiently defocused to be illegible. Line drawings presented randomly to the right or the left visual field showed greater amplitude over the right central regions than over the left, but there were no asymmetries in the parietal regions. Neville also reported preliminary results from an experiment wherein the same stimuli could be perceived according to their alphabetical or spatial nature. This was an elegant design analogous to that used by Wood et al. (1971) in the analysis of auditory speech perception. There were only small asymmetries in the letter task, but there was a very prolonged right frontal positive wave during the spatial task. We have also recorded potentials during a spatial-orientation task and have found a prolonged frontal positivity (Stuss et al. 1982). In our recordings, this is associated with a parietal negativity and there is no obvious hemispheric asymmetry.

Hink, Kaga, and Suzuki (1980) recorded the evoked potentials to Japanese words presented in ideographic (*kanji*) or phonetic (*kana*) script. The kanji words evoked larger responses than the kana words. Furthermore, the scalp distributions of the responses were different; the kana words evoked responses with relatively greater amplitudes over the left parietal regions than the kanji words. These results suggest more involvement of the left parietal cortex in the processing of the kana words.

Visual-Probe Studies
Linguistic information is initially processed in a more discrete fashion in the visual system than in the auditory system, since the visual input is usually broken up by saccadic eye movements. The linguistic processing may, nevertheless, be continuous because of the long duration of the visual image. Thus, the evoked potentials to discrete visual stimuli may still not tap language processes. Several studies have therefore recorded the evoked potentials to irrelevant visual stimuli while subjects were engaged in ongoing language tasks. Galin and Ellis (1975) recorded flash-evoked potentials from left and right parietal and temporal electrodes using a vertex reference. Eye movements were monitored, and

trials in which there were blinks or movements of greater than 6° were excluded from the averaging. It was found that the amplitude of the flash-evoked potential was smaller over the left hemisphere when the subjects were engaged in writing down remembered information and smaller over the right hemisphere when they were engaged in constructing a complex geometric pattern. An incoming stimulus will synchronously activate fewer neuronal processes when these are engaged in some ongoing task. The asymmetry of the probe response therefore suggests an inverse asymmetry of the neuronal processes underlying the different cognitive modes; verbal processing involves a busy left hemisphere and spatial processing a busy right hemisphere. The asymmetries are somewhat difficult to interpret, however, because of the activity of the vertex reference. Furthermore, it is possible that there might have been some interaction with the ongoing motor responses. The evoked-potential asymmetry might have reflected task differences in the motor activity of the hands rather than an asymmetry of cognitive mode. Mayes and Beaumont (1977) replicated this experiment using slightly different electrode locations and adding further conditions wherein the subject was required to make mental rather than motor responses. They found no definite task-related asymmetries, although differences between the motor and mental conditions and between the verbal and spatial tasks were common to both hemispheres. Since the same sensory modality was used for both the task and the probe, it is also possible that the asymmetry noted by Galin and Ellis was more related to sensory than to cognitive processes. Papanicolaou (1980) used visual probes during an auditory task. He recorded flash-evoked potentials from frontal, temporal, and parietal electrodes in reference to linked ears. The subject listened to a meaningless sequence of words in order to detect words belonging to a particular category such as "vegetables" (a semantic task), words containing particular stop consonants (a phonetic task), or words having different frequency of intensity characteristics (an acoustic task). In a control condition the subject was asked to ignore the auditory stimuli and attend to the flashes. The amplitude of the N1 component was significantly reduced over the left hemisphere and enhanced over the right when the responses obtained during the semantic and phonetic tasks were compared with those obtained during the acoustic or the control task. These results are particularly intriguing because of the well-designed ongoing tasks. Unfortunately there was no control of the direction of gaze or of the accommodation—a failing common to the previous studies as well.

Subtle changes in direction of gaze can greatly alter the scalp distribution of the visual evoked potential.

Semantic Matching Tasks

Thatcher has evolved a "background-information probe" paradigm to record the evoked potentials to both informative stimuli and irrelevant probes (Thatcher 1977a,b; Thatcher and Maisel 1979). The basic task is "delayed matching to sample." A variable number of random-dot displays are followed by the informative standard stimulus. The standard is followed by a second series of random-dot displays which lead to a test stimulus that matches or mismatches the standard. The evoked potentials to the random-dot displays may probe such ongoing cerebral processes as rehearsal between the standard and test stimuli. The evoked potentials to the standard and test stimuli will reflect the cerebral processes specifically related to information processing. Extensive scalp-distribution studies of these evoked potentials have been recorded using various matching tasks. Of particular interest are the ERP findings during semantic matching. In this task the standard word was synonymous (*small*), antonymous (*large*), or neutral (*down*) with respect to the test word (*little*). The very large amount of data generated in these experiments has been evaluated in several principal-component analyses. These have been performed both across all conditions in one electrode derivation and across all electrodes within one stimulus condition. The resultant "electrotopography" is very complex and difficult to understand. There appear to be asymmetric ERP waveforms related to semantic processing. These asymmetries are most apparent in the response to the test stimulus. The most prominent asymmetry is in the late positive waves having peak latencies between 300 and 400 msec, which tend to be larger over the left hemisphere.

Thatcher (1977b) found some early differences between the evoked potentials to matching and mismatching test stimuli in the stimuli letter-matching variant of his paradigm, although these were overshadowed by larger differences in the late positive waves, which were larger for the matching test stimuli. Boddy (1981) and Boddy and Weinberg (1981) evaluated the ERPs during a semantic priming task. The subjects were given a category ("Is it a fish?") and then presented with a word that was either in or out of that category. They found that the N1–P2 amplitude of the response to the test stimulus was larger when the stimulus was in the category than when it was out. Comparisons between lateral electrodes did not reveal any significant hemispheric asymmetries in this effect. The results, however, nicely demonstrate early access to selective semantic processing.

Goto et al. (1979, 1980) used Thatcher's delayed-semantic-matching paradigm with *kanji* (ideographic) and *kana* (phonetic) Japanese words and observed a large late positive wave (P650) in the response to the test word. This component was larger over the right hemisphere for the *kanji* words and larger over the left hemisphere for the *kana* words. There were no obvious asymmetries in earlier components of the response.

Expectancy Waves

Goto et al. (1979) also recorded a slow negative baseline shift between the standard and test stimuli. This contingent negative variation (CNV) was resolved when a match or a mismatch was detected. No CNV asymmetry was reported. Several other researchers have recorded CNVs in paradigms involving expectancies for linguistic or nonlinguistic information. Butler and Glass (1974) reported that the CNV preceding numerical information was greater over the left hemisphere. However, the asymmetry also occurred when the numbers presented at S2 were uninformative. Marsh, Poon, and Thompson (1976) found that the CNV preceding the reading of a visually presented word at S2 was greater in amplitude over the left hemisphere, whereas the CNV preceding a visuospatial judgment was not asymmetrical. The differences were small, and it is difficult to determine whether there were asymmetrical artifacts from the speech muscles. Donchin, Kutas, and McCarthy (1977) studied the CNV preceding tasks requiring either structural or functional matching of visual stimuli. In one condition ("fixed") the same type of matching was used throughout a block of trials; in a second condition ("mixed") either of the two types of matching could be required, depending on the information provided by the warning stimulus. The CNV was generally found to be larger over the left hemisphere, and this asymmetry was greater when the type of task was not known until S1. We found no clear asymmetry of the CNV preceding perceptual tasks involving different amounts of linguistic and spatial processing (Stuss and Picton 1978). In summary: There appear to be no clear relations between the CNV and the linguistic or spatial nature of the task that it precedes. There are several components in the CNV (Picton and Stuss 1980), and relations to language may occur only for some of these components. Furthermore, the actual expectancy strategy of the subject may vary greatly in the interval preceding a task. This strategy will have to be controlled and monitored more closely.

Dimensions of Meaning

The relationship of ERPs to the semantic meanings of words presented visually was studied using Osgood's analysis of semantic meaning by Chapman et al. (1977, 1978, 1980). Much of the semantic variance among different words can be distributed along three independent dimensions of meaning: evaluative (good-bad), potency (strong-weak), and activity (fast-slow). When two lists of relatively pure examples of the positive and negative poles of each of the three dimensions (neutral on the other two dimensions) were presented to 12 subjects (Chapman et al. 1977), the subjects attended to the words in order to pronounce them out loud about 2 seconds later. ERPs were recorded from "CPZ" (1/3 of distance from Cz to Pz) referred to a linked earlobe reference. Visual analysis of the waveforms of three subjects to the positive and negative poles of the evaluative (E) dimension revealed ERP differences that were consistent across the two lists. Subtraction of these waveforms yielded a difference template to which the ERPs to E+ and E− stimuli were correlated in all 12 subjects. The ERPs to E+ and E− words were significantly different in their correlation to this standard-difference template in 5 of the 12 subjects at $P < 0.01$ and in 8 of the 12 subjects at $P < 0.05$, "supporting the conclusion that differences on the evaluative dimension of visually presented words result in detectable and statistically reliable differences in brain responses." Backward classification of the ERPs based on their similarity to or dissimilarity from the template resulted in 71% correct classification across all 12 subjects. Results for the two other dimensions were not as strong (potency 69%, activity 56%).

Chapman et al. (1978) also analyzed a different set of ERP data recorded from different subjects tested in an identical paradigm. They standardized the ERPs for each subject separately, performed a principal-component analysis on the standardized waveforms from all 10 subjects, and submitted the component scores to a multiple discriminant analysis into the six semantic classes. The accuracy of classification of the ERP into one of the six semantic categories was 56.7%. Three cross-validation procedures were used. In the "jackknifed" procedure, each case was left out of the discriminant analysis and then classified according to the discriminant function derived from the other subjects' data. This procedure yielded a classification accuracy of 42%.

Validation of the data from one list to the other gave an accuracy of 40%. Finally, data from the multiple discriminant function analysis was used to predict the results of one new subject. The accuracy of semantic-group classification for this subject's ERPs was 42%. The introduction of a semantic judgment ("rate the word" rather than "say the word") gave strikingly similar results (Chapman et al. 1980).

These results are complex and difficult to evaluate. The physiological meaning of standardized ERP waveforms is difficult to understand. The relation between the components derived from such standardized waveforms and the simple averaged ERP is not clear. The significance of the discriminant analyses is probably overestimated because the same experimental variance is used to determine both the components and the discriminant functions. The results appear to be replicable across different groups of subjects, although there have not been direct comparisons among the components derived in the different studies. The cross-validation techniques used so far have shown that the discriminations between the ERP waveforms are significant. The most stringent validation procedure would be to use the component loadings derived from one group of subjects to obtain component scores in a separate group of subjects for evaluation using multiple discriminant analysis. ERPs to words with different connotations appear to be discriminable along the Osgood dimensions. Whether these ERP effects are universal across many different subjects and across words with meanings outside the poles of the three dimensions is yet to be determined. The physiological basis of the ERP differences (how "meaning" occurs in the brain) remains unknown.

Late Positive Waves
Visually presented words have a remarkable ability to evoke the late positive wave or P3 component in the event-related potential. Saccadic eye movements during reading evoke late positive waves in the occipital regions with peak latencies between 200 and 400 msec (Barlow 1971; Kurtzberg and Vaughan 1979).

Shelburne (1972, 1973) recorded evoked potentials to individual letters making up either a three-letter word or a three-letter nonsense syllable. There were no hemispheric asymmetries noted for any of the evoked-potential components. The third letter allowed a decision about the type of stimulus to be made and evoked a large late positive wave (P3). Since the stimuli presented and the decisions entailed were linguistic in nature for both the words and the nonsense stimuli, one would not have expected a difference in hemispheric asymmetry between them. Nevertheless, there was also no underlying hemispheric asymmetry common to both types of linguistic stimuli.

When target words are detected in a train of other words, the latency of the late positive wave in the target-evoked potential varies with how difficult it is to discriminate the target (compare the difficulty of discriminating *Nancy* in a list of names, girls' names in a list of names, and synonyms of the word *prod* in a list of words) (Kutas, McCarthy, and Donchin 1977). The late positive wave appears to be generated at the end of stimulus evaluation and to be relatively unaffected by manipulations of response selection. It may therefore be used as an index of stimulus-evaluation time (Donchin 1981). Polich and his colleagues (Polich 1982; Polich et al. 1983) have recorded ERPs during a matching task wherein words could be compared phonologically (*moose-juice*), orthographically (*some-home*), or on the basis of both cues (*cake-bake*). The late positive wave was delayed when there was a conflict between the phonological and orthographic cues. This suggests that there is ongoing interaction between the analysis of the two codes, and not just a response-selection conflict after independent or sequential analyses of the codes.

Friedman et al. (1975a) found that the words making up a sentence, when presented separately, each evoked a P3 response, regardless of whether the words were redundant in that they were always present ("*the* heel *is on the* shoe") or informative ("the *wheel* is on the axle"). The more informative the word, the longer the latency of the P3 component. The largest P3 wave occurred at the end of the sentence, regardless of whether that word carried the main information content of the sentence. This was interpreted as representing "syntactic closure." There were no significant hemispheric asymmetries for these P3 components.

Kutas and Hillyard (1980a) also recorded late positive waves following visual words presented in a sentence context. They found, however, a definite asymmetry, with the late positivity of greater amplitude over the left temporo-parietal region than over the right. Their sentences were much less repetitive than those used by Friedman et al. and may have therefore been more able to engage normal language processes.

Late Negative Waves

The most striking finding in the work of Kutas and Hillyard (1980b,c), however, is the large late negative wave, with a peak latency of 400 msec, that was recorded following a semantically incongruous word occurring at the end of a sentence ("I take coffee with cream and dog"). This N400 wave was larger over the posterior scalp and somewhat greater in amplitude over the right hemisphere than over the left. It is possible that this asymmetry may be related to a simultaneous slow positive wave of greater amplitude over the left hemisphere like that recorded to the previous words in the sentence. On the other hand, physically anomalous words ("I take coffee with cream and SUGAR") elicited a late positive wave rather than a late negativity. In their initial report of the N400 wave Kutas and Hillyard raised the possibility that this represented the special "reprocessing required by the semantically anomalous word from subjects who were unable to reconcile themselves to nonsense."

Somewhat similar late negative waves have been recorded in response to other meaningful stimuli. Symmes and Eisengart (1971) presented photographic slides to children between the ages of 5 and 11 years. The slides elicited a late negative wave with a peak latency of 500 msec and with a frontocentral scalp distribution. Blank or defocused slides did not elicit this component, which suggested that it is related to the "perceptual integration of meaningful stimuli." However, as Symmes and Eisengart point out, it is difficult to rule out a relationship to the contrast pattern of the focused stimuli. Courchesne (1978) recorded large frontal negative waves from children in response to novel stimuli, but not in response to other improbable or meaningful stimuli. He suggested that this component of the response is elicited by stimuli for which there is no perceptually available category, and that it may there-fore reflect "the perception of attention-getting or interesting events that require further detailed processing or assessment." Adult subjects do not generate this response to novel stimuli (Courchesne, Hillyard, and Galambos 1975). Focused complex stimuli elicit large negative waves with peak latencies of about 500 msec in infants and P350 components in adults (Schulman-Galambos and Galambos 1978). It is probable that the criteria for eliciting the late negative wave change with maturation. In infants this wave may occur in response to any complex stimulus, in older children in response to novel stimuli, and in adults in response to stimuli that after initial evaluation are assessed as requiring further processing (often linguistic in nature).

We recorded a late negative wave that is quite similar in morphology to the N400 in response to pictures that subjects were asked to name (Stuss et al. 1982). Different pictures from the Boston Naming Test were presented after a warning stimulus, and the subject gave a delayed response after another cue stimulus. Our late negative wave is somewhat more frontal than that recorded by Kutas and Hillyard, but it is otherwise strikingly similar in its morphology, its latency, and its slight right-hemispheric predominance. We also recorded this negative wave when a subject was presented with the written names of the pictures he had previously seen. Results similar to ours were obtained with words by Neville, Kutas, and Schmidt (1982); however, they recorded a negative wave that was somewhat greater in amplitude over the left frontal scalp regions. It is possible that the scalp-distribution differences may have been related to differences in the manner in which the stimulus was presented. Neville et al. presented half-field vertically written words, whereas we presented full-field horizontally written words. Another possible explanation of these differences in scalp distribution may be the variability of the preferred cognitive strategy employed by the subjects in linguistic tasks as revealed by their different ear dominances on dichotic digit tests (Bakker et al. 1980). These results initially suggested to us that the N400 wave reflected a cerebral process involved when stimuli required immediate linguistic processing. A contextually appropriate word occurring in a sentence would be handled by smoothly and continuously updating the available information through the cerebral mechanisms reflected in the lateralized P3 wave. An anomalous word requiring immediate and extensive processing of both the word and its context, an isolated word that must be read, or an unpredictable picture that must be named would elicit an N400 wave.

Polich and his colleagues (personal communication) presented subjects with a series of seven words wherein the first six belonged to one category and the last word either did or did not belong to that category. A late negative wave occurred in response to the last word when it did not belong to the same category as the previous words. Polich suggested therefore that the N400 might represent a mismatch-detection process based on the semantic category of the stimulus. It would then be functionally similar to the earlier N2 wave that occurs during the detection of a physical mismatch. There are probably many different negative waves that occur as a visual stimulus is perceptually processed, each negative wave reflecting in its timing and its scalp distribution the evaluation of different aspects of the stimulus (Ritter et al. 1982; Harter, Aine, and Schroeder 1982).

Fischler et al. (1982) recorded evoked potentials during visually presented sentences of the form "a sparrow / is not / a vehicle." The sentences were either affirmative ("is") or negative ("is not") and were either true or false. An N400 wave was evoked by the last word of the sentence when it was either affirmative and false or negative and true. Because of this second finding, the N400 could not reflect a semantic mismatch of the sentence with previous knowledge or expectancy. They suggested rather that the N400 reflected a semantic mismatch among the components of the sentence. Some recent work by Kutas and Hillyard (personal communication) also sheds light on this problem. They presented sentences the final word of which, although always semantically appropriate, was more or less expected from everyday usage: "Most students prefer to work during the day/night/summer/evenings/week" (Bloom and Fischler 1980). The less predictable the word on the basis of the sentence context (the lower its "cloze" probability), the larger the late negative wave at 400 msec it evoked. This fits well with the results of Fischler et al., since the final category given in their sentences was not predictable if the proposition was affirmative and false or negative and true. The N400 wave therefore appears to represent the cerebral processing of an informative stimulus whose meaning cannot be predicted before it occurs and to differ from the late positive wave that occurs in the processing of stimuli of predictable character but unpredictable occurrence (for example, targets occurring randomly in a train of standard stimuli).

Most studies of this late negative wave have used linguistic paradigms. We have also used a complex spatial orientation task wherein a subject determined whether two rotated versions of the same shape were identical or mirror images (Stuss et al. 1982). We recorded an N400 wave in this task, followed by prolonged parietal negativity and frontal positivity. A more simple nonlinguistic task, the detection of one of two possible shapes, elicited a P3 wave and no N400. It appears, then, that the N400 is not specifically linguistic, although it is most usually evoked in linguistic situations. It probably reflects some cerebral process invoked to monitor the perceptual evaluation of complex unpredictable stimuli and/or to provide a context wherein such stimuli might be evaluated.

Summary
There appear to be some definite ERP manifestations of language processing in the visual modality. As in the auditory modality, the evidence for localization is not consistent, although there is some suggestion of left-hemispheric specialization for the phonological interpretation of visual stimuli. The clearest and most exciting findings, however, concern

the structure and timing of the cerebral events that occur during language perception. The ERPs have shown early semantic access in the processing of words (Boddy and Weinberg 1981) and have demonstrated specific cerebral processes that can be invoked by semantic incongruities (Kutas and Hillyard 1980b). It will probably be more fruitful in the near future to use ERPs to study the "how" rather than the "where" of language in the brain.

Event-Related Potentials and Language Disorders

There are two different reasons for studying the event-related potentials of patients with language disorders. The first aims at describing the pathology using neurophysiological measures; the second aims at understanding the neurophysiology using the different kinds of pathology as experimental tools (cf. Shagass 1972; Picton 1980). In the first approach, the ERPs may provide clinical information that is objective, discriminative, and predictive. Information of this kind would be particularly helpful in the evaluation of patients with disorders such as childhood dyslexia. In these patients behavioral testing may be difficult, and even when possible it may give culturally biased results. Because there are no objective indices, diagnoses are often made more by intuition than by fact. Testing is often not possible until the child has been severely affected by the inability. It would be very helpful to make the diagnosis of dyslexia before a child goes to school. Behavioral examination may be helpful in guiding therapy, but this is normally done in a symptomatic manner without regard to etiology. In the second approach, the ERPs of patients with language disorders may allow insight into the anatomy and physiology of both normal and abnormal language processes. Language and its ERP concomitants can break down only in certain ways. Many studies have used both approaches. This is possible, but one must be cautious that the logic not become circular. An ERP measure cannot, in the same study, be validated both as a marker of a particular type of pathology and as an index of the type of behavior represented in patients with that pathology.

One of the most important methodological requirements in both approaches is an accurate definition of the patient groups (Denckla 1978; Hughes 1978). The patients must fulfill well-defined criteria on neurological evaluation, on neuropsychological testing, and on neuroradiological assessment. Furthermore, the patient groups must be reasonably homogeneous on other variables. There must be no obvious disparities within the group in such parameters as age, sex, and educational experience. The normal subjects must be reasonably well matched to the patients and should not be characterized by risk factors that could suggest subclinical disorder. It is only through such well-defined and homogeneous groups that one can hope to find neurophysiological indices of pathology; otherwise, within-group differences may distort or even cancel any of the between-group effects.

Once significant differences have been obtained between normal subjects and patients with a language disorder, the two experimental approaches diverge. If the goal is to establish some diagnostic measure, the investigator should assess patients with other disorders to make sure that the findings have some diagnostic specificity and are not just indicative of "patienthood" and must evaluate further subgroups of patients with the disorder to see whether the measure is related to the disorder in general or to one particular variant thereof. Finally, the results must be evaluated in other patients with different degrees of abnormality to determine the relationship of the neurophysiological measure to the behavioral manifestations. This step-by-step development of neurophysiological diagnostic rules has been used by Duffy et al. (1980b) in evaluating children with dyslexia. It differs from the "neurometric" approach of John (1977), wherein a large heterogeneous group of patients is evaluated using many neurophysiological measures, which are then submitted to multivariate cluster analysis. This technique has been applied to learning-disabled children and has demonstrated distinct neurophysiological subgroups. The difficulty is that the analysis is data-driven and will demonstrate subgroups in all but the most homogeneous sets of data. The results must therefore be validated by some other criteria based on etiology or response to treatment.

If one is using the neurophysiological results from patients to determine more about language, the further steps are different from when one is seeking a diagnostic measure. One must become more specific about the language functions disrupted in the patients with the abnormal ERP measures. This can be done by studying patients with different clinical disorders who have overlapping language dysfunctions. One must also make the anatomical basis of the abnormalities more specific. Do patients with similar lesions of the opposite hemisphere have homologous ERP abnormalities? Do patients with lesions in approximately the same location but without the language disorder show different waveforms?

In addition to these complex issues of experimental design, multiple technical difficulties must be overcome in recording ERPs in patients. Patients are just not the same as the college students that are the typical subjects in ERP experiments. Indeed, these subjects are not even appropriate normal subjects for most ERP studies of patients with language disorders, because of their age and their educational experience. Artifacts from the eyes, the scalp muscles, and the tongue are particular problems in patients. More time is required to explain the paradigm to patients, and the time taken for averaging is longer because of the higher levels of background EEG noise and the larger number of trials that must be rejected because of artifacts.

The paradigms used to study the ERPs of patients with language disorders vary in the amount of patient involvement required and in the linguistic content of the stimuli or the task. One type of paradigm uses a nonlinguistic stimulus and records from a passive subject. The goal of this approach is to find some simple neurophysiological marker of diagnosis or prognosis. The very simplicity of the paradigm makes it easy to administer in patients and facilitates obtaining extensive normative data. The main disadvantage is that a simple paradigm may not be able to demonstrate abnormalities that are evident only in conditions requiring complex brain functions. Even when abnormalities are demonstrated, it is possible that they are related only obliquely to the language disorder. Liberson (1966) found that patients with aphasia could have normal or abnormal evoked potentials to stimulation of the median nerve. Those with normal responses were less severely affected and responded better to speech therapy. These results were probably more related to the extent and the location of the lesion causing the aphasia than to the language disorder itself. Patients with larger and/or more parietal lesions would be more likely to have abnormal somatosensory evoked potentials and would also be less likely to improve. Other experimental designs require the subject to perceive and respond to linguistic stimuli while the ERPs are recorded. The ERPs are averaged separately, depending upon the accuracy and the speed of the response. The advantage of this approach is that it will show the ERP changes that occur as a patient demonstrates abnormal language function. The major disadvantage of this technique is that it may not provide any more information than is available in the behavioral responses. Otto et al. (1976) found a greater than normal positivity in the ERPs recorded from children with aphasia and dyslexia during visual matching of pictures or letters. These results were attributed to "extensions of the P300 wave reflecting the uncertainty of aphasics and dyslexics in processing stimulus information." However, such ERP abnormalities could have also occurred when the patients were certain of their responses, indicating some abnormality of visual matching unrelated to uncertainty. This would require a more extensive evaluation of the ERPs in relation to the behavioral responses. Within the range of these extremes several other kinds of paradigms have been used. Subjects may be asked to make behavioral responses but not to linguistic stimuli, or they may be requested to perceive linguistic stimuli such as speech sounds or written words but not to make any specifically linguistic response to the stimuli. In the first case the assumption is that a language disorder may be more likely to show up when the subject is relating stimulus to response than when he is just perceiving the stimulus. In the second case it is hypothesized that certain language functions may be unavoidably initiated by linguistic stimuli even when no linguistic response is required.

The psychological concomitants of language disorder must always be considered in evaluating the ERPs of these patients. Patients will tend to be more anxious and less certain during the paradigms. They may generally be less motivated and less attentive than normal subjects. The effects of these psychological processes on the ERPs must always be assessed and if possible differentiated from the effects specifically due to the language disorder.

Studies of Dyslexia

Conners (1970) made the first extensive recordings of the flash-evoked potential in dyslexic children. He recorded from left and right occipital and parietal regions using a vertex reference. The basic paradigm required that the children look at a 1-per-1.6-sec train of flashes and press a telegraph key in response to occasional (one in five) dim flashes. Only the evoked potentials to the bright flashes were recorded. Children with reading disorders showed a significant decrease in the amplitude of the positive wave at about 140 msec and in that of the negative wave at about 200 msec in the parietal regions, particularly on the left. These changes were significantly related to the learning quotient, the reading ability, and the verbal ability of the dyslexic child. Furthermore, an attenuation of the left parietal response was found in all the dyslexic members of a family. This study is impressive for the number of patients evaluated. The heterogeneity of the patients, however, makes it difficult to determine whether the abnormalities detected were specifically related to dyslexia. Conners himself raises the possibility that some of his poor readers may have suffered from "generalized diminished attentiveness, low arousal, or other motivational factors," which could have affected the evoked potentials in both hemispheres rather than just in the left hemisphere. Although the subjects were asked to detect the occasional dim flashes, the speed and accuracy of their performance were not assessed. A major shortcoming of this study is the lack of any recordings from normal children. Normal children can certainly have asymmetrical visual evoked potentials. Furthermore, the location of the reference electrode at the vertex makes it impossible to determine whether the abnormalities noted were located at the parietal or at the vertex regions of the scalp (Kooi 1972).

Several other studies have recorded the evoked potentials of dyslexic patients to simple visual stimuli. Shields (1973) found that the visual evoked potentials of children with learning disabilities were significantly larger and later than those of normal children. She recorded from left and right central regions and used a variety of visual stimuli: "light flashes, pictures, designs, words, and nonsense words." Preston, Guthrie, and Childs (1974) recorded the evoked potentials to flashes and to the word *cat* from dyslexic children, age-matched normal children, and younger normal children matched to the reading levels of the dyslexic children. The dyslexic children had significantly smaller amplitudes in the left parietal region, but the confessed technical problems encountered in the study make the data very unconvincing. Sobotka and May (1977) replicated the study of Conners using both normal and dyslexic children. Both groups of children had flash-evoked potentials that were greater in amplitude over the right hemisphere. The dyslexic children had significantly greater amplitudes than the normal children at both parietal and occipital locations. The interpretation of these findings is difficult because of the active vertex reference electrode. Weber and Omenn (1977) found no consistent asymmetries of the visual or auditory evoked potentials in the dyslexic members of families with reading disabilities. Duffy et al. (1980a) found small differences in the late components of the flash-evoked potential between normal and dyslexic children, but these differences were at the midline parietal regions and were of questionable significance.

Two early studies of the ERPs in dyslexic children used combinations of stimuli. Shipley and Jones (1969) compared the responses to simultaneous flashes and clicks with the responses obtained when the stimuli were presented separately. They found that the bimodal responses of dyslexic children were smaller than the sum of the separate responses, whereas in normal children the bimodal responses tended to be larger. These results suggest a failure of intermodality integration in the dyslexic children, although the statistical significance of the results is not clear. Ross, Childers, and Perry (1973) recorded the responses to binocular and dicoptic red and green stimuli. They found an asymmetry in the response, but this asymmetry is very difficult to evaluate because of their unusual recording montage.

There have been several studies of the visual evoked potentials to words in dyslexic patients. Preston et al. (1977) studied the evoked potentials to three-letter words and to unpatterned flashes in adult dyslexic subjects. The subjects were required to count the appearances of a designated target word in the train of presented words, and the evoked potentials were averaged over targets and nontargets. During the flash conditions the subjects counted all stimuli. Normal subjects showed significantly larger left parietal late positive waves in response to the words but not to the flashes (Preston 1979). This asymmetry was significantly less in the dyslexic patients. Matsumiya (1976) reported that dyslexic children had visual evoked potentials to meaningful words that were more asymmetrical than the responses of normal children, but insufficient data allow no meaningful interpretation of these results. Symann-Louett et al. (1977) compared the evoked potentials of normal and dyslexic children to visually presented common words. They recorded the responses from left and right occipital and parietal regions and from electrodes over Wernicke's area and an equivalent region of the right hemisphere, using linked earlobes as a reference. The dyslexic children had significantly fewer components in the evoked potential occurring before 200 msec at the left parietal region and over Wernicke's area. The significance of this finding is, however, questionable; the fact that 2 out of 21 comparisons were significant at $P < 0.05$ is more probably due to chance than to a valid difference. The lack of any striking differences between the normal and dyslexic children may have been related to the absence of any discrimination task to ensure linguistic involvement. Shelburne (1978) recorded the evoked potentials to visually presented letters that formed words or nonsense CVC trigrams. He found that, in comparison with normal children, dyslexic children had no significant late positive wave in the response to the final and informative letter of the trigram. This finding correlated with the poor performance of the dyslexic children on the discrimination task. Unfortunately, the evoked potentials were not analyzed separately for correct and incorrect discriminations. Bakker et al. (1980) found that dyslexic children had significantly longer latencies in the parietal regions for the P310 and N440 components of the response to words. Since the subjects "read the flashed word aloud upon appearance," it is impossible to rule out the possibility of contaminating speech-muscle potentials. These would probably have occurred at different latencies in the two groups of children.

Several groups have studied the evoked potentials of dyslexic children to speech sounds. Duffy et al. (1980a) required normal and dyslexic children to discriminate between the words *tight* and *tyke* and count the number of times that one of these equiprobable stimuli occurred. Very significant differences between the two groups of children were found in the parietal and posterior temporal regions at multiple latencies. These differences were highly accurate in discriminating normal and dyslexic children (Duffy et al. 1980b). Unfortunately, it is difficult to determine what these differences actually represent in terms of the ERP waveform, since only their significance is mapped over the scalp. The differences were most apparent in the 150–186msec region of the response and were far more striking for these speech-evoked potentials than they were for click-evoked potentials. Fried et al. (1981) recorded the evoked potentials to spoken words (*do* and *go*) and strummed chords presented in random order to normal children and to dyslexic children classified as "dysphonetic," "dyseidetic," or "alexic" (Boder 1973). Normal readers and those dyslexic children with visuospatial deficits showed significant waveform differences between the responses to words and the responses to chords, particularly over the left hemisphere. These differences were mainly related to latency changes but were measured using correlation techiques. Dysphonetic children did not have this asymmetry.

There have been many studies of the contingent negative variation in dyslexic children. Fenelon (1968, 1978) found that the CNV in dyslexic children was significanly lower than that of normal children. This difference was most apparent in the left parietal area and when the S1 and S2 were auditory, although these effects were not significant on the statistical analysis. Cohen (1980) used a visual S1 and an auditory S2. The CNV was significantly lower in amplitude in dyslexic children than in age-matched normal children. There were no significant differences in the auditory evoked potentials; the dyslexic children had significantly later and smaller visual evoked potentials. There were no significant differences between the two groups in reaction time, which suggested that the ERP differences were not related to an impairment of attention. The findings were therefore interpreted as indicating a dysfunction in visual processing and in the integration of visual and auditory information. There is no evidence that such abnormalities are specific to dyslexia, however, and it is possible that the dyslexic patients were more anxious and/or less attentive in the experimental environment than the normal children.

In summary: There are no consistent or specific ERP findings from patients with dyslexia. The most promising findings have been obtained with paradigms that require some degree of linguistic processing from the subjects. Future studies should probably concentrate on active linguistic paradigms. It might be possible, for example, to record the ERPs during reading itself (Kurtzberg and Vaughan 1979). There is some evidence that different subgroups of dyslexic children (Boder 1973; Doehring and Hoshko 1977) may show different ERP findings. Any future study of ERPs in dyslexic children must therefore pay careful attention to these different types of dyslexia.

Studies of Aphasia
Pace et al. (1979) recorded the auditory evoked potentials to common words in two groups of brain-damaged subjects: three left-hemisphere-damaged fluent aphasics and three right-hemisphere-damaged subjects. Each trial consisted of paired stimuli: a category name (furniture or animal) followed in 4–6 seconds by a high- or a low-frequency member of the presented category or of another category. Sixteen word pairs were presented. Subjects responded by pressing one key if the words were related and another if they were not. The evoked potentials from T3, T4, and P3 electrode sites were averaged over 16 trials for each category, membership frequency, match-mismatch sequence, and electrode site. A principal-component analysis of these results yielded 12 components. Significant differences between the patient groups were reported for 3 of these components. Unfortunately, the significance of these data is extremely difficult to evaluate. The number of subjects was small and the statistical analysis of the data overwhelming. Since the number of trials in an average was small and there was no control for scalp-muscle or ocular potentials, it is impossible to know to what extent these factors were related to unaveraged EEG noise or extracranial artifacts.

Szirtes et al. (1980) studied the auditory evoked potentials in 6 right-hemisphere-damaged patients, 8 normal control subjects, and 12 left-hemisphere-damaged (aphasic) patients. One-second pre- and post-stimulus epochs were averaged from F7, F8, C3, and C4 as well as from temporal electrodes place between T3 and T5 (W1) and T4 and T6 (W2), using a linked-ear reference. Subjects were requested to listen to one- and two-syllable words presented under one of four conditions: left monaural, right monaural, dichotic, and diotic. Although no artifact controls were mentioned, the "artifact-free" trials for each subject were grouped for averaging under the four conditions as well as for a grand average. The amplitudes of N1 and P2 were smaller in the aphasic patients. In addition, the aphasic patients appeared to have a very small slow negative wave (probably an auditory sustained potential) or none at all. No asymmetries were reported. Fourier analysis did not appear to add any further information. There was no documentation of the type of aphasia or the size and location of the lesion.

Three projects have analyzed visual ERPs in aphasic patients.

In the first, Neville and her colleagues (Neville et al. 1979; Neville 1980) studied three patients who had a clinically and neurologically defined syndrome: alexia without agraphia. These patients, having left occipital lesions that also presumably involved the splenium of the corpus callosum, were compared with normal control subjects and with one patient with a large left occipital lesion that presumably spared the splenium (he could read normally). Words and pictures were presented to the left and right visual fields. Recordings were taken from left and right occipital, parietal, and central regions using a linked-mastoid reference. The ERPs of the alexic patients had two major differences from those of the normal subjects. First, both types of stimulus information in the normal subjects evoked a large N150 over the occiput contralateral to stimulus presentation. In the alexic patients, the responses to visual stimuli presented to the left visual field (going to the intact right hemisphere) were dependent on content. Line drawings evoked an essentially normal N1. The words, which were reported accurately, did not evoke an N1 over the intact right occipital lobe but did evoke apparently normal N1s from more anterior C4 and P4 leads. This was interpreted as reflecting an anterior relay of the visual language information for processing in the right hemisphere or for transfer via intact callosal areas to the language-dominant left hemisphere. The second major finding concerned the P3 (325 msec) wave. The P3 was large and symmetrical in normal subjects but asymmetrical at parietal leads (left greater than right) for the aphasic patients. If a P3 of greater amplitude reflects increased processing, this result may corroborate the second hypothesis, suggesting a more anterior callosal transfer for subsequent left-hemisphere processing. The patient who had a similar left occipital lesion but who was not alexic (presumably because of an intact splenium) had a normal N1 distribution with left-visual-field presentation, suggesting that left occipital damage alone could not explain the absent right occipital N1 in alexic patients. These results are complex and (as the authors themselves note) will be easily understood only after additional studies of similar patients. The report does, however, demonstrate how precisely defined patients and vigorously controlled experimental paradigms may be used to delineate language processes in patients with brain damage.

In the second study using visual ERPs in aphasic patients, Thatcher and April (1976) recorded the ERPs of normal subjects and two aphasic patients (one Broca's aphasic and one Wernicke's aphasic) during performance of visual matching tasks. Matching was done between letters and word meanings. The word pairs were simpler for the aphasics than for the normal subjects. Recordings were taken from many scalp locations in reference to linked earlobes. Normal subjects had large late positive waves (300–500 msec) to the first letter and the matching second letter, both of which were significantly more positive than the response to a mismatched second letter. Similar results occurred when the responses to antonyms and synonyms (match) were compared with the responses to neutral words (mismatch). In the severe Broca's aphasic, the late positive component was enhanced bilaterally but was not large in the left inferior frontal and left temporal regions. In the Wernicke's aphasic, the late positive component was absent in the anterior electrodes in the delayed semantic matching task. This result was difficult to interpret because the patient was unable to perform the task correctly. Goto et al. (1979) also found reduced late positive waves in aphasic patients during a similar semantic matching task. Both of these studies were limited by the small number of aphasic patients tested.

In an attempt to obtain an objective index of verbal discrimination ability in aphasic patients, Burian et al. (1972) used a CNV paradigm that allowed comparison of meaningful and meaningless words. The meaningful word (*mama*) was presented auditorially and was followed in 1.5 seconds by a flash. Randomly interspersed with these trials was a meaningless phonetic equivalent (*fama*), which was not followed by a flash. Development of a CNV to the meaningful word and not to the meaningless word could be objective evidence of correct word discrimination. Only 7 of 11 aphasic patients developed a CNV, but no correlation was made with the patients' ability to comprehend. A single-case study of aphasia using essentially the same paradigm (Gloning and Burian 1975) addressed the issue of different types of perseveration in aphasia. The aphasic patient developed a CNV to a meaningful word but not to a meaningless word, demonstrating intact discrimination. When the flash was shifted the follow to meaningless and not the meaningful word, the subject verbalized the change but continued to develop a CNV only after the meaningful word. Perseveration of the CNV indicated a failure of an extinction mechanism, rather than a defect in mnestic control. Such CNV results are difficult to interpret. There were no controls for possible EOG contamination. Furthermore, the CNV may be increased when a greater amount of effort is devoted to a task, decreased by anxiety, and prolonged by uncertainty (Picton and Low 1971; Low and Swift 1971; Jacobson and Gans 1981).

In summary: There have been no consistent ERP findings in patients with aphasia. Very few studies have been done. Those that have been reported are characterized by too few patients, inadequate artifact control, little if any anatomical documentation, and insufficient correlation with abnormal language performance. This is unfortunate, since the neurophysiological study of patients with well-documented brain damage and well-described language abnormalities could allow a deeper understanding of both language and neurophysiology.

Conclusions

Very little is known about the neurophysiology of language. It is, however, a field with great promise, and much more research is needed. We would not have been so exhaustively critical of past research if we felt that further study was unjustified. The following four principles for future research are derived from our analysis of the facts and the flaws of the literature to date.

• The ERPs should be recorded while the subject is involved in language processing. This processing must be clearly defined by the experimental paradigm and closely monitored by behavioral responses. Only then will the complex relations between neurophysiological measurement and language performance become accessible. Some of the inconsistencies in the literature may be related to the variety of perceptual strategies that subjects can employ to perform the same task.

• The ERPs must be recorded with obsessive regard for the technical aspects of analyzing electrical fields at a distance from their generation.

All possible sources of noncerebral potentials must be controlled and monitored. Extensive scalp-distribution studies are imperative, since there is far more information in the EEG than that which can be seen in one or two channels of recording. Averaging must be adequate to provide clear recordings, and replications must be performed to ensure their reliability.

• The research must address specific questions and must be clear about which questions are being asked. The initial questions should concern the relations between ERP components and language processes. These may be followed by questions about the structure of language processing as revealed by ERP measures with defined relations to particular aspects of language.

• The neurophysiological evaluation of patients with language disorders will demand that both their behaviors and their lesions be as accurately and extensively delineated as their ERPs.

Acknowledgments

We appreciate the financial support of the Medical Research Council and the Ontario Mental Health Foundation. Francine Sarazin assisted in the preparation of the references.

References

Anderson, S. W. 1977. Language-related asymmetries of eye-movement and evoked potentials. In *Lateralization in the Nervous System*, S. Harnad, R. W. Doty, L. Goldstein, J. Jaynes, and G. Krauthamer, eds. New York: Academic.

Bakker, D. J., R. Licht, A. Kok, and A. Bouma. 1980. Cortical responses to word-reading by right- and left-eared normal and reading-disturbed children. *Journal of Clinical Neuropsychology* 2: 1–12.

Baribeau-Braun, J. 1977. Investigation du moment des premiers effets d'attention sélective sur les potentials évoqués auditifs. Thesis, University of Ottawa.

Barlow, J. S. 1971. Brain information processing during reading: Electrophysiological correlates. *Diseases of the Nervous System* 32: 668–672.

Bickford, R. G. 1972. Physiological and clinical studies of microreflexes. *Electroencephalography and Clinical Neurophysiology Supplement* 31: 93–108.

Bloom, P. A., and I. Fischler. 1980. Completion norms for 329 sentence contexts. *Memory and Cognition* 8: 631–642.

Blumhardt, L. D., G. Barrett, and A. M. Halliday. 1977. The asymmetrical visual evoked potential to pattern reversal in one half field and its significance for the analysis of visual field defects. *British Journal of Opthalmology* 61: 454–461.

Boddy, J. 1981. Evoked potentials and the dynamics of language processing. *Biological Psychology* 13: 125–140.

Boddy, J., and H. Weinberg. 1981. Brain potentials, perceptual mechanisms and semantic categorisation. *Biological Psychology* 12: 43–61.

Boder, E. 1973. Developmental dyslexia: A diagnostic approach based on three atypical reading-spelling patterns. *Developmental Medicine and Child Neurology* 15: 663–687.

Brooker, B. H., and M. W. Donald. 1980a. The search for scalp-recordable speech potentials. *Progress in Brain Research* 54: 782–789.

Brooker, B. H., and M. W. Donald. 1980b. Contribution of the speech musculature to apparent human EEG asymmetries prior to vocalization. *Brain and Language* 9: 226–245.

Brown, W. S., and D. Lehmann. 1979. Linguistic meaning-related differences in ERP scalp topography. In *Human Evoked Potentials: Applications and Problems*, ed. D. Lehmann and E. Callaway. New York: Plenum.

Brown, W. S., D. Lehmann, and J. T. Marsh. 1980. Linguistic meaning-related differences in evoked potential topography: English, Swiss-German and imagined. *Brain and Language* 11: 340–353.

Brown, W. S., J. T. Marsh, R. E. Ponsford, and L. E. Travis. 1982. Language laterality, sex and stuttering: ERPs to contextual meaning. Unpublished.

Brown, W. S., J. T. Marsh, and J. C. Smith. 1973. Contextual meaning effects on speech-evoked potentials. *Behavioral Biology* 9: 755–761.

Brown, W. S., J. T. Marsh, and J. C. Smith. 1976. Evoked potential waveform differences produced by the perception of different meanings of an ambiguous phrase. *Electroencephalography and Clinical Neurophysiology* 41: 113–123.

Brown, W. S., J. T. Marsh, and J. C. Smith, 1979. Principal component analysis of ERP differences related to the meaning of an ambiguous word. *Electroencephalography and Clinical Neurophysiology* 46: 709–714.

Buchsbaum, M., and D. Drago. 1977. Hemispheric asymmetry and the effects of attention on visual evoked potentials. *Progress in Clinical Neurophysiology* 3: 243–253.

Buchsbaum, M., and P. Fedio. 1969. Visual information and evoked responses from the left and right hemispheres. *Electroencephalography and Clinical Neurophysiology* 26: 266–272.

Buchsbaum, M., and P. Fedio. 1970. Hemispheric differences in evoked potentials to verbal and non-verbal stimuli in the left and right visual fields. *Physiology and Behavior* 5: 207–210.

Burian, K., G. F. Gestring, K. Gloning, and M. Haider. 1972. Objective examination of verbal discrimination and comprehension in aphasia using the contingent negative variation. *Audiology* 11: 310–316.

Butler, S. R., and A. Glass. 1974. Asymmetries in the CNV over left and right hemispheres while subjects await numeric information. *Biological Psychology* 2: 1–16.

Chapman, R. M., H. R. Bragdon, J. A. Chapman, and J. W. McCrary. 1977. Semantic meaning of words and average evoked potentials. *Progress in Clinical Neurophysiology* 3: 36–47.

Chapman, R. M., J. W. McCrary, J. A. Chapman, and H. R. Bragdon. 1978. Brain responses related to semantic meaning. *Brain and Language* 5: 195–205.

Chapman, R. M., J. W. McCrary, J. A. Chapman, and J. K. Martin. 1980. Behavioral and neural analyses of connotative meaning: Word classes and rating scales. *Brain and Language* 11: 319–339.

Cohen, J. 1980. Cerebral evoked responses in dyslexic children. *Progress in Brain Research* 54: 502–506.

Cohn, R. 1971. Differential cerebral processing of noise and verbal stimuli. *Science* 172: 599–601.

Conners, C. K. 1970. Cortical visual evoked response in children with learning disorders. *Psychophysiology* 7: 418–428.

Courchesne, E. 1978. Neurophysiological correlates of cognitive development: Changes in long-latency event-related potentials from childhood to adulthood. *Electroencephalography and Clinical Neurophysiology* 45: 468–482.

Courchesne, E., S. A. Hillyard, and R. Galambos. 1975. Stimulus novelty, task relevance and the visual evoked potential in man. *Electroencephalography and Clinical Neurophysiology* 39: 131–143.

Curry, S. H., J. F. Peters, and H. Weinberg. 1978. Choice of active electrode site and recording montage as varables affecting CNV amplitude preceding speech. In *Multidisciplinary Perspectives in Event-Related Brain Potential Research*, ed. D. A. Otto. Washington: U.S. Environmental Protection Agency.

Davis. A. E., and J. A. Wada. 1977. Spectral analysis of evoked potential asymmetries related to speech dominance. *Progress in Clinical Neurophysiology* 3: 127–139.

Davis, H. 1976. Principles of electric response audiometry. *Annals of Otology, Rhinology and Laryngology Supplement* 28: 4–96.

Deecke, L., B. Grözinger, and H. H. Kornhuber. 1976. Voluntary finger movement in man: Cerebral potentials and theory. *Biological Cybernetics* 23: 99–119.

Denckla, M. B. 1978. Critical review of "Electroencephalographic and Neurophysiological Studies in Dyslexia." In *Dyslexia: An Appraisal of Current Knowledge*, ed. A. L. Benton and D. Pearl. New York: Oxford University Press.

Doehring, D. G., and I. M. Hoshko. 1977. Classification of reading problems by the Q-technique of factor analysis. *Cortex* 13: 281–294.

Donald, M. W. 1979. Current theories of transient evoked potentials: limits and alternative approaches. *Progress in Clinical Neurophysiology* 6: 187–199.

Donchin, E. 1981. Surprise! . . . Surprise? *Psychophysiology* 18: 493–513.

Donchin, E., M. Kutas, and G. McCarthy. 1977. Electrocortical indices of hemispheric utilization. In *Lateralization of the Nervous System*, ed. S. Harnad et al. New York: Academic.

Donchin, E., and E. F. Heffley. 1978. Multivariate analysis of event-related potential data: A tutorial review. In *Multidisciplinary Perspectives in Event-Related Brain Potential Research*, ed. D. A. Otto. Washington: U.S. Environmental Protection Agency.

Donchin, E., W. Ritter, and W. D. McCallum. 1978. Cognitive psychophysiology: The endogenous components of the ERP. In *Event-Related Brain Potentials in Man*, ed. E. Callaway et al. New York: Academic.

Duffy, F. H., M. B. Denckla, P. H. Bartels, and G. Sandini. 1980a. Dyslexia: Regional differences in brain electrical activity by topographic mapping. *Annals of Neurology* 7: 412–420.

Duffy, F. H., M. B. Denckla, P. H. Bartels, G. Sandini, and L. S. Kiessling. 1980b. Dyslexia: Automated diagnosis by computerized classification of brain electrical activity. *Annals of Neurology* 7: 421–428.

Ertl, J., and E. W. P. Schafer. 1967. Cortical activity preceding speech. *Life Sciences* 6: 473–479.

Ertl, J., and E. W. P. Schafer. 1969. Erratum. *Life Sciences* 8: 559.

Fenelon, B. 1968. Expectancy waves and other complex cerebral events in dyslexic and normal subjects. *Psychonomic Science* 13: 253–254.

Fenelon, B. 1978. Hemispheric effects of stimulus sequence and side of stimulation on slow potentials in children with reading problems. In *Multidisciplinary Perspectives in Event-Related Brain Potential Research*, ed. D. A. Otto. Washington: U.S. Environmental Protection Agency.

Fischler, I., P. A. Bloom, D. A. Childers, S. E. Roucos, and N. W. Perry. 1982. Potentials related to sentence verification: Lexical versus sentential processes and the N400. (In press)

Fried, J., P. E. Tanguay, E. Boder, C. Doubleday, and M. Greensite. 1981. Developmental dyslexia: Electrophysiological evidence of clinical subgroups. *Brain and Language* 12: 14–22.

Friedman, D., R. Simson, W. Ritter, and I. Rapin. 1975a. Cortical evoked potentials elicited by real speech words and human sounds. *Electroencephalography and Clinical Neurophysiology* 38: 13–19.

Friedman, D., R. Simson, W. Ritter, and I. Rapin. 1975b. The late positive component (P 300) and information processing in sentences. *Electroencephalography and Clinical Neurophysiology* 38: 255–262.

Gaillard, A. 1978. *Slow Brain Potentials Preceding Task Performance.* Institute for Perception, TNO, Soesterberg.

Galambos, R., P. Benson, T. S. Smith, C. Schulman-Galambos, and H. Osier. 1975. On hemispheric differences in evoked potentials to speech stimuli. *Electroencephalography and Clinical Neurophysiology* 39: 279–283.

Galin, D., and R. R. Ellis. 1975. Asymmetry in evoked potentials as an index of lateralized cognitive processes: Relation to EEG alpha asymmetry. *Neuropsychologia* 13: 45–50.

Genesee, F., J. Hamers, W. E. Lambert, L. Mononen, M. Seitz, and R. Starck. 1978. Language processing strategies of bilinguals: A neurophysiology study. *Brain and Language* 5: 1–12.

Geschwind, N., and W. Levitsky. 1968. Left-right asymmetries in temporal speech region. *Science* 161: 186–187.

Glaser, E. M., and D. S. Ruchkin. 1976. *Principles of Neurobiological Signal Analysis.* New York: Academic.

Gloning, K., and K. Burian. 1975. Perseveration in aphasia as demonstrated by LERA. *Revue de Laryngologie* 96: 207–209.

Goto, H., T. Adachi, T. Utsonomiya, and I. C. Chen. 1979. Late positive component (LPC) and CNV during processing of linguistic information. In *Human Evoked Potentials Applications and Problems*, ed. D. Lehmann and E. Callaway. New York: Plenum.

Goto, H., T. Adachi, T. Utsonomiya, and I. C. Chen. 1980. Late positive component (LPC) during semantic information processing in Kanji and Kana words. In *Evoked Potentials*, ed. C. Barber. Baltimore: University Park Press.

Grabow, J. D., A. E. Aronson, K. P. Offord, D. E. Rose, and K. L. Greene. 1980a. Hemispheric potentials evoked by speech sounds during discrimination tasks. *Electroencephalography and Clinical Neurophysiology* 49: 48–58.

Grabow, J. D., A. E. Aronson, D. E. Rose, and K. L. Greene. 1980b. Summated potentials evoked by speech sounds for determining cerebral dominance for language. *Electroencephalography and Clinical Neurophysiology* 49: 38–47.

Grabow, J. D., and F. W. Elliot. 1974. The electrophysiological assessment of hemispheric asymmetries during speech. *Journal of Speech and Hearing Research* 17: 64–72.

Greenberg, H. J. and J. T. Graham. 1970. Electroencephalographic changes during learning of speech and nonspeech stimuli. *Journal of Verbal Learning and Verbal Behavior* 9: 274–281.

Grözinger, B., H. H. Kornhuber, and J. Kriebel. 1977. Human cerebral potentials preceding speech production, phonation, and movements of the mouth and tongue, with reference to respiratory and extracerebral potentials. *Progress in Clinical Neurophysiology* 3: 87–103.

Grözinger, B., H. H. Kornhuber, and J. Kriebel. 1979. Participation of mesial cortex in speech: Evidence from cerebral potentials preceding speech production in man. *Experimental Brain Research Supplement* 11: 189–192.

Grözinger, B., H. H. Kornhuber, J. Kriebel, J. Szirtes, and K. T. P. Westphal. 1980. The *Bereitschaftspotential* preceding the act of speaking. Also an analysis of artifacts. *Progress in Brain Research* 54: 798–804.

Haaland, K. Y. 1974. The effect of dichotic, monaural and diotic verbal stimuli on auditory evoked potentials. *Neuropsychologia* 12: 345–399.

Harter, M. R., C. Aine, and C. Schroeder. 1982. Hemispheric differences in ERP measures of selective attention. In *Proceedings of the Sixth International Conference on Event-Related Potentials of the Brain*, ed. R. Karrer, J. Cohen, and P. Tueting. In press.

Hillyard, S. A., T. W. Picton, and D. Regan. 1978. Sensation, perception and attention: Analysis using ERPs. In *Event-Related Brain Potentials in Man*, ed. E. Callaway et al. New York: Academic.

Hillyard, S. A., and D. L. Woods. 1979. Electrophysiological analysis of human brain function. In *Handbook of Behavioral Neurology, volume 2*, ed. M. S. Gazzaniga. New York: Plenum.

Hink, R. F., S. A. Hillyard, and P. J. Benson. 1978. Event-related potentials and selective attention to acoustic and phonetic cues. *Biological Psychology* 6: 1–16.

Hink, R. F., K. Kaga, and J. Suzuki. 1980. An evoked potential correlate of reading ideographic and phonetic Japanese scripts. *Neuropsychologia* 18: 455–464.

House, J. F., and P. Naitoh. 1979. Lateral-frontal slow potential shifts preceding language acts in deaf and hearing adults. *Brain and Language* 8, no. 3: 287–302.

Hughes, J. R. 1978. Electroencephalographic and neurophysiological studies in dyslexia. In *Dyslexia: An Appraisal of Current Knowledge*, ed. A. L. Benton and D. Pearl. New York: Oxford University Press.

Jacobson, G. P., and D. P. Gans. 1981. The contingent negative variation as an indicator of speech discrimination difficulty. *Journal of Speech and Hearing Research* 24: 345–350.

John, E. R. 1977. *Neurometrics: Clinical Applications of Quantitative Electrophysiology*. New York: Wiley.

Kinsbourne, M. 1972. Eye and head turning indicate cerebral lateralization. *Science* 176: 539–541.

Kooi, K. A. 1972. Letter to the editor. *Psychophysiology* 9: 154.

Kooi, K. A., and R. E. Marshall. 1979. *Visual Evoked Potentials in Central Disorders of the Visual System*. New York: Harper & Row.

Kurtzberg, D., and H. G. Vaughan, Jr. 1979. Maturation and task specificity of cortical potentials associated with visual scanning. In *Human Evoked Potentials. Applications and Problems*, ed. D. Lehmann and E. Callaway. New York: Plenum.

Kutas, M., and S. A. Hillyard. 1980a. Reading between the lines: Event-related brain potentials during natural sentence processing. *Brain and Language* 11: 354–373.

Kutas, M., and S. A. Hillyard. 1980b. Reading senseless sentences: Brain potentials reflect semantic incongruity. *Science* 207: 203–205.

Kutas, M., and S. A. Hillyard. 1980c. Event-related brain potentials to semantically inappropriate and surprisingly large words. *Biological Psychology* 11: 99–116.

Kutas, M., G. McCarthy, and E. Donchin. 1977. Augmenting mental chronometry: The P300 as a measure of stimulus evaluation time. *Science* 197: 792–795.

Lawson, E. A., and A. W. K. Gaillard. 1981. Evoked potentials to consonant-vowel syllables. *Acta Psychologica* 49: 17–25.

Ledlow, A., J. M. Swanson, and M. Kinsbourne. 1978. Reaction times and evoked potentials as indicators of hemispheric differences for laterally presented name and physical matches. *Journal of Experimental Psychology. Human Perception and Performance* 4: 440–454.

Leissner, P., L.-E. Lindholm, and I. Petersen. 1970. Alpha amplitude dependence on skull thickness as measured by ultrasound technique. *Electroencephalography and Clinical Neurophysiology* 29: 392–399.

Levy, R. S. 1977. The question of electrophysiological asymmetries preceding speech. *Studies in Neurolinguistics* 3: 287–318.

Levy, R. S. 1980. A note on prespeech slow potentials with special reference to Brooker's and Donald's (1980) inaccuracies in describing Levy (1977). *Brain and Language* 11: 204–208.

Liberson, W. T. 1966. Study of evoked potentials in aphasics. *American Journal of Physical Medicine* 45: 135–142.

Low, M. D., and M. Fox. 1977. Scalp-recorded slow potential asymmetries preceding speech in man. *Progress in Clinical Neurophysiology* 3: 104–111.

Low, M. D., and S. J. Swift. 1971. The contingent negative variation and the "resting" DC potential of the human brain: Effects of situational anxiety. *Neuropsychologia* 9: 203–208.

Low, M. D., J. A. Wada, and M. Fox. 1973. Electroencephalographic localization of the conative aspects of language production in the human brain. *Transactions of the American Neurological Association* 98: 129–133.

McAdam, D. W., and H. A. Whitaker. 1971. Electrocortical localization of language function—Reply to Morrell and Huntington. *Science* 174: 1360–1361.

McCallum, W. C., and S. H. Curry. 1980. The form and distribution of auditory evoked potentials and CNVs when stimuli and responses are lateralized. *Progress in Brain Research* 54: 767–775.

Marsh, G. R., L. W. Poon, and L. W. Thompson. 1976. Some relationships between CNV, P300, and task demands. In *The Responsive Brain*, ed. W. C. McCallum and J. R. Knot. Bristol: Wright.

Marslen-Wilson, W., and L. G. Tyler. 1980. The temporal structure of spoken language understanding. *Cognition* 8: 1–71.

Matsumiya, Y. 1976. The psychological significance of stimuli and cerebral evoked response asymmetry. In *Behavior Control and Modification of Physiological Activity*, ed. D. I. Mostofsky. Englewood Cliffs, N.J.: Prentice-Hall.

Matsumiya, Y., V. Tagliasco, C. T. Lombroso, and H. Goodglass. 1972. Auditory evoked response: Meaningfulness of stimuli and interhemispheric asymmetry. *Science* 175: 790–792.

Mayes, A., and G. Beaumont. 1977. Does visual evoked potential asymmetry index cognitive activity? *Neuropsychologia* 15: 219–256.

Megala, A. L., and T. J. Teyler. 1978. Event-related potentials associated with linguistic stimuli: Semantic vs. lower-order effects. In *Multidisciplinary Perspectives in Event-Related Brain Potential Research*, ed. D. A. Otto. Washington: U.S. Environmental Protection Agency.

Michalewski, H. J., H. Weinberg, and J. V. Patterson. 1977. The contingent negative variation (CNV) and speech production: Slow potentials and the area of Broca. *Biological Psychology* 5: 83–96.

Molfese, D. L. 1977. The ontogeny of cerebral asymmetry in man: Auditory evoked potentials to linguistic and non-linguistic stimuli in infants and children. *Progress in Clinical Neurophysiology* 3: 188–204.

Molfese, D. L. 1978a. Neuroelectrical correlates of categorical speech perception in adults. *Brain and Language* 5: 25–35.

Molfese, D. L. 1978b. Left and right hemisphere involvement in speech perception: Electrophysiological correlates. *Perception and Psychophysics* 23: 237–243.

Molfese, D. L. 1980a. Hemispheric specialization for temporal information: implications for the perception of voicing cues during speech perception. *Brain and Language* 11: 285–299.

Molfese, D. L. 1980b. The phoneme and the engram: Electrophysiological evidence for the acoustic invariant in stop consonants. *Brain and Language* 9: 372–376.

Molfese, D. L., R. B. Freeman, and D. S. Palermo. 1975. The ontogeny of brain lateralization for speech and nonspeech stimuli. *Brain and Language* 2: 356–368.

Molfese, D. L., and T. M. Hess. 1978. Hemispheric specialization for VOT perception in the preschool child. *Journal of Experimental and Child Psychology* 26: 71–84.

Molfese, D. L., and V. J. Molfese. 1979a. VOT distinctions in infants: Learned or innate? *Studies in Neurolinguistics* 4: 225–240.

Molfese, D. L., and V. J. Molfese. 1979b. Hemisphere and stimulus differences as reflected in the cortical responses of newborn infants to speech stimuli. *Developmental Psychology* 15: 505–511.

Molfese, D. L., and V. J. Molfese. 1980. Cortical responses of preterm infants to phonetic and nonphonetic speech stimuli. *Developmental Psychology* 16: 574–581.

Molfese, D. L., and V. J. Molfese. 1982. Hemispheric specialization in infancy. In *Manual Specialization and the Developing Brain: Longitudinal Research*, ed. G. Young. New York: Academic. In press.

Molfese, V. J., Molfese D. L., and C. Parsons. 1982. Hemispheric processing of phonological information. In *Language Functions and Brain Organization*, ed. S. Segalowicz. New York: Academic. In press.

Morrell, L. K., and D. A. Huntington. 1971. Electrocortical localization of language production. *Science* 174: 1359–1360.

Morrell, L. K., and J. G. Salamy. 1971. Hemispheric asymmetry of electrocortical response to speech stimuli. *Science* 174: 164–166.

Neville, H. J. 1974. Electrographic correlates of lateral asymmetry in the processing of verbal and nonverbal auditory stimuli. *Journal of Psycholinguistic Research* 3: 151–163.

Neville, H. J. 1980. Event-related potentials in neuropsychological studies of language. *Brain and Language* 11: 300–318.

Neville, H. J., M. Kutas, and A. Schmidt. 1982. Event-related potential studies of cerebral specialization during reading: A comparison of normally hearing and congenitally deaf adults. In *Proceedings of the Sixth International Conference on Event-Related Slow Potentials of the Brain*, ed. R. Karrer et al. In press.

Neville, H. J., E. Snyder, R. Knight, and R. Galambos. 1979. Event related potentials in language and non-language tasks in patients with alexia without agraphia. In *Human Evoked Potentials Applications and Problems*, ed. D. Lehmann and E. Callaway. New York: Plenum.

Otto, D. A., K. Houck, H. Finger, and S. Hart. 1976. Event-related slow potentials in aphasic, dyslexic and normal children during pictorial and letter-matching. In *The Responsive Brain*, ed. W. C. McCallum and J. R. Knott. Bristol: Wright.

Pace, S. A., D. L. Molfese, A. L. Schmidt, W. Mikula, and C. Ciano. 1979. Relationships between behavioral and electrocortical responses of aphasic and non-aphasic brain-damaged adults to semantic materials. In *Evoked Brain Potentials and Behavior*, ed. H. Begleiter. New York: Plenum.

Papanicolaou, A. C. 1980. Cerebral excitation profiles in language processing: The photic probe paradigm. *Brain and Language* 9: 269–280.

Papanicolaou, A. C., H. M. Eisenberg, and R. S. Levy. 1982. Evoked potential correlates of left hemisphere dominance in covert articulation: An auditory probe paradigm. In press.

Picton, T. W. 1980. The use of human event-related potentials in psychology. In *Techniques of Psychophysiology*, ed. P. H. Venables and T. Martin. New York: Wiley.

Picton, T. W., and P. G. Fitzgerald. 1982. A general description of the human auditory evoked potentials. In *Bases of Auditory Brainstem Responses*, ed. E. J. Moore. New York: Grune & Stratton.

Picton, T. W., S. A. Hillyard, H. I. Krausz, and R. Galambos. 1974. Human auditory evoked potentials. 1: Evaluation of components. *Electroencephalography and Clinical Neurophysiology* 36: 179–190.

Picton, T. W., and M. D. Low. 1971. The CNV and semantic content of stimuli in the experimental paradigm: Effects of feedback. *Electroencephalography and Clinical Neurophysiology* 31: 451–456.

Picton, T. W., and D. T. Stuss. 1980. The component structure of the human event-related potentials. *Progress in Brain Research* 54: 17–49.

Picton, T. W., D. L. Woods, D. T. Stuss, and K. B. Campbell. 1978. Methodology and meaning of human evoked potential scalp distribution studies. In *Multidisciplinary Perspectives in Event-Related Brain Potential Research*, ed. D. A. Otto. Washington: U.S. Environmental Protection Agency.

Pinsky, S. D., and D. W. McAdam. 1980. Electroencephalographic and dichotic indices of cerebral laterality in stutterers. *Brain and Language* 11: 374–397.

Pisoni, D. B. 1977. Identification and discrimination of the relative onset time of two component tones: Implications for voicing perception in stops. *Journal of the Acoustical Society of America* 61: 1352–1361.

Polich, J. 1982. Cognition and ERPs: The relation of post-stimulus negativities to information processing. Unpublished.

Polich, J. M., G. McCarthy, W. S. Wang, and E. Donchin. 1983. When words collide: Orthographic and phonological interference during word processing. *Biological Psychology* 16: 155–180.

Preston, M. S. 1979. The use of evoked response procedures in studies of reading disability. In *Evoked Brain Potentials and Behavior*, ed. H. Begleiter. New York: Plenum.

Preston, M. S., J. T. Guthrie, and B. Childs. 1974. Visual evoked responses (VERs) in normal and disabled readers. *Psychophysiology* 11: 452–457.

Preston, M. S., J. T. Guthrie, I. Kirsch, D. Gertman, and B. Childs. 1977. VERs in normal and disabled adult readers. *Psychophysiology* 14: 8–14.

Ritter, W., Simson, H. G. Vaughan, and M. Macht. 1982. Manipulation of event-related potential manifestations of information processing stages. *Science* 215: 1413–1415.

Roemer, R. A., and T. J. Teyler. 1977. Auditory evoked potential asymmetries related to word meaning. *Progress in Clinical Neurophysiology* 3: 48–59.

Rösler, F., and D. Manzey. 1981. Principal components and varimax-rotated components in event-related potential research: Some remarks on their interpretation. *Biological Psychology* 13: 3–26.

Ross, J. J., D. G. Childers, and N. W. Perry. 1973. The natural history and electrophysiological characteristics of familial language dysfunction. In *The Disabled Learner: Early Detection and Intervention*, ed. P. Satz and J. J. Ross. University of Rotterdam.

Ruchkin, D. S., and S. Sutton. 1979. Latency characteristics and trial by trial variation of emitted potentials. In *Cognitive Components in Cerebral Event-Related Potentials and Selective Attention*, ed. J. E. Desmedt. Basel: Karger.

Rugg, M. D., and J. G. Beaumont. 1978. Visual evoked potentials to visual spatial and verbal stimuli: Evidence of differences in cerebral processing. *Physiological Psychology* 6: 501–504.

Rugg, M. D., and J. G. Beaumont. 1979. Late positive component correlates of verbal and visuospatial processing. *Biological Psychology* 9: 1–11.

Schulman-Galambos, C., and R. Galambos. 1978. Cortical responses from adults and infants to complex visual stimuli. *Electroencephalography and Clinical Neurophysiology* 45: 425–435.

Schwartz, M. 1976. Averaged evoked responses and the encoding of perception. *Psychophysiology* 13: 546–553.

Seitz, M. R. 1976. The effects of response requirements and linguistic context on the averaged electroencephalic responses to clicks. *Human Communication* 1: 49–57.

Seitz, M. R., and B. A. Weber. 1974. Effects of response requirements on the location of clicks superimposed on sentences. *Memory and Cognition* 2: 43–46.

Shagass, C. 1972. *Evoked Brain Potentials in Psychiatry*. New York: Plenum. Pp. 220–222.

Shelburne, S. A. 1972. Visual evoked responses to word and nonsense syllable stimuli. *Electroencephalography and Clinical Neurophysiology* 32: 17–25.

Shelburne, S. A. 1973. Visual evoked responses to language stimuli in normal children. *Electroencephalography and Clinical Neurophysiology* 34: 135–143.

Shelburne, S. A. 1978. Visual evoked potentials to language stimuli in children with reading disabilities. In *Multidisciplinary Perspectives in Event-Related Brain Potential Research*, ed. D. A. Otto. Washington: U.S. Environmental Protection Agency.

Shibasaki, H., G. Barrett, E. Halliday, and A. M. Halliday. 1980. Components of the movement-related cortical potential and their scalp topography. *Electroencephalography and Clinical Neurophysiology* 49: 213–226.

Shields, D. T. 1973. Brain responses to stimuli in disorders of information processing. *Journal of Learning Disabilities* 6: 37–41.

Shipley, T., and R. W. Jones. 1969. Initial observations on sensory interaction and the theory of dyslexia. *Journal of Communication Disorders* 2: 295–311.

Shucard, D. W., K. R. Cummins, D. G. Thomas, and J. L. Shucard. 1981. Evoked potentials to auditory probes as indices of cerebral specialization of function—Replication and extension. *Electroencephalography and Clinical Neurophysiology* 52: 389–393.

Shucard, D. W., and D. G. Thomas. 1977. Auditory evoked potentials as probes of hemispheric differences in cognitive processing. *Science* 197: 1295–1298.

Simson, R., H. G. Vaughan, and W. Ritter. 1977. The scalp topography of potentials in auditory and visual discrimination tasks. *Electroencephalography and Clinical Neurophysiology* 42: 528–535.

Sobotka, K. R., and J. G. May. 1977. Visual evoked potentials and reaction time in normal and dyslexic children. *Psychophysiology* 14: 18–24.

Stuss, D. T., F. F. Sarazin, E. E. Leech, and T. W. Picton. 1983. Event-related potentials during naming and mental rotation. *Electroencephalography and Clinical Neurophysiology* 56: 133–146.

Stuss, D. T., and T. W. Picton. 1978. Neurophysiological correlates of human concept formation. *Behavioral Biology* 23: 135–162.

Sutton, S., M. Braren, J. Zubin, and E. R. John. 1965. Evoked potential correlates of stimulus uncertainty. *Science* 150: 1187–1188.

Symann-Louett, N., G. G. Gascon, Y. Matsumiya, and C. T. Lombroso. 1977. Wave form difference in visual evoked responses between normal and reading disabled children. *Neurology* 27: 156–159.

Symmes, D., and M. A. Eisengart. 1971. Evoked response correlates of meaningful visual stimuli in children. *Psychophysiology* 8: 769–778.

Szirtes, J., V. Diekmann, A. Rothenberger, and R. Jurgens. 1980. Fourier analysis of acoustic evoked potentials in healthy, aphasic and right hemisphere damaged subjects. *Progress in Brain Research* 54: 496–501.

Szirtes, J., and H. G. Vaughan, Jr. 1977. Characteristics of cranial and facial potentials associated with speech production. *Electroencephalography and Clinical Neurophysiology* 43: 386–396.

Tanguay, P., J. M. Taub, D. Doubleday, and D. Clarkson. 1977. An interhemispheric comparison of auditory evoked responses to consonant-vowel stimuli. *Neuropsychologia* 15: 123–131.

Taub, J. M., P. E. Tanguay, and D. Clarkson. 1976. Electroencephalographic and reaction time asymmetries to musical chord stimuli. *Physiology and Behavior* 17: 925–930.

Teyler, T. J., R. A. Roemer, T. F. Harrison, and R. F. Thompson. 1973. Human scalp-recorded evoked-potential correlates of linguistic stimuli. *Bulletin of the Psychonomic Society* 1: 333–334.

Thatcher, R. W. 1977a. Evoked-potential correlates of hemispheric lateralization during semantic information-processing. In *Lateralization in the Nervous System*, ed. S. A. Harnad et al. New York: Academic.

Thatcher, R. W. 1977b. Evoked-potential correlates of delayed letter matching. *Behavioral Biology* 19: 1–23.

Thatcher, R. W., and R. S. April. 1976. Evoked potential correlates of semantic information processing in normals and aphasics. In *The Neuropsychology of Language*, ed. R. W. Rieber. New York: Plenum.

Thatcher, R. W., and E. B. Maisel. 1979. Functional landscapes of the brain: An electrotopographic perspective. In *Evoked Brain Potentials and Behavior*, ed. H. Begleiter. New York: Plenum.

Wada, J. A. 1949. A new method for the determination of the side of cerebral speech dominance. A preliminary report on the intracarotid injection of sodium amytal in man. *Medical Biology* 14: 221–222.

Walter, W. G., R. Cooper, V. J. Aldridge, W. C. McCallum, and A. L. Winter. 1964. Contingent negative variation: An electric sign of sensori-motor association and expectancy in the human brain. *Nature* 203: 380–384.

Weber, B. A., and G. S. Omenn. 1977. Auditory and visual evoked responses in children with familial reading disabilities. *Journal of Learning Disabilities* 10: 153–158.

Weinberg, H., W. G. Walter, R. Cooper, and V. J. Aldridge. 1974. Emitted cerebral events. *Electroencephalography and Clinical Neurophysiology* 36: 449–456.

Wolpaw, J. R., and J. K. Penry. 1975. A temporal component of the auditory evoked response. *Electroencephalography and Clinical Neurophysiology* 39: 609–620.

Wolpaw, J. R., and C. C. Wood. 1982. Scalp distribution of human auditory evoked potentials. I. Evaluation of reference electrode sites. *Electroencephalography and Clinical Neurophysiology* 54: 15–24.

Wood, C. C. 1975. Auditory and phonetic levels of processing in speech perception: Neurophysiological and information processing analyses. *Journal of Experimental Psychology: Human Perception and Performance* 104: 3–20.

Wood, C. C., W. R. Goff, and R. S. Day. 1971. Auditory evoked potentials during speech perception. *Science* 173: 1248–1251.

Wood, C. C., and J. R. Wolpaw. 1982. Scalp distribution of human auditory evoked potentials. II. Evidence for multiple sources and involvement of auditory cortex. *Electroencephalography and Clinical Neurophysiology* 54: 25–38.

Zimmerman, G., and J. R. Knott. 1974. Slow potentials of the brain related to speech processing in normal speakers and stutterers. *Electroencephalography and Clinical Neurophysiology* 37: 599–607.

Chapter 15

Neural Models and Very Little About Language

James A. Anderson

In this chapter I briefly review a theoretical approach to brain organization that may have certain applications to language. I also present some new material. The overall problem concerning us here is how some general models of brain function might be applied to higher-level cognitive activities, among which (at least in humans) are the complex of issues related to language behavior.

How do we study a system as complicated as the brain? There is no obvious best approach. Traditionally, there have been two groups who try to understand the brain. First, we have those who might call themselves neuroscientists. They have close ties to biology and medicine. Their academic affiliations are usually with a medical school or biology department. They tend to be very fact-oriented and somewhat anti-theory, partly because theory has a poor record of performance in neuroscience. Second, we have a group of scientists who study psychology. Their primary emphasis is on the study of behavior by itself, though there is always the realization that it is the biological structure that is giving rise to the behavior. Psychologists often talk as if they were operating at a different "level" than neurobiologists, and lawful relations in psychology can seem to have relatively little connection to the underlying biology. Most psychologists believe, however, that when "true" psychological theories arise they will turn out to be "physiologically reasonable" when all the facts are known. Much of the material presented in this book falls at this particular interface.

There is a third group, of very recent genesis, which I will argue has a lot to teach us about the brain, though many of its most prominent practitioners might deny this. I refer to those who are concerned with formal models of intellectual function, particularly those who make computer models of cognition in the field now called "artificial intelligence." In many ways, these are the inheritors of the approach taken by many distinguished theoreticians of the past, such as Warren McCulloch, Walter Pitts, Norbert Wiener, or John von Neumann. They work at a level of abstraction far removed from the biology and even, in many cases, the psychology of actual humans. Many of them are interested in making machines do things that previously could only be done by humans, and they really don't care how people do them. However, others take inspiration (sometimes rather remote) from the details of human perception and cognition. What can we learn from the experience of this very intelligent and active though sometimes exasperating group? They found very quickly that problems of cognition are much more difficult than they had first thought and that many obvious approaches simply and demonstrably do not work. It is in this emphasis on performance that we can learn the most from them. Researchers in cognitive science and artificial intelligence place a great emphasis on models that work. They must actually do what they claim to do and must be demonstrated and studied in performing systems. This kind of abstract experimentation, I feel, is very important. Until one has tried to make a model of cognition actually do something cognitive, it is hard to appreciate how difficult and subtle it is. For this reason, I feel that any successful attack on brain organization must make use of information from all three areas, all of which offer insights and factual knowledge that can be useful to all the others. Furthermore, theories must be checked in ways used by each group. This is obviously difficult, but I suspect that it can indeed be done. It should be kept in mind as a desirable goal.

Let me start with the first area, neuroscience, from which we take inspiration for our models. I will briefly mention a couple of key assumptions that we must make to start.

First, real neurons tend to be graded in their activity. That is, instead of being on or off, they respond to stimuli in most (though not all) cases as continuous-valued voltage-to-frequency converters. Neurons act much more simply in this respect than one might expect, turning voltage or current into firing frequency as a relatively straightforward transducer might. There are important nonlinearities, but they are often tractable, and a modest degree of linearity coupled with some simple nonlinearities is sometimes an adequate approximation of neuron activity.

Second, we have a complicated set of concepts that arise when we consider the behavior of groups of single neurons. In mammals, particularly in cerebral cortex, a neuron is a member of a large set of neurons with similar connection patterns and morphologies. It must be always kept in mind that there are not a few but millions of cells in the mammalian nervous system. Many of the most prominent aspects of brain organization may arise from the need to make efficient and productive use of large numbers of elementary entities. Brain tissue carries a high biological cost in terms of energy consumption, potential for damage, and required support facilities, and it must earn its keep.

The Selectivity of Single Neurons

There are two extreme positions taken on neuronal selectivity. One claims great specificity and importance for single neurons. We look at the physiological data and observe that single neurons respond to only a small number of stimuli; that is, they have considerable selectivity. We then conjecture that when a neuron is active it signals very precise information about the sensory input. This point of view is stated most clearly by Horace Barlow (1972), who formulates his ideas in the form of "dogmas." One of these relates directly to linguistics. Dogma 2 states: "The sensory system is organized to achieve as complete a representation of the sensory stimulus as possible with the minimum number of active neurons." Dogma 4 states: "Perception corresponds to the activity of a small selection from the very numerous high-level neurons, each of which corresponds to a pattern of events of the order of complexity of the events symbolized by a word." Dogma 5 states: "High impulse frequency in such neurons corresponds to high certainty that the trigger feature is present." (Barlow 1972, p. 371). Therefore, Barlow argues, single neurons are very important, very specific, and signal certainty with increased activity.

There are many virtues to this position. It is easy to understand and it makes good intuitive sense. Cells do something. When a cell fires, something specific and important happens, both because few or no other cells are firing and because the cell is "meaningful." The mental image, at least to me, is that of an enormous bank of green lights (part of the Hollywood set for a submarine, say, or the control room of a nuclear reactor), and when one of the greens turns to red the operator knows that there is trouble at valve 42-B. In psychology this model typically is called the "grandmother cell" or "yellow Volkswagen detector" hypothesis.

Many severe problems arise immediately from this model, often problems that are of a psychological cast. How can such selectivity develop? There are some very powerful learning assumptions concealed. Also, it is hard to make a system of this kind work in a practical sense. There are too many potential connections between cells to function well in large systems. I strongly suspect that models based on this kind of selectivity face the kind of exponential retrieval catastrophes shown by most information-retrieval systems of any complexity. Some of the associative network models proposed in artificial intelligence are faced with the same problem: a combinatorial explosion of possible connections between items.

Another set of criticisms also hold. How does this system generalize? Does our yellow Volkswagen detector also respond to chartreuse Volkswagens? How about the green Volkswagen detector? And here is where Barlow found it necessary to modify his original model. At the end of his paper, he essentially said that not single cells but small groups of selective cells (what he called a "college of cardinals") were what was really significant. The conceptual simplicity of the system crumbles with this assumption. Perception no longer corresponds to active single cells (a single red light), but rather to groups of active single cells (a syndrome of red lights). The groups can be more selective than the single cells; the class of yellow Volkswagens might be the simultaneous activation of the "yellow" cell and the "Volkswagen" cell.

Distribution

Models in which it is essential to consider the simultaneous activities of single elementary entities are called "distributed." It is possible to build models of this type, but they have pronounced and somewhat unfamiliar properties. The strongest and most extreme statement of distribution in the classical literature is the following, due to Karl Lashley:

It is not possible to demonstrate the isolated localization of a memory trace anywhere within the nervous system. . . . the characteristics of the nervous network are such that when it is subject to any pattern of excitation, it may develop a pattern of activity . . . by spread of excitations, much as the surface of a liquid develops an interference pattern of spreading waves. . . . There is no great excess of cells which can be reserved as the seat of special memories. The same neurons which retain the memory traces of one experience must also participate in countless other activities. . . . Recall involves the synergic action or some kind of resonance among a very large number of neurons. . . . (Lashley 1950, pp. 477–480)

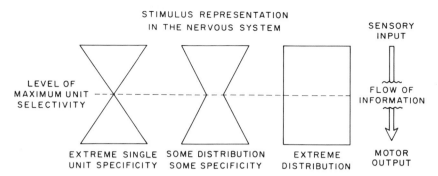

Figure 15.1

Although we know much more about the details of cortical physiology than Lashley did, and we know that some of these statements are too strong, still they emphasize some of the features that might be found in distributed representations of the nervous system: the nonlocalizability of complex events, the lack of privileged sites for specific memories, the importance of the pattern of simultaneous activities of large numbers of cells, and the possibility of resonancelike effects.

The actual nervous system seems to show an intermediate selectivity. Cells clearly are very selective, but not as selective as a "single neuron–single word" model would hold. Yet they are not completely distributed either. Perhaps we have the situation diagrammed in figure 15.1, where the nervous system is biased toward selectivity but displays a modest amount of distribution. A small percentage of a lot of neurons responding to a complex stimulus is still a lot of neurons. There must be some kind of optimality of stimulus coding going on here that we do not understand at present.

Parallelism

The brain is certainly highly parallel in its architecture. A system that is parallel at one level may be serial at another. At the level of motor neurons there is parallelism, since any complex movement is composed of the simultaneous discharge of many motor neurons. Yet if this pattern is used as the unit of description then patterns follow one another in series. Different areas of the brain project to one another over parallel pathways, and processing may proceed serially, from one step to another, with parallelism at each level.

Even in conventional digital computers, often considered highly serial, there is parallelism at the level of the single word. Single machine instructions are simultaneous patterns of ones and zeros, usually based on a single computer word. The more elements composing a single word, the more powerful the resulting instructions, which, however, are typically executed in series by the programs. The evolution of microcomputers over the past decade illustrates this. From simple 4-bit microprocessors, the next step was to 8-bit series, which are the central processing units of most of the current personal computers. The next step, to 16-bit words, is in progress, and several companies have already announced microprocessors with 32-bit words—machines that contain nearly the power of a mainframe on a single chip. Much of the increase in power and speed of such machines is due to the increase in word size.

The degree of parallelism of the nervous system—that is, the effective size of the word—is enormous. The power of the resulting "elementary" operations may also be intense, but it may not be easily understandable in terms of the simple logical operations that characterize current digital computers. Although the brain's "machine instructions" may be very powerful, they are also very slow and noisy, leading to quite different strategies for computation than conventional computers, even though the result may be superficially quite similar.

Association

We assume that what is of importance in the operation of the system are patterns of activity shown by many neurons in a set of neurons. Such activity patterns are denoted as vectors. Vectors of this type are called *state vectors*. It is possible to develop mathematical structures that let these activity patterns act as fundamental primitive entities. We have moved away from individual neurons, which are small parts of these elementary units. I feel that understanding the lawful behavior of state vectors is the key to developing a theory that will be both physiologically reasonable and cognitively interesting. I predict that, when a satisfactory brain theory arises, it will be in the form a calculus of such state vectors.

A key psychological aspect of memory that inspires these models is association. Association is not a logical process; rather, associations are formed because of contiguity, happenstance, similarity, or other capricious events external to the actual association that is formed. By using association, it is possible to build structures than can be used to perform the kinds of operations we call "cognitive." This is an approach widely used in cognitive science, though details of the associative process differ among models. Notice that the impetus for making a "general-purpose" associator is not from neuroscience, but from psychology.

Most neuroscientists are convinced that precisely specified changes in synaptic coupling store memory. In general, it is impossible for mammals to form new neurons after birth. Large changes in dendritic branching and formation of new synapses are possible in immature organisms but rare in adults. Therefore, change in strength of preexisting synapses is the most likely candidate for learning in adults. My own candidate for one modifiable structure in adult mammalian neocortex is the dendritic spines. These structures seem to respond to experience in some degree, and have electrical properties as well as anatomies that seem highly suitable for modification. In the invertebrate *Aplysia* there is no doubt that synaptic change underlies a simple kind of learning (Kandel 1979).

In any case, detailed change in the coupling between cells seems to be the best mechanism for most learning. The most common departure point for further development seems to be the following suggestion made by Donald Hebb in 1949: "When an axon of cell A is near enough to excite a cell B and repeatedly or persistently takes part in firing it, some growth process or metabolic change takes place in one or both cells such that A's efficiency as one of the cells firing B is increased." (Hebb 1949, p. 62) This suggestion predicts that cells will tend to become correlated in their discharges, and a synapse acting like this is sometimes called a correlational synapse. In recent years a number of models have been put forth that build on this idea of learning. (See Kohonen 1977 for a review of this literature.) The particular model described here uses the notation of Anderson 1972 and has also been discussed by Cooper (1974). Although experimental evidence to support this attractive notion was lacking for many years, psysiologists have recently found data in several systems that seem to be most reasonably explained by some kind of Hebbian synapse (Levy and Steward 1979; Rauschecker and Singer 1979, 1981).

If we assume that synaptic change is related to the magnitude of both presynaptic and postsynaptic activity, then, we can immediately show that the system can act as a general-purpose associator.

Suppose we have two sets of N neurons, α and β, which are completely convergent and divergent, that is, every neuron in α projects to every

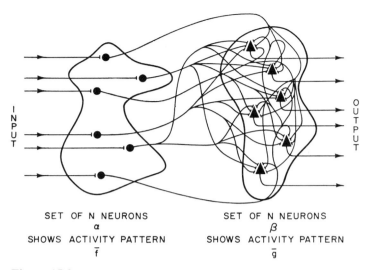

SET OF N NEURONS   SET OF N NEURONS
        α   β
SHOWS ACTIVITY PATTERN   SHOWS ACTIVITY PATTERN
        f̄   ḡ

Figure 15.2

neuron in β (figure 15.2). A neuron j in α is connected to a neuron i in β by way of a synapse with strength $A(i, j)$. Our first basic assumption is that we are primarily interested in the behavior of the set of simultaneous individual neuron activities in a group of neurons. We stress pattern of individual neuron activities, because our current knowledge of cortical physiology suggests that cells are highly individualistic; cell properties differ from cell to cell. We represent these large patterns as state vectors with independent components. We also assume that these components can be positive or negative. This can occur if the relevant physiological variable is considered to be the deviation around a nonzero spontaneous activity level.

Suppose that a pattern of activity **f** occurs in α and another pattern of activity **g** occurs in β. Suppose that for some reason we wish to associate these two arbitrary patterns. We assume a detailed rule of synaptic modification: To associate patterns **f** in α and **g** in β we need to change the set of synaptic weights according to the product of presynaptic activity at a junction with the activity of the postsynaptic cell. This information is locally available at the junction. If $f(j)$ is the activity of cell j in α and $g(i)$ is that of cell i in β, then the change in synaptic strength is given by

$$A(i, j) = f(j)\, g(i).$$

This defines an $n \times n$ matrix $\mathbf{A}$ of the form $\mathbf{A} = gf^T$. Suppose $\mathbf{f}$ is normalized, so that the inner product $[\mathbf{f},\mathbf{f}] = 1$, and the above equality holds. If pattern of activity $\mathbf{f}$ recurs in α, since β is connected to α a pattern of activity will also appear in β. If the activity of a neuron is a simple weighted sum of its inputs, as is found in the *Limulus* eye (Knight, Toyoda, and Dodge 1970) and most simple variants of integrator models for neurons, then we can compute the pattern on β from the matrix $\mathbf{A}$ and from $\mathbf{f}$ as follows:

Pattern on $\beta = \mathbf{A}\mathbf{f}$

$$= \mathbf{g}\mathbf{f}^T\mathbf{f}$$

$$= \mathbf{g}[\mathbf{f},\mathbf{f}]$$

$$= \mathbf{g}.$$

Suppose that instead of having only one association we have K of them, $(\mathbf{f}_1,\mathbf{g}_1)$, $(\mathbf{f}_2,\mathbf{g}_2)$, ... $(\mathbf{f}_K,\mathbf{g}_K)$, each associated with an incremental matrix of the form $\mathbf{A}_i = \mathbf{g}_i\mathbf{f}_i^T$. Since there are only N synapses in the system, the same synapses must participate in storing all the associations: that is, they are used over and over again. This is an inexorable consequence of distributed systems, and it has very powerful consequences.

Suppose that the overall connectivity is given by

$$\mathbf{A} = \sum_i \mathbf{A}_i .$$

Just as before, what happens when pattern of activity $\mathbf{f}$ appears on α? Let us do a very simple calculation for the case when the $\mathbf{f}$s are orthogonal, that is, when the inner product $[\mathbf{f}_i,\mathbf{f}_j]$ equals 0. In the case of orthogonal $\mathbf{f}$s, when $\mathbf{f}_i$ appears on α, then

Pattern on $\beta = \mathbf{A}\mathbf{f}_i$

$$= \mathbf{A}_i\mathbf{f}_i + \sum_{i \neq j} \mathbf{A}_j\mathbf{f}_i$$

$$= \mathbf{g}_i + \sum_{i \neq j} \mathbf{g}_j[\mathbf{f}_j,\mathbf{f}_i]$$

$$= \mathbf{g}_i.$$

For orthogonal $\mathbf{f}$s the system associates random vectors perfectly.

The capacity of the system is related to N. In the case of association of vectors with orthogonal inputs, there is a maximum of N different inputs. Since there are a very large number of neurons in the brain, this is not as serious a limitation as it might seem. Many of the most powerful generalizing, averaging, and analyzing properties of these models follow from a lack of orthogonality at the input. The system is formally a content-addressable memory, in that the specific input must be provided to evoke the association. There is no indexing or referencing. At this level, the activities of a single cell or synapse or even of sizable numbers of synapses is of little importance to the overall functioning of the system. This property has been extensively demonstrated in computer ablation simulations (Gordon 1983; Wood 1978, 1983).

Prototype Formation

The physical world contains objects that, for reasons of utility and convenience, people have decided to classify in groups. For example, there are a great many different physical objects that are called "tables." There seem to be no absolute defining properties of "tableness" that are shown by tables and by nothing else. What is required by most complicated high-level cognitive models is an elementary entity that consists of some things that are called "ideas" or "concepts." What seems to be required is some kind of equivalence class containing examples of items that are somehow like other members of the class even though they may not be identical to them.

There are several aspects to this problem, the simplest of which we call "abstraction." Let us make an observation about the behavior of our simple associator. Suppose we have a number of examples of a category. Suppose we call the category name the state vector $\mathbf{g}$ and the different examples $\mathbf{f}_1, \mathbf{f}_2, \ldots \mathbf{f}_k$. Each incremental matrix is given as before, and we arrive at an overall connectivity matrix given by

$$\mathbf{A} = \mathbf{g} \sum_i \mathbf{f}_i^T.$$

This expression contains the sum of the **f**s. This term acts like an average-response computer. If the **f**s are correlated, as they will be in many cases of interest, then the central tendency of the **f**s will be extracted. The associated response to the central tendency will be the most vigorous. The amplitude of the response will give a measure of the distance from the central tendency. This result is very similar to a view held by many psychologists (see, for example, Rosch and Lloyd 1978), in which most natural categories are held to contain a "prototype"

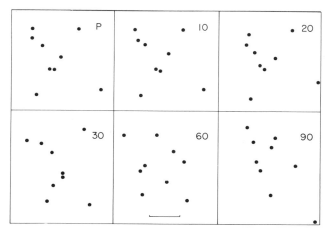

Figure 15.3

and particular examples are judged to be better or poorer category members depending on their distance from the prototype.

Posner and Keele (1968, 1970) have demonstrated what appears to be this process taking place in the perception of patterns of dots. Their experiment is quite simple. The experimenters generate prototype patterns of nine dots. Then noisy examples of the prototype are made by moving dots in random directions and at random distances. Subjects learn to classify distortions of a single prototype together in an equivalence class simply by being told that they are correct when this is what they do. After they learn to do this, they are tested with new patterns. Posner and Keele found, and Knapp (1979) has replicated, that subjects respond most strongly to the prototype they never saw. The primary experimental variable is the "level of distortion," which describes the average distance a given dot will move when an example is being generated from a particular prototype. Figure 15.3 shows a prototype and several examples at different levels of distortion generated from that prototype (Knapp and Anderson, in press). In fact, in our experiments, we structured the stimuli so that the prototype could effectively never appear as a stimulus, yet subjects responded, in many conditions, as if they were more certain they had seen the prototype than what they actually had seen.

It is not difficult to make the neural model I have described form a prototype. All that is required is some reasonable spatial representation

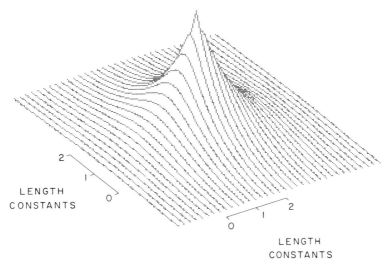

LENGTH
CONSTANTS

LENGTH
CONSTANTS

Figure 15.4

of the stimulus. Suppose each dot gives rise to a spatially distributed pattern of activity on some spatial representation of the visual field. (There are a large number of these spatial maps known in the cerebral cortex, so this is not an unreasonable assumption.) Assume that there is an exponential decay of activity from a central location. Figure 15.4 shows the pattern of activity arising from a single dot. If a number of examples of the same prototype are seen and are represented in a memory of the kind we have been discussing, the memory representation contains the sum of all examples. If the examples are all different, then the sum, with appropriate parameters, will contain a significant component at the prototype location, where the distributions all overlap. Figure 15.5 shows what these sums look like when four examples, equally spaced from the prototype and each other, are stored at various separations. Note the great enhancement at the prototype location, even for quite large separations. When this simple model is analyzed quantitatively, the experimental data, for a number of variations of this experiment, can be fitted quite accurately. An exponential distribution is, in fact, the simple function giving the best fit to the data.

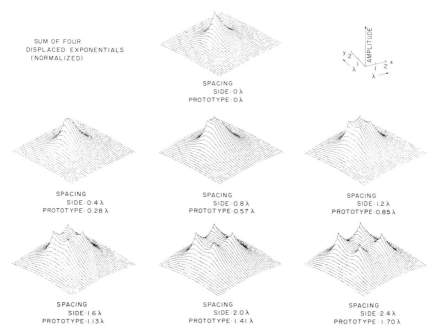

SUM OF FOUR
DISPLACED EXPONENTIALS
(NORMALIZED)

SPACING
SIDE : 0 λ
PROTOTYPE : 0 λ

SPACING
SIDE : 0.4 λ
PROTOTYPE : 0.28 λ

SPACING
SIDE : 0.8 λ
PROTOTYPE : 0.57 λ

SPACING
SIDE : 1.2 λ
PROTOTYPE : 0.85 λ

SPACING
SIDE : 1.6 λ
PROTOTYPE : 1.13 λ

SPACING
SIDE : 2.0 λ
PROTOTYPE : 1.41 λ

SPACING
SIDE : 2.4 λ
PROTOTYPE : 1.70 λ

Figure 15.5

Feature Analysis and Categorization

The simple kind of averaging abstraction just described is not adequate to explain many categorization and classification phenomena. It is necessary to invent a stronger categorizer.

A word that is used often by both psychologists and those interested in language is *feature*. What seems to be meant by *feature* is a kind of perceptual atom out of which complex stimuli are made. Those interested in pattern recognition also speak about features in roughly the same way, and with the same rationale.

Suppose we have a perceptual task requiring the classification of an external stimulus into one or another of a set of categories. Recognizing a set of noises as a phoneme, or a set of marks as a particular letter, is a good example. The ability to categorize appropriately is what gives language in particular and human cognition in general much of its power.

Those interested in language, psychology, and pattern recognition soon realized, with some compelling experimental evidence, that a very useful strategy for this task would be the following: Complex events never recur exactly, and may be embedded in a highly variable context. Suppose, however, that there were certain aspects of the particular item that were always present. If we looked for these aspects alone and ignored the other information as irrelevant, we could make efficient categorizations based on a much simplified set of information. These exceptionally useful aspects of the stimulus set are called "features." In linguistics, where this approach first appeared, a feature might correspond to something like the presence or absence of vocal-cord vibration, or of energy in certain frequency bands, or of a noiselike character to the stimulus, or of a number of other acoustic aspects of the input (often somewhat ill defined). It has so far, as a matter of fact, been impossible to physically characterize, in terms of unambiguous frequency or time patterns, any of the features that seem to be present in speech. This is due partly to the extremely powerful context effects present in speech, where the identical physical stimulus may be characterized in different ways depending on the surrounding phonemes; conversely, a phoneme may have the same name when it has widely differing physical properties. Another reason seems to be that phonetic features often seem to have their natural definition in terms of their production—where the tongue happens to be or whether or not the vocal cords are vibrating—and a simple change in production may give rise to very complicated changes in acoustic waveform.

Models for Feature Analysis

Neuronal selectivity as discussed above is particularly relevant here. Features, as elementary aspects of the stimulus, might seem particularly likely to correspond to the known single-unit selectivities found in the nervous system. One obvious example concerns recognition of written letters. Numerous feature sets have been published suggesting that letters are composed of oriented line segments like the cells known to exist in profusion in area 17 of visual cortex. However, if one looks carefully at published data on the selectivity of single cells, one rarely seems to find cells responding to stimuli with either the generality or the invariant properties expected of true "feature detectors." Typical area 17 cells will respond to a number of aspects of the visual stimulus. It seems clear that features must correspond, to be useful, to somewhat more complex and abstract aspects of the stimulus.

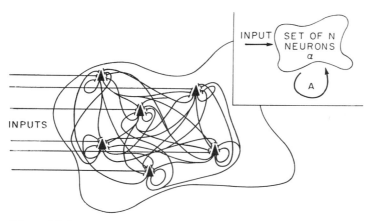

Figure 15.6

My feeling is that a "single cell–single feature" model of perception must fail for the same reasons that "yellow Volkswagen detectors" make an inferior psychological model of perception. What alternatives are there?

A Distributed Model for Features

Since I have been emphasizing the centrality of state vectors for models of nervous-system organization, can I suggest a model based on state vectors for feature analysis? The answer is Yes, and the resulting analysis sheds interesting light on single-cell selectivity.

Consider a system very much like the simple associator just discussed. Now, however, consider only a single set of neurons, and let this set connect to itself by way of a set of lateral interconnections. This system is illustrated by figure 15.6, in which α is a set of N neurons and every neuron in α is connected to every other neuron in α through a learning matrix of synaptic connectivities, A.

As a matter of anatomical fact, cortex contains an extensive set of connections of exactly this kind: the recurrent collateral system of cortical pyramids. Because of this extensive series of lateral connections, one pyramid can influence others up to several millimeters away. The effect of the collaterals is not simple lateral inhibition, since evidence (Szentagothai 1978; Shepherd 1979) suggests the effect is excitatory. Let us make the assumption that this set of lateral interconnections is capable of the kind of modification that gave rise to the associative model. When a pattern of activity learns itself, the synaptic increment is given by

$$A(i, j) = A(j, i) \propto f(i) f(j).$$

Since **A** now is a symmetric matrix, the eigenvectors are orthogonal and the eigenvalues are real. The more interesting result is that the matrix **A** is related closely to the sample covariance matrix. This means that the kinds of results obtained in principal-component analysis hold and, as a result, the eigenvectors of the resulting matrix with the largest positive eigenvalues contain the largest amount of the variance of the system.

If a set of items are to be discriminated, then these eigenvectors are the most useful for making discriminations among members of the set; this is what makes principal-component analysis so useful.

Remember, however, that this is a feedback system. The coefficient of feedback, the amount of feedback, is set by the eigenvalue. Positive feedback will cause a relative enhancement of the eigenvectors with large positive eigenvalues. Since these are also the most useful parts of the stimulus for making discriminations among members of the set, we have a weighting that favors the most "useful" eigenvectors. If we call these particular eigenvectors (the ones with large positive eigenvalues) "features," then we have a system that does pretty much what we want feature analysis to do.

We have here a fine demonstration of the powers of a parallel system. If this kind of weighting is desired, the resulting system, both theoretically and in computer simulation, is tremendously robust. However, it is not an analytical decomposition. At no point are the features themselves ever available, only a weighted sum. The system is fast, but not as flexible as a more analytical system.

Microfeatures and Macrofeatures

We now have a system where things like "features" are patterns of activity rather than selective single cells. Yet clearly there must be a close relationship between the kinds of selectivities found in the single elementary units that make up the patterns of activities and the selectivities of the features. To avoid confusion, I shall coin two ugly new words from a nice but ambiguous old one: *feature.*

Let me define a microfeature as the kind of selectivity shown by single cells. The microfeatures of visual cortex seem to be oriented line segments. There are very many of them in visual cortex, and they change from cell to cell. When neuroscientists talk about features, they are talking about microfeatures.

Let me define a macrofeature as the pattern of activity that corresponds to a particular perceptual feature. However, this entity is a pattern of activity, a vector. Perception takes place by analysis in terms of macrofeatures. The vectors are made up of components that respond to microfeatures. When psychologists, linguists, and cognitive scientists talk about features, they are talking about macrofeatures.

The difference between microfeatures and macrofeatures is the difference between a scalar and a vector. When cognitive scientists discuss "higher-level" features, they usually seem to mean features that are more complex that "simple features," which often seem to show a strong flavor of the physical stimulus. High-level features are farther along in a processing hierarchy. This may or may not correspond to a scalar-vector distinction, depending on the particular processing model being discussed.

Feedback and Saturation

Let me make some further modifications of the feedback system I have just presented. Two aspects of the model just presented should be carefully separated: the learning and the dynamics. The feedback takes place in real time; macrofeature analysis is a dynamic process, provoked by a stimulus input. Learning—change in strength of connection—takes place on a different time scale. Once the connections have been formed for a particular stimulus set, learning can cease and the feedback system will continue to analyze the inputs in terms of the inputs. Learning can be slow and painful, but the analysis, once learned, can be very fast.

There are many variants of both processes we have investigated. Let me describe a particularly simple approach. Suppose the response to the input pattern and the output of the feedback system interact significantly and have comparable time courses. We chose a very simple dynamics for our first modeling efforts.

Let $x(t)$ denote the state vector at time t. Integer values are taken by t in the simulations, since these were computer models. Assume that there is no significant delay; thus the activity at time $t+1$ is assumed to be the sum of activity at time t and the activity of the feedback matrix on the activity at time t. Therefore we have the simple expression

$$\mathbf{x}(t+1) = \mathbf{x}(t) + \mathbf{A}\,\mathbf{x}(t)$$

$$= (\mathbf{I}+\mathbf{A})\mathbf{x}(t),$$

where $\mathbf{I}$ is the identity matrix. This is potentially a positive feedback system.

The "Brain State in a Box"

However, this system may not be stable. All positive values of the eigenvalue lead to activity growing without bound. We must break with the linearity of dynamics we have assumed up to now. We need to keep the system from blowing up.

The simplest way of containing the activity is to observe that neurons have limits on their activities: They cannot fire faster than some frequency or slower than zero. Suppose we incorporate this in our model. A particular state vector is a point in a high-dimensional space. Putting limits on firing rate corresponds to putting allowable state vectors in a hypercube; this is what leads to the nickname for this model.

The qualitative dynamics of the system are straightforward. Suppose we start off with an activity pattern receiving positive feedback. The vector lengthens until it reaches a wall of the box, that is, a firing limit for one component. The vector tries to lengthen but cannot escape. It heads for a corner, where it will remain if the corner is stable (not all corners are stable). If we allow decay there are stable points outside of corners, but many components of the state vector will still saturate. An example of a two-dimensional system of this type is shown in figure 15.7, where a number of starting points and their trajectories under the influence of the feedback system are drawn.

Categorial Perception

The dynamics of this model form a categorial perceiver, since the model takes a set of initial points in a region of space and lumps them together in a single corner. We have produced a classification algorithm, in which the state vector representing the categorization is the classification. Previously, the behavior of the system was described in contexts where it seemed to act somewhat like the strong form of "categorial perception" that seems to be characteristic of consonant perception.

Let me briefly describe a large computer simulation of this algorithm that I performed with Michael Mozer, using the traditional material:

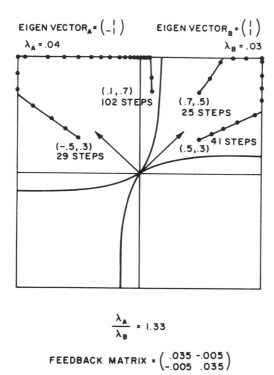

EIGEN VECTOR$_A$ = $\begin{pmatrix} 1 \\ -1 \end{pmatrix}$ EIGEN VECTOR$_B$ = $\begin{pmatrix} 1 \\ 1 \end{pmatrix}$

λ_A = .04 λ_B = .03

(.1, .7)
102 STEPS

(.7, .5)
25 STEPS

(-.5, .3)
29 STEPS

(.5, .3) 41 STEPS

$$\frac{\lambda_A}{\lambda_B} = 1.33$$

FEEDBACK MATRIX = $\begin{pmatrix} .035 & -.005 \\ -.005 & .035 \end{pmatrix}$

Figure 15.7

capital letters (Anderson and Mozer 1981). We coded the letters into 117 element vectors. The coding was straightforward and was meant to be as much like the way the visual system codes stimuli as possible. Part of the vector was a point-for-point excitatory mapping of the visual stimulus with an accompanying inhibitory surround. The letters were placed on a 7 × 7 matrix and drawn in what seemed to be a reasonable way. The surround could extend into a 9 × 9 grid, making 81 matrix elements. The other 36 elements were coded by looking for oriented line segments. The presence and orientation of line segments in these nine regions were coded by the response of orientation-selective units to the presence or absence of a line segment oriented at 0°, 45°, 90°, and 135°. Vectors were normalized, and the mean vector across the stimulus set was subtracted. Figure 15.8 shows four examples of the stimulus codings we used. In the learning phase of the simulation, letter vectors were presented at random. The feedback matrix was applied

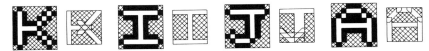

Figure 15.8
Four sample letter codings.

according to the previous equation for seven iterations. The final state, x, was then learned according to the outer-product learning rule, $\mathbf{A} \propto \mathbf{x}\mathbf{x}^T$, with one modification: We assumed that it was necessary to limit learning at some point. Our limitation assumption (there are many other possibilities) notes the final value of the cell's firing rate after seven iterations. If it is at a limit, then there is no change in the synaptic strengths of that cell. There is absolutely no instruction in this procedure. The system does not know whether classification was fast or slow, or correct or incorrect, or anything else. It seems obvious that even the most elementary feedback along these lines should improve results. The behavior of the system is very insensitive to the learning parameter. Simulations were made where the learning parameter varied over two orders of magnitude with no essential difference in behavior.

The same pattern of results we have seen before appeared here. At first, the system miscategorized different letters into particular corners, forming clumps of letters. With more trials, the clumps separated, until at 5,000 presentations only W and N, whose codings were extremely similar in our simulation, were classified together. At 10,000 learning trials, W and N were still classified together, but the simulation was halted. Table 15.1 shows the categorizations as they evolved with the number of presentations.

One of the best features of this system is that when we study the dynamics of classification we find that the speed of categorization increases as the system learns more. The system gets both faster and more accurate with experience.

Many of the most significant aspects of behavior of the system should appear in the structure of the eigenvectors and the eigenvalues. Because of their hypothesized closeness to the factors of factor analysis, we should be able to represent the letters as appropriately weighted sums of the first few eigenvectors and expect to find behavior very similar to that when all the eigenvectors are present. This is so.

Since the eigenvectors represent macrofeatures, one might wonder if there was any obvious "interpretation" of the eigenvectors with largest

Table 15.1
Equivalence classes where more than one letter coding is categorized in the
same final state ("clumps"), as a function of learning trials.

No. of trials	Different letters, categorized identically
1,000	B,E,G,O
	A,I
	C,K,N,W,X
	D,M,Q,S,T,V,Y
	F,P,R
2,000	B,E
	G,O,S,V
	N,W
	K,X
	Q,Y
3,000	F,P
	G,S
	N,W
4,000	F,P
	G,S
	N,W
5,000	N,W
6,000	N,W
7,000	N,W
8,000	N,W
9,000	N,W
10,000	N,W

Source: Anderson and Mozer 1981.

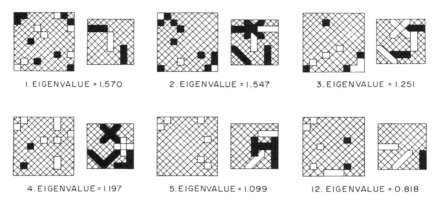

1. EIGENVALUE = 1.570   2. EIGENVALUE = 1.547   3. EIGENVALUE = 1.251

4. EIGENVALUE = 1.197   5. EIGENVALUE = 1.099   12. EIGENVALUE = 0.818

Figure 15.9
Six "feature" vectors.

positive eigenvalue. Six eigenvectors are presented schematically in figure 15.9. Close scrutiny certainly did not reveal any obvious interpretations to us, at least not anything that looked like the lines and corners of published feature sets. Occasionally one came across a feature that seemed to "mean" something; the eigenvector with twelfth largest eigenvalue means roughly "presence of a horizontal bar in the middle coupled with relative lack of horizontal letters in lower right corner," a feature that discriminated A and F from E and G.

Cell Selectivity

Since we have been emphasizing that cell selectivity in a distributed system is not sufficient to explain perceptual selectivity, it is of some interest to look at the selectivities that the model neurons develop in the simulation.

Because any single letter involves a large-scale pattern of activation of vector elements, there is little selectivity for a stimulus. The output is a fully limited state, where each element can only be at the positive or the negative limit, so there is no selectivity at the single-unit level at the output. The output vectors are fairly evenly divided between plus and minus, as a matter of fact. If we look at the initial response of the single elements to features, however, a somewhat different picture emerges. (We look at initial response before feedback has had a chance to modify the picture. This might correspond to an "experimental" situation where feedback was disabled or where only part of a stimulus was presented, which would not allow the powerful cooperative effects on which the categorization depends to occur.)

There was indeed some selectivity at the single-element level. There was also variation from macrofeature to macrofeature. Out of 117 elements, a given feature might have large responses from only two or three elements. Another macrofeature might give rise to large responses from five or six. All features had large numbers of small elements, which, in aggregate, would be significant in their effects. Single neurons could respond strongly to more than one macrofeature. The picture of initial cell responses that emerged was, somewhat to our suprise, that of a modest amount of single-cell selectivity. This is not inconsistent with the picture seen in the actual nervous system, but it would be premature to do more than observe the result with interest. (Details of this simulation are presented at tedious length in Anderson and Mozer 1981.)

Numerical Ablations

One of the most obvious characteristics of distributed models is that they are highly "damage-resistant." Because different items affect the values of many different memory storage elements, loss of one or a few of the storage elements usually will not disrupt the system. This has been demonstrated in considerable detail in a series of computer simulations by Wood (Wood 1980, 1983). One of Wood's findings was that the effects of ablation (that is, removing some of the elements, or setting some of the synaptic strengths in the association matrix to zero) could be paradoxical. Although the statistical predictions are clear-cut, for an individual association, a great deal of representation might happen to be in a particular location. Loss of that location could have a serious effect on one association in a set yet leave the others untouched. Wood commented that this kind of behavior shown by what is formally a highly distributed model suggested that the hard and fast distinction between distributed and localized models might not be so obvious as it seems, since a distributed system could show either pattern, or both simultaneously.

The simulation described above, in which model neurons seemed to develop a degree of selectivity, supports Wood's observation for a bigger system. Loss of a single neuron can be critical for the behavior of a macrofeature in some cases and not in others. Behavior of the system with damage is not easily predictable, even though the statistical behavior may be clear.

Because of the interest shown at the conference in the question of the behavior of these distributed models with damage, I thought it would be useful to look at the classification behavior of our categorizor when it was subjected to damage.

An Ablation Simulation

The nervous system operates in a noisy environment. One of the reasons for the utility of categorization is that it allows the system to ignore "irrelevant" differences between similar stimuli in order to form an equivalence class. We should expect this kind of behavior to be even more resistant to harm than, say, association where there can be slow changes in the direction of a vector with damage. It is possible that the categories might show no change even though there was considerable internal disruption.

A 50-dimensional system was presented with codings representing the 26 letters. The codings were a subset of the 117-dimensional system described earlier, the 36 orientation-selective elements concatenated with 14 elements that did not receive a stimulus input. The system learned the letters for 5,000 presentations, by which time a number of very stable categories had been formed (that is, all the letters were in corners, either by themselves or with a small number of other letters). The simulation was done five times. The final number of categories varied from 12 to 22. Perfect categorization was not particularly wanted. The stimulus set was designed to have some letters very similar to each other (correlation as high as 0.85) and some nearly orthogonal to each other, for testing purposes.

The correlation matrix was 90% connected. That is, during learning a random 10% of the connections were permanently set to zero, corresponding to a lack of connection between the cells involved. After 5,000 presentations, learning had essentially ceased and the categories were extremely stable. An additional 45,000 learning trials showed no significant changes in the corners or the categories, though there were occasional minor switches involving a change of sign of one or two elements. There was an average 0.99 correlation between the corners at 5,000 and 50,000 learning trials, and most corners did not change at all.

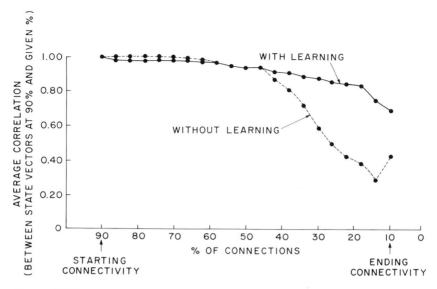

Figure 15.10

All five simulations, which used different seeds for the random number generator, behaved similarly. The graphs and exact numbers are for one simulation, but the others did not differ in any essential respect. For the ablation, I took the matrix at 5,000 trials and started setting matrix elements to zero at random—that is, metaphorically breaking connections between the involved cells. I preferred this to removing model neurons because it is not obvious what is the best way to interpret the resulting reduced-dimensionality state vector. Physiologically, this would correspond to loss of synaptic connection at random without death of the presynaptic or the postsynaptic cell. This could occur by ablation if two regions were connected through a third region and the third region was ablated; some degenerative processes could have this result as well. Starting at 90%, the number of connections was reduced in increments of 4% to a final value of 10%. The curve labeled "without learning" in figure 15.10 shows the average value of correlation between the corners at 90% and the corners at the various intermediate stages. Remarkably, there was no change in categorization until over 20% of the matrix elements had been removed; thus an "outside observer" might be unaware of a rather large amount of damage. There was, however, an increase in the average number of iterations required to

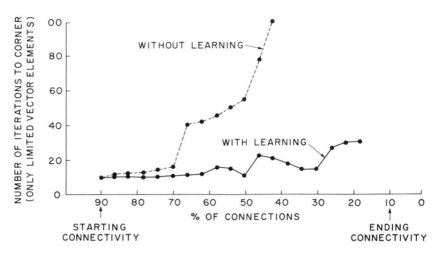

Figure 15.11

reach a corner, from 9 at 90% connectivity to 15 at 70% connectivity. The curve labeled "without learning" in figure 15.11 shows the average number of iterations required to reach a corner.

The system initially responded to diffuse damage by slowing down, but otherwise it performed correctly. One might conjecture that the system would be progressively less reliable in the presence of noise. Our simulations seemed to show this as well, though no quantitative measures of this were made. From 70% to about 50% connectivity errors in categorization started to occur, though the average correlation between corners was still 0.94 at 50% connectivity. Over 50 iterations were now required to reach a corner. All vector elements took longer to saturate, not just a critical few. Below about 50% connectivity, the system deteriorated rapidly. Even around 10% connectivity there was still a significant positive average correlation between the corners in the damaged and the undamaged system, but I suspect that the 10% system would not be of much use for categorization.

Learning While an Ablation Takes Place

It is part of the received wisdom of ablation studies, and it is taught to undergraduates in Psychology 1 as a fact, that slow ablations are much less damaging than rapid ablations. Since the brain is an adaptive system, it can presumably adapt to slow damage or change continuously, whereas sudden change is much more difficult to handle. The above study corresponds to rapid damage, since the matrix elements did not change between disconnections.

It seemed interesting to see whether the model would behave the way the biological system did when confronted with a gradual ablation. I used the same starting matrix as above, and after each 4% ablation (the exact matrix elements removed during the ablation were the same in the two studies) the system learned for 1,000 trials with the same parameters as it did when the categories were first being formed.

As I expected, the learning system was more resistant to damage than the system with no learning. From 90% to 70% connectivity there were now a few changes in corners (in one sense, learning allowed change where there was none before), and the average correlation between corners with 90% connectivity to 70% connectivity became 0.98. There was little difference in corner correlation at 50% connectivity between the learning and nonlearning cases; each had an average correlation of 0.94. However, when connectivity dropped below about 50%, the adaptive system did much better, as can be seen from the graph labeled "with learning" in figure 15.10. Even at 10% connectivity there was still a correlation of 0.69 between the corners in the damaged and undamaged systems, a remarkable amount of compensation. Even more striking was the number of iterations required to reach a corner. As can be seen from the graph labeled "with learning" in figure 15.11, there was almost no increase in number of iterations required until around 50% connectivity. Really serious slowdowns did not start until about 30% connectivity. Even this simple model system with limited learning potential showed a considerable increase in damage resistance when learning was allowed, and an even more striking retention of response speed.

Detailed Modeling Assumptions

These state-vector models are a long way from single-unit neurophysiology, yet I would like to believe that they have not entirely lost contact with neuroscience. Certainly the models I have suggested have wide applications. I would like to discuss very briefly the most recent applications of these models to the retrieval of information from a distributed data base.

Because these applications are so far afield, I would like to discuss some of the details of the simulations, in particular some of the assumptions about noise and some observations about robustness of the simulations. I want to emphasize that the models are not modified severely for a new application, but that their behavior has been reinterpreted. This is a critical point. I believe that neocortex is pretty much neocortex in that the basic operations and connections, viewed sharply as operations, are quite similar from place to place. Yet activity in visual cortex is interpreted as visual perception, whereas the identical operations in a part of the cortex devoted to language might have a linguistic interpretation. However, the permutations of the state vectors could be the same.

Thanks to grants from the National Science Foundation, with assistance from the Digital Equipment Corporation and the Alfred P. Sloan Foundation, we have a DEC VAX 11/780 computer in the Center for Cognitive Science at Brown University. This computer has provided many hours of computation for studying the "brain state in a box" model. For the first time, we have been able to carry out an extensive study of parameters and variations on the same basic theme. Let me state two qualitative conclusions:

- Almost all variants of the basic model work, more or less.
- The most critical modeling assumption seems to be the one that I suspect has had a major effect on the design of the real nervous system: how to turn learning off when the system has learned "enough."

As a neuroscientist, I feel that the question of the stability of the nervous system in both the short and the long term may be the most difficult problem that neocortex must face. When I consider the potential for feedback and for instability, it strikes me as nearly miraculous that there are no equivalents of seizures in normal, healthy individuals. It must be emphasized that small deviations from normality—fevers, small brain injuries, some minor biochemical abnormalities—can totally destabilize the system. The system is stable, but barely so.

Some kind of gain control for learning is required. Here again—and this is encouraging—the simulations indicate that most of the reasonable ways that one can invent for terminating or slowing learning work after a fashion. Our first suggestion (Anderson et al. 1977) was that, when the cell reached a firing limit, it simply no longer learned, and that none of the synpases associated with that cell (neither the rows nor the columns of the connectivity matrix) were modified. This assumption was rather crude, but it was easy to program and it worked surprisingly well. The most successful variant I have tested is a slight extension of this. I simply let the amount of learning (the learning parameter) be different for each cell. I let this parameter be a function of how close the final cell activity (before limiting) was to a threshold. Below this threshold the cellular learning parameter is positive; above it it is negative. This assumption has the effect of equalizing activities across the set of cells. In simulations it worked very well and was easy to program. This local learning limitation is also agreeable philosophically. However, the matrix is not generally symmetric anymore. We have recently had some success with the Widrow-Hoff procedure, an error-correcting technique that is stable and computationally convenient.

Let me make two more points about the simulations. One of the trickiest parts of the feedback model to interpret is the term for the feedback of a cell onto itself. These terms correspond to the diagonal matrix elements. Direct synaptic self-connection is called an "autapse" and is known to exist. However, numerous other more complex circuits of this kind are known. Possibly the most famous is the Renshaw cell system found in spinal motor neurons, where the firing of a motor neuron excites an inhibitory interneuron which then turns off the motor neuron. This important system seems to serve a gating and stabilizing function, and this kind of inhibitory feedback circuitry may be common. The conclusion is that self-feedback is tricky business and may involve a number of special circuits. A formal problem we face in our models is that we are storing $f(i)f(j)$, which in the case of self-feedback is always positive $[f(i)f(i)]$. In most recent simulations, including those in this paper, I have required that this term be zero, which requires all the categorization properties of the system to be performed with the cross-connections.

A third technical point involves the presence of noise. This is a critical aspect of a neurally based model, though I think its importance is often not realized by those who make cognitive models. The brain is a very noisy environment for many reasons, some of which follow:

- There is actual quantization noise at synapses as well as from the use of pulse-code modulation in the cell.
- Unrelated and spontaneous activities are ubiquitous. This may not be noise in the sense of a random process, but from the point of view of a learning system it has the same effect.
- There is noise from the outside environment. Identical objects are seen in different contexts, and so on.
- There is "connection noise." A cell receives connections from a subset of other cells, not from all of them.

• Presentation history is random. Since the learning matrix mixes events up and responds differently at different times, the internal context is different for different histories of presentation.

These are not minor considerations. All of our models, from the beginning, have been chosen to be optimal or close to optimal in the presence of noise of the kinds described above.

All current simulations make use of a significant amount of noise. In a typical learning run, around half the matrix elements would be held at zero (i.e., the model nervous system would be partially connected); the presentation of the stimuli to be learned would be random; and the vectors actually learned would be corrupted by varying amounts of additive, zero-mean, Gaussian noise, often quite large (that is, the average values of noise would be comparable to stimulus input values).

The simulations show a very striking, nearly universal result: The system learns better in the presence of noise. Adding noise during learning is helpful. Adding noise during categorization has the expected result of causing confusions. There are good reasons why this result is not surprising. One of the things that is happening is that noise eliminates the many metastable solutions that appear in the noise-free case. Also, it adds eigenvectors with nonzero eigenvalues in all dimensions, which has important effects on corner stability. Random connectivity makes model neurons more individualistic in their behavior—a physiologically appealing result.

Retrieval of Information from a Distributed Data Base

In 1977, Donald Norman asked me if distributed systems could form relational triples. I wasn't aware of it at the time, but this is the essential minimum required to construct "semantic networks" (Norman and Rumelhart 1975; Brachman 1979). The basic notion is that we have two concepts (ill-defined but useful elementary entities), which are linked together by a "relation." This is usually presented in the form

Node (Concept) $\longrightarrow$ Link $\longrightarrow$ Node (Concept).

My first idea was that the most reasonable realization of this structure would be to assume that the nodes were state vectors and the links were associative matrices. This idea did not work.

We really have only two types of elements in the simplest model we have worked with: state vectors and associative matrices. Most links, as usually described, are quite complicated, with elaborate structure.

They were very conceptlike themselves, so let us let them be concepts. Now a relational triple looks as follows.

Node $\xrightarrow{}$ Association $\xrightarrow{}$ Relational $\xrightarrow{}$ Association $\xrightarrow{}$ Node
(Concept)   matrix   link   matrix   (Concept)

This is a pentuple rather than a triple, but it lets us use our model. Note particularly that there need be only one association matrix: that is, we can have a relation between different parts of a state vector using the same sets of connections (forming a quadruple instead of a triple or a pentuple, I suppose).

For now, let us look at the minimal space-domain realization of a model system of this kind. The more interesting and difficult problems will arise when we expand our horizons to include significant temporal-domain behavior (an obvious necessity for any complex cognitive behavior). (See Anderson and Hinton 1981 and Hinton and Anderson 1981 for reviews of applications of some of the ideas arising from parallel, distributed models to cognition.)

Take a simple assertion of the kind one might find in a data base. The following simulations were performed using a 50-dimensional system, because I had the programs already written for 50 dimensions. The use of these particular examples is due to Hinton (1981). Let us consider

Frank father Jane.

All three words are denoted by state vectors. To maintain the space-domain representation, let us partition the state vector:

Locations 1–16   17–32   33–48   49–50

         Frank   father   Jane   0

Each one of these words is given a set of numerical values in the appropriate locations. I used Walsh functions for the actual vectors, since they were easy to remember and type into the computer.

For the first set of relations we had five simple assertions to be learned, coded as above:

Frank father Jane.
Jane mother Herb.
Jane sister Alice.
Alice sister Jane.
Herb father Ted.

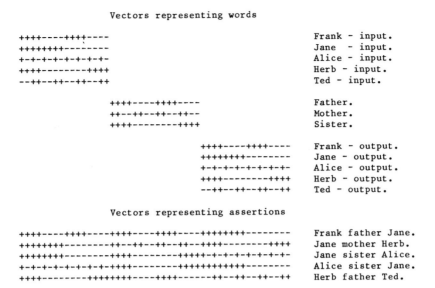

Vectors representing words

```
++++----+++++----                                    Frank - input.
++++++++--------                                     Jane  - input.
+-+-+-+-+-+-+-+-                                      Alice - input.
++++--------++++                                      Herb  - input.
--++--++--++--++                                      Ted   - input.

                ++++----+++++----                    Father.
                ++--++--++--++--                      Mother.
                ++++--------++++                      Sister.

                                ++++----+++++----    Frank - output.
                                ++++++++--------      Jane  - output.
                                +-+-+-+-+-+-+-+-      Alice - output.
                                ++++--------++++      Herb  - output.
                                --++--++--++--++      Ted   - output.
```

Vectors representing assertions

```
++++----+++++----+++++----+++++----++++++++-------   Frank father Jane.
++++++++--------++--++--++--++--++++----------++++    Jane mother Herb.
++++++++--------+++++-------++++-+-+-+-+-+-+-+-       Jane sister Alice.
+-+-+-+-+-+-+-+-++++--------++++++++++++++-------     Alice sister Jane.
++++--------++++++++----+++++------++--++--++--++     Herb father Ted.
```

Figure 15.12

These five relations contained a total of 15 words. We assumed that subject and object appeared in different vector locations. More realistic assumptions will follow further work. After a reasonable amount of experimentation (2 VAX CPU hours and a weekend), a scheme emerged. Figure 15.12 shows the patterns of 1, -1, and 0 that we used to code the words and assertions. First, we taught the matrix the words involved. That is, we simply viewed these 15 words as any stimulus pattern to be learned. After around 1,500 random presentations (100 per word), the resulting WORDS matrix categorized every word into a different corner. (Since there was noise in the system and 30% of the interconnections were set to zero, all 50 elements responded to each input.) Second, once we had the word matrix, we could proceed to teach the system the desired relations between words. Using WORDS as a starting point, we now taught it the five relations, simply presented as patterns. After 400 learning trials (with a low learning coefficient) we had a system that correctly classified all five patterns into different corners. The final corners did not exactly reproduce the shape of the input pattern of signs.

Since the function of the resulting matrix, RELATIONS, was to be a data base, we had to ask it questions to see how well it worked. The actual information was spread all over the matrix and was not accessible in any obvious form.

One of the most striking formal aspects of all these matrix models is that they are profoundly reconstructive. That is, they will reconstruct missing portions of the input state vector. This is essential behavior for a neural system, since state vectors are in no sense abstract patterns or representations but are physical entities that drive other regions of the brain and eventually drive muscle tissue by means of neural interconnections.

Our approach was to ask the matrix a question by requiring it to reconstruct the missing portion of the relation. Suppose we wanted to know who Frank is the father of. We would construct a state vector called "Frank father ? " with these values:

Locations 1–16   17–32   33–48   49–50

          Frank   father   0   0

We would then use this vector as the input to the "brain state in a box" iteration and inspect the final state. If the system operated correctly, we should find in locations 33–48 the identical values that would be found if the entire assertion was present.

There were five learned assertions, each with three terms, for a total of 15 possible questions formed in the above manner. When these queries were put to the system, there were 11 completely correct results, 2 near misses (off by 2 out of the 16 values), and 2 completely spurious results. Figure 15.13 shows the stimulus set corresponding to the questions and the final corner corresponding to the final state of the system.

Hinton (1981) constructed a set of assertions that are especially difficult for correlational-type models to handle. The model has sufficient noise coupled with simple nonlinearities to suggest that it might be able to handle this set, so I went through roughly the same learning procedure with the stimulus set

Zero same Zero.
Zero different One.
One different Zero.
One same One.

After a little fiddling with parameters, the system answered 9 out of the 12 possible queries perfectly and was partially correct (2 out of 16

Query Vectors

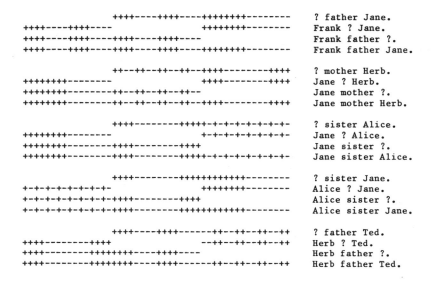

```
                  ++++----++++----++++++++--------      ? father Jane.
++++----++++----                  ++++++++--------      Frank ? Jane.
++++----++++----++++----++++----                        Frank father ?.
++++----++++----++++----++++----++++++++--------        Frank father Jane.

                  ++--++--++--++--++++--------++++      ? mother Herb.
++++++++--------                  ++++--------++++      Jane ? Herb.
++++++++--------++--++--++--++--                        Jane mother ?.
++++++++--------++--++--++--++--++++--------++++        Jane mother Herb.

                  ++++--------+++++-+-+-+-+-+-+-        ? sister Alice.
++++++++--------                  +-+-+-+-+-+-+-        Jane ? Alice.
++++++++--------++++--------++++                        Jane sister ?.
++++++++--------++++--------+++++-+-+-+-+-+-+-          Jane sister Alice.

                  ++++--------++++++++++++--------      ? sister Jane.
+-+-+-+-+-+-+-                    ++++++++--------      Alice ? Jane.
+-+-+-+-+-+-+-++++--------++++                          Alice sister ?.
+-+-+-+-+-+-+-++++--------++++++++++++++--------        Alice sister Jane.

                  ++++----++++------++--++--++--++      ? father Ted.
++++--------++++                  --++--++--++--++      Herb ? Ted.
++++--------+++++++++---++++----                        Herb father ?.
++++--------+++++++++---++++------++--++--++--++        Herb father Ted.
```

Response Vectors

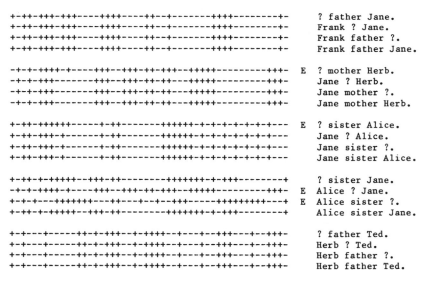

```
+-++-+++-+++----++++----++--+-------+++--------+-        ? father Jane.
+-++-+++-+++----++++----++--+-------+++--------+-     Frank ? Jane.
+-++-+++-+++----++++----++--+-------+++--------+-     Frank father ?.
+-++-+++-+++----++++----++--+-------+++--------+-     Frank father Jane.

-+-+-++++-+----+++--+++-++-+++--++++---------+++-  E  ? mother Herb.
-+-+-+++------+++-+++-++-++--++++---------+++-        Jane ? Herb.
-+-+-+++------+++-+++-++-++--++++---------+++-        Jane mother ?.
-+-+-+++--------+++-+++-++-+--++++---------+++-       Jane mother Herb.

+-++-+++++-----+-++-------+++++-+-+-+-+-+-+---   E    ? sister Alice.
+-++-+++-+------+-++-------+++++-+-+-+-+-+-+---       Jane ? Alice.
+-++-+++-+------+-++-------+++++-+-+-+-+-+-+---       Jane sister ?.
+-++-+++-+------+-++-------+++++-+-+-+-+-+-+---       Jane sister Alice.

+-++-+-+++++--+++-++-------+++++++-+-+++--------+      ? sister Jane.
-+-+-++++-+----+++--+++-++-+++--++++---------+++-  E  Alice ? Jane.
+-+-+---+++++++--++-+----+-+-+++-----+++++++++---+ E  Alice sister ?.
+-++-+-+++++--+++-++-------+++++++-+-+++--------+      Alice sister Jane.

+-+---+-----++-+-+++-+-+++++--+---+--++++---+--+++-   ? father Ted.
+-+---+-----++-+-+++-+-+++++-+---+--++++---+--+++-   Herb ? Ted.
+-+---+-----++-+-+++-+-+++++--+---+--++++---+--+++-  Herb father ?.
+-+---+-----++-+-+++-+-+++++--+---+--++++---+--+++-  Herb father Ted.
```

E = Error in inference.

Figure 15.13

Vectors representing words

```
++++++++--------                                               one - input.
++--++--++--++--                                               zero - input.
        --++--++--++--++                                       same.
        ++++----++++----                                       different.
                ++++++++--------                               one - output.
                ++--++--++--++--                               zero - output.
```

Assertions to be learned

```
++++++++----------++--++--++--++++++++++--------               One same one.
++++++++--------++++----++++----++--++--++--++--               One different zero.
++--++--++--++----++--++--++--++++--++--++--++--               Zero same zero.
++--++--++--++--++++----++++----++++++++--------               Zero different one.
```

Final states of the possible queries

```
--++--++--++--++--++--++--++--++++++++++--------+-             ? same one.
--++--++--++--++--++--++--++--++++++++++--------+-             one ? one.
++++++++----------++++----++++++-++-+++--++-++++   E           one same ?.
--++--++--++--++--++--++--++--++++++++++--------+-             One same one.

-++++++++-+------++--+++--++--++--+---++-++---+---++           ? different zero.
-++++++++-+------++--+++--++--++--+---++-++---+---++           one ? zero.
-++++++++-+------++--+++--++--++--+---++-++---+---++           one different ?
-++++++++-+------++--+++--++--++--+---++-++---+---++           One different zero.

-+--++--++--++--+---+++--+--++--++--++--++--++---+             ? same zero.
-+--++--++--++--+---+++--+--++--++--++--++--++---+             zero ? zero.
+-------+-++++++--++---+--++--++-+++--+--+++-+++--   E         zero same ?
-+--++--++--++--+---+++--+--++--++--++--++--++---+             Zero same zero.

--++--++-+++--++++++--+--+++-----++-++-+--+-+----+-            ? different one.
--++--++-+++--++++++--+--+++-----++-++-+--+-+----+-            zero ? one.
--++--++-+++--++++++-- -+++------+--+-+--+-+----+-  E          zero different ?
--++--++-+++--++++++--+--+++-----++-++-+--+-+----+-            Zero different one.
```

E = error in final state.

Figure 15.14

vector elements wrong) in one more. Figure 15.14 presents the words and the set of assertions that the matrix learned, as well as the final corners attained by the final system.

Conclusions

What I have tried to show is that it is possible to apply a simple model to a wide range of effects that seem initially rather far removed from the starting point. The models use state vectors as basic entities, and are formally parallel, distributed, and associative. I emphasize that the most testable and interesting applications of a model with a base in neuroscience may not be in neuroscience at all but in psychology and cognitive science. This makes more sense than it might seem to at first, because it is very difficult to understand a system if we do not know what the system is used for. Since at least one prominent use of mammalian neocortex seems to be cognition, there might reasonably be expected to be a close connection between some general organizing principles and cognitive models. I have presented very simple, almost skeletal models for association, prototype formation, categorization, and retrieval of information from a distributed data base.

These models ignore many details of cortical physiology and anatomy; this is obvious. Yet, at the same time, they show what I consider surprisingly rich behavior, which has occasional similarities—sometimes qualitative and sometimes quantitative—to the kind of behavior we seem to show.

Acknowledgments

This work was supported by a grant administered by the Memory and Cognition Section of the National Science Foundation. The computing facilities were provided and supported by grants from the National Science Foundation, the Alfred P. Sloan Foundation, and Digital Equipment Corporation.

References

Anderson, J. A. 1972. A simple neural network generating an interactive memory. *Mathematical Biosciences* 14: 197–220.

Anderson, J. A., and G. Hinton. 1981. Models of information processing in the brain. In Hinton and Anderson 1981.

Anderson, J. A., and M. Mozer. 1981. Categorization and selective neurons. In Hinton and Anderson 1981.

Anderson, J. A., J. W. Silverstein, S. A. Ritz, and R. S. Jones. 1977. Distinctive features, categorical perception, and probability learning: Some applications of a neural model. *Psychological Review* 84: 413–451.

Barlow, H. B. 1972. Single units and sensation: A neuron doctrine for perceptual psychology? *Perception* 1: 371–394.

Bienenstock, E. L., L. N. Cooper, and P. W. Monro. 1982. A theory for the development of neuron selectivity: Orientation selectivity and binocular interactions in visual cortex. *Biological Cybernetics* 2: 32–48.

Brachman, R. 1979. On the epistemological status of semantic networks. In *Associative Networks: Representation and Use of Knowledge by Computers*, ed. N. V. Findler. New York: Academic.

Cooper, L. N. 1974. A possible organization of animal memory and learning. In *Proceedings of the Nobel Symposium on Collective Properties of Physical Systems*, ed. B. Lundquist and S. Lundquist. New York: Academic.

Gordon, B. 1983. Confrontation naming: Computational model and disconnection simulation. In *Neural Models of Language Processes*, ed. M. A. Arbib, D. Caplan, and J. C. Marshall. New York: Academic.

Hebb, D. O. 1949. *The Organization of Behavior*. New York: Wiley.

Hinton, G. E. 1981. Implementing semantic networks in parallel hardware. In Hinton and Anderson 1981.

Hinton, G., and J. A. Anderson, eds. 1981. *Parallel Models of Associative Memory*. Hillsdale, N.J.: Erlbaum.

Kandel, E. 1979. Small systems of neurons. In *The Brain*. San Francisco: Scientific American/Freeman.

Knapp, A. 1979. Determinants of Typicality Structures. Thesis, Center for Neural Sciences, Brown University.

Knapp, A., and J. A. Anderson. 1983. A signal-averaging theory of categorization. In press.

Kohonen, T. 1977. *Associative Memory: A System Theoretic Approach*. Berlin: Springer.

Knight, B. W., J.-I. Toyoda, and F. A. Dodge, Jr. 1970. A quantitative description of the dynamics of excitation and inhibition in the eye of *Limulus*. *Journal of General Physiology* 56: 421–437.

Lashley, K. S. 1950. In search of the engram. In *Physiological Mechanisms in Animal Behavior*. New York: Academic.

Levy, W. B., and O. Steward. 1979. Synapses as associative memory elements in the hippocampal formation. *Brain Research* 175: 233–245.

Norman, D., and D. Rumelhart. 1975. *Explorations in Cognition*. San Francisco: Freeman.

Posner, M. I., and S. W. Keele. 1968. On the genesis of abstract ideas. *Journal of Experimental Psychology* 77: 353–363.

Posner, M. I., and S. W. Keele. 1970. Retention of abstract ideas. *Journal of Experimental Psychology* 83: 304–308.

Rauschecker, J. P., and W. Singer. 1979. Changes in the circuitry of the cat's visual cortex are gated by post-synaptic activity. *Nature* 280: 58–60.

Rauschecker, J. P., and W. Singer. 1981. The effects of early visual experience on the cat's visual cortex and their possible explanation by Hebb synapses. *Journal of Physiology* 310: 215–239.

Rosch, E., and B. B. Lloyd, eds. 1978. *Cognition and Categorization*. Hillsdale, N.J.: Erlbaum.

Shepherd, G. 1979. *The Synaptic Organization of the Brain*, second edition. New York: Oxford University Press.

Szentagothai, J. 1978. Specificity versus (quasi) randomness in cortical connectivity. In *Architectonics of the Cerebral Cortex*, ed. M. A. B. Brazier and H. Petsche. New York: Raven.

Wood, C. 1978. Variations on a theme by Lashley: Lesion experiments on the Neural Models of Anderson, Silverstein, Ritz, and Jones. *Psychological Review* 85: 582–591.

Wood, C. 1983. Implications of simulated lesion experiments for the interpretation of lesions in real nervous systems. In *Neural Models of Language Processes*, ed. M. A. Arbib et al. New York: Academic.

Contributors

André Ali-Cherif
C.H.U. de la Timone
and EHESS, Marseille

James A. Anderson
Brown University

Anna Basso
Università di Milano

Alain Bonafe
C.H.U. de Purpan, Toulouse

David Caplan
Montreal Neurological Institute
Centre Hospitalier Côte des Neiges
and McGill University

Dominique Cardebat
C.H.U. de Purpan, Toulouse

Peter D. Eimas
Brown University

Albert Galaburda
Charles A. Dana Research Institute
Beth Israel Hospital
and Harvard Medical School, Boston

Merrill F. Garrett
Max Planck Institute for Psycholinguistics
and Massachusetts Institute of Technology

Norman Geschwind
Harvard Medical School,
Beth Israel Hospital, Boston,
and Massachusetts Institute of Technology

Yves Joanette
Centre Hospitalier Côte des Neiges
and Université de Montréal

Mary-Louise Kean
University of California, Irvine

Andrew Kertesz
St. Joseph's Hospital, London, Ontario
and University of Western Ontario

Andre Roch Lecours
Centre Hospitalier Côte des Neiges
and Université de Montréal

Sylvia Moraschini
Università di Milano

John Morton
MRC Cognitive Development Unit, London

Jean-Luc Nespoulous
Université de Montréal
and Centre Hospitalier Côte des Neiges

Fernando Nottebohm
Rockefeller University

Loraine Obler
Boston University School of Medicine
and Boston V.A. Medical Center

Terence W. Picton
University of Ottawa
and Ottawa General Hospital

Michael Poncet
C.H.U. de la Timone
and EHESS, Marseille

Michele Puel
C.H.U. de Purpan, Toulouse

Andre Rascol
C.H.U. de Purpan, Toulouse

Donald T. Stuss
University of Ottawa
and Ottawa General Hospital

Edgar Zurif
Brandeis University
and Boston V.A. Medical Center

Index